T0211781

Multiantenna Systems for MIMO Communications

Multiantenna Systems for MIMO Communications

Franco De Flaviis, Lluis Jofre, Jordi Romeu, and Alfred Grau

ISBN: 978-3-031-00408-7   paperback

ISBN: 978-3-031-01536-6   ebook

DOI: 10.1007/978-3-031-01536-6

A Publication in the Springer series

SYNTHESIS LECTURES ON ANTENNAS #5

Lecture #5

Series Editor: Constantine A. Balanis, Arizona State University

Series ISSN

ISSN 1932-6076   print
ISSN 1932-6084   electronic

Multiantenna Systems for MIMO Communications

Franco De Flaviis
University of California, Irvine

Lluis Jofre
Universitat Politècnica de Catalunya

Jordi Romeu
Universitat Politècnica de Catalunya

Alfred Grau
University of California, Irvine

SYNTHESIS LECTURES ON ANTENNAS #5

ABSTRACT

Advanced communication scenarios demand the development of new systems where antenna theory, channel propagation and communication models are seen from a common perspective as a way to understand and optimize the system as a whole. In this context, a comprehensive multiantenna formulation for multiple-input multiple-output systems is presented with a special emphasis on the connection of the electromagnetic and communication principles.

Starting from the capacity for a multiantenna system, the book reviews radiation, propagation, and communication mechanisms, paying particular attention to the vectorial, directional, and time-frequency characteristics of the wireless communication equation for low- and high-scattering environments. Based on the previous concepts, different space—time methods for diversity and multiplexing applications are discussed, multiantenna modeling is studied, and specific tools are introduced to analyze the antenna coupling mechanisms and formulate appropriate decorrelation techniques. Miniaturization techniques for closely spaced antennas are studied, and its fundamental limits and optimization strategies are reviewed. Finally, different practical multiantenna topologies for new communication applications are presented, and its main parameters discussed.

A relevant feature is a collection of synthesis exercises that review the main topics of the book and introduces state-of-the art system architectures and parameters, facilitating its use either as a text book or as a support tool for multiantenna systems design.

KEYWORDS

multiantenna systems, MIMO, channel coding, radio channel, antenna modeling, antenna diversity techniques, decorrelation networks

Preface

Since the last decade, multiantenna wireless communication systems have gained a strong interest in both the academic and industrial sectors. Also known in the literature as multiple-input multiple-output (MIMO) [1] wireless communication systems, they use multiport antennas at the transmitter and/or the receiver to increase the transmission rate and the strength of the received signal, as compared with traditional single-input single-output systems, which use one transmit antenna and one receive antenna. Most importantly, these gains come with no additional increase in bandwidth or transmission power [1], which are scarce resources; rather, they come at the cost of system complexity.

This book provides a multidisciplinary approach to the analysis and design of multiantenna systems for MIMO communications, including principles from systems theory, coding theory, radio channel theory, and antenna theory. Note that the term multiantenna system is used in a wide sense because we use it to refer to any multiport antenna system in which each port can be associated with distinct and physically separated antennas [2], distinct polarizations [3,4], or different radiation patterns [5,6]. Based on the previous concepts, different space–time methods for diversity and multiplexing applications are discussed. Also, the modeling of multiantenna systems is studied and specific tools are introduced to analyze the antenna coupling mechanisms and different decorrelation techniques. Miniaturization and mutual coupling for closely spaced antennas are studied and its fundamental limits and optimization strategies reviewed. Finally, different practical multiantenna topologies for new communication applications are presented and its main parameters discussed.

Organization of the Book

In Chapter 1, we review the principles and fundamental concepts of multiantenna communication systems. In Chapter 2, we discuss the properties of the radio channel. We also review specific channel models for the analysis and simulation of MIMO systems. Chapter 3 reviews some of the fundamental MIMO channel coding techniques and the figures of merit that are used to assess the performance of MIMO communication systems from a coding perspective. In Chapter 4, we explain the modeling of a multiantenna system, and give performance measures involving antenna parameters. Chapter 4 also reviews the fundamental limitations imposed by antennas. In Chapter 5, design criteria and procedures for optimal performance of the multiantenna systems within a MIMO communication system are exposited. Chapter 6 concludes this book, presenting some particular design examples and discussing the performance of different multiantenna system topologies.

Acknowledgments

The authors thank Pete Balsells, founder of Balseal Engineering, who has built bridges between California and Catalonia through the creation of the Balsells Fellowship and the California Catalonia Engineering Innovation Program, and Dr. Roger H. Rangel, chair of the Mechanical and Aerospace Engineering Department of the Henry Samueli School of Engineering at the University of California at Irvine (UCI), who has made these bridges work and has made possible the fruitful collaboration between UCI and the Universitat Politècnica de Catalunya (UPC), which has resulted in this book.

Finally, the authors thank Javier Rodriguez De Luis, Marc Jofre, Stan Dossche, and Santiago Capdevila for their help in revising the book, testing some of the examples, and editing the figures.

Contents

CHAPTER 1

Principles of Multiantenna Communication Systems

1.1 INTRODUCTION TO WIRELESS COMMUNICATIONS SYSTEMS

Originally, wireless communications were based on analog systems [7]. In 1837, Morse developed the first digital communication system, known as telegraphy. As with any other digital communications system, telegraphy provided a series of advantages over its analog counterpart, summarized below:

1. The transmitted signal can be regenerated at greater distance compared with analog systems, thus making the system less expensive.
2. The signal fidelity can be better controlled because at each regeneration point the effects of noise can be reduced.
3. The redundancy on the message can be removed using source coding techniques, thus providing channel bandwidth savings.

These advantages were initially investigated by R.V.L. Hartley and H. Nyquist, who are considered among the fathers of modern digital communications. Nyquist (1928) investigated the problem of determining the maximum signaling rate that can be used over a telegraph channel of a given bandwidth B without intersymbol interference [8]. His studies led him to conclude that the maximum pulse rate is $2B$ pulses per second. He also determined that this rate can be achieved using an optimum pulse shape given by $g(t) = \frac{\sin 2\pi Bt}{2\pi Bt}$ [7]. As a result of these studies, the Nyquist sampling theorem was formulated in Ref. [8]. The Nyquist theorem allows us to determine the minimum sampling rate at which an analog signal needs to be sampled (digitized) so that it can be correctly reconstructed. The minimum sampling rate is known as the Nyquist frequency. For a signal with maximum spectral content, or bandwidth, B, the sampling rate f_s must satisfy

$$f_s > 2B. \tag{1.1}$$

Hartley (1928) considered the same problem as Nyquist when multiple amplitude levels for transmissions were used [9]. Both developed pulse code modulation (PCM). Later, in 1948, Shannon [7] established the mathematical basis of a new technological field called *information theory* and derived the fundamental limits for digital communication systems. Shannon extended the studies of Nyquist and Hartley by considering the effects of noise and the statistical structure of the original message. In particular, he investigated how to quantify the information content of a source, adopting a logarithmic measure for it. He also demonstrated that the transmitted power (P^S), the bandwidth used during the communication (B), and the additive noise associated with the channel, can be put into a single parameter, called the channel capacity (C).

1.2   CAPACITY IN A SINGLE-INPUT/SINGLE-OUTPUT COMMUNICATION SYSTEM

Based on Shannon's theorem, capacity is a measure of the maximum transmission rate for reliable communication on a given channel—reliable in the sense that it is possible to transmit at such a speed with an arbitrarily small error probability. In other words, what Shannon's theorem says is that the main limitation that noise causes in a communication channel is not on the quality of the communication itself but on its speed. Interestingly, the theorem does not tell us how to build such reliable systems; instead, it gives us an upper bound for its performance. To achieve or approach these limits, channel coding techniques need to be used. Capacity is therefore the signaling rate for the best possible encoding technique.

The capacity of a particular wireless communication system depends on many factors, such as the noise, the number of antennas being used in the transmitter and receiver, and the number of reflections or multipath components generated in the propagation channel. Some of these parameters are fixed, whereas others can be engineered to maximize the performance of the communication system.

Let us consider first the simple case where we have one antenna in the transmitter and one antenna in the receiver, also called a single-input single-output (SISO) communication system. If we assume an additive white Gaussian noise interference (AWGN), the capacity of the channel, C, can be expressed as [7]:

$$C = \log_2\left(1 + \rho_0\right) \qquad \text{bits/s/Hz (bps/Hz)}, \qquad (1.2)$$

where $\rho_0 = \frac{P^{D,av}}{N_0}$ is the average signal-to-noise ratio (SNR) at the receiver, N_0 is the noise power of the additive noise, and $P^{D,av}$ is the average received power. The capacity, as defined in Eq. (1.2), is also known as the spectral efficiency.

For many years, wireless communications systems have been based on SISO configurations and line-of-sight (LOS) propagation conditions. To improve the performance of SISO systems, engineers traditionally used several techniques to combat multipath, such as temporal diversity techniques or the use of directional antennas instead of omnidirectional antennas. Temporal diversity techniques will be described later in this chapter; for now, we will give some hints on the benefits of directional antennas. From an antenna perspective, the proper use of directional antennas increases the SNR of a SISO system and thus increases its capacity or coverage range. This happens because directional antennas efficiently focus the radiated energy in a desired direction by narrowing its main beam pattern. Let us define G^T as the gain, in linear scale, of the transmitter antenna and G^R as the gain of the receiver antenna and assume that the directions of maximum radiation from these two antennas are pointing to each other. Then the SNR using directional antennas, ρ_{0_G}, can be expressed as [10]

$$\rho_{0_G} = G^T \rho_0 G^R. \tag{1.3}$$

Another important parameter that we need to consider for the antennas is the antenna noise temperature, T^A. This parameter is defined as the temperature of a hypothetical resistor at the input of an ideal noise-free receiver that would generate the same output noise power per unit bandwidth as that at the antenna output at a specified frequency. If $G^A(\theta, \phi)$ is the gain pattern of a particular (receiving) antenna, then T^A is normally given by

$$T^A = \int\int_{4\pi} T^A(\theta, \phi) G^A(\theta, \phi) \sin(\theta) d\theta d\phi, \tag{1.4}$$

where $T^A(\theta, \phi)$ is the noise temperature distribution seen by the antenna. Notice that if $T^A(\theta, \phi)$ is not uniform, then directional antennas can be used to reduce the noise due to the use of narrower beams. However, because $T^A(\theta, \phi)$ is normally uniform in nature, then independently of the directivity of the antenna being used, the noise power will remain the same [10]. In other words, the SNR of a SISO system will improve because of the increase in received power as a result of a higher directivity of the antennas but not because of the possible reduction in noise power as a result of using narrower beams. The capacity of a SISO communication system using directional antennas at both the transmitter and receiver can then be expressed as

$$C = \log_2\left(1 + \rho_{0_G}\right) = \log_2\left(1 + G^T \rho_0 G^R\right). \tag{1.5}$$

Obviously, an increase in the directivity of the antennas implies that the coverage is reduced in the directions where the energy is not focused, as compared with an omnidirectional antenna. In SISO systems, omnidirectional antennas are typically used for broadcasting where the transmitter does not know where the receivers are located, whereas directional antennas are used in radio links.

On the other hand, notice that directivity is proportional to the antenna aperture size; thus, the use of directional antennas is limited in practical compact wireless communication systems [10].

1.3   IMPAIRMENTS OF THE WIRELESS CHANNEL

Some of the desired attributes of a wireless communication system are high spectral efficiency (capacity) and data rates, high quality of service, low bit error rate (BER), wide coverage range, and low deployment, maintenance, and operation costs. All these requirements need to be guaranteed, sometimes simultaneously, over a wireless channel, which by nature is very hostile, especially in non–line-of-sight (NLOS) scenarios. In the last few decades, NLOS propagation scenarios have become common in wireless communication as a result of the massive introduction of radio communication in urban environments. This has raised the necessity to develop new wireless communication architectures that not only work well in LOS scenarios but also in NLOS conditions.

In addition to the typical impairments of the wireless channel in LOS propagation conditions, which include, among others, noise, co–channel interference, path losses, and scarce available bandwidth, in NLOS propagation conditions, there exist other detrimental impairments such as signal fading, which consists of severe fluctuations in the received signal level. This last one has traditionally been one of the most difficult impairments to combat and is caused by one of the

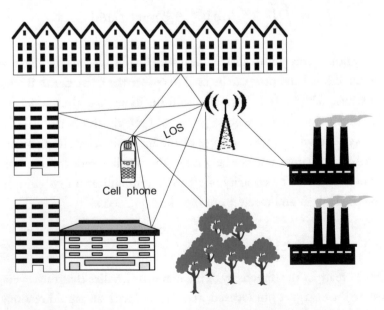

FIGURE 1.1: Representation of a typical wireless propagation channel with randomly localized objects in between the transmitter and the receiver.

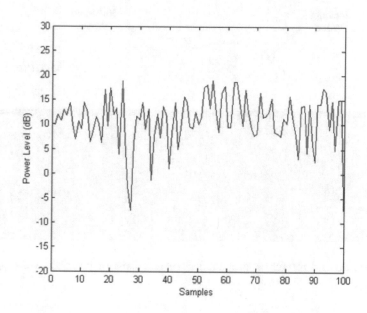

FIGURE 1.2: Representation of a temporal response of the received signal on a channel with multipath propagation, and not LOS, causing signal fading.

most important propagation phenomena arising from the scattering nature of the propagation environments, known as multipath. Multipath is originated in scattered environments (NLOS propagation conditions) when multiple replicas of the transmitted signals arrive at the receiver as a result of the different interactions that the electromagnetic waves undergo with the objects present in the propagation environment, as shown in Fig. 1.1. At the receiver, the destructive addition of multipath signals causes sudden declines of signal power, as shown in Fig. 1.2. Therefore, the SNR is degraded, resulting in a poor or nonexistent communication link between the transmitter and the receiver.

Notice that in SISO systems, narrow radiation beams have also been used to help reduce the effects of multipath and interference in NLOS scenarios. However, the use of directional antennas has limited applicability, especially in cellular communications and broadcasting.

1.3.1 Propagation Electromagnetic Phenomena

To provide a simple explanation of how multiple copies of the transmitted signal can appear at the receiver, in this section, we describe some electromagnetic propagation phenomena that an electromagnetic wave undergoes [10,11]:

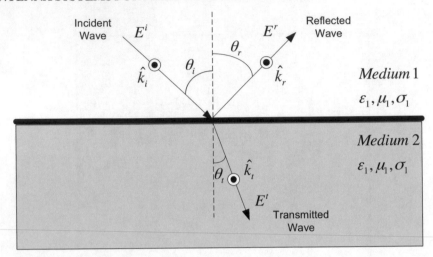

FIGURE 1.3: Reflection and refraction propagation mechanisms of polarized waves (perpendicular polarization shown).

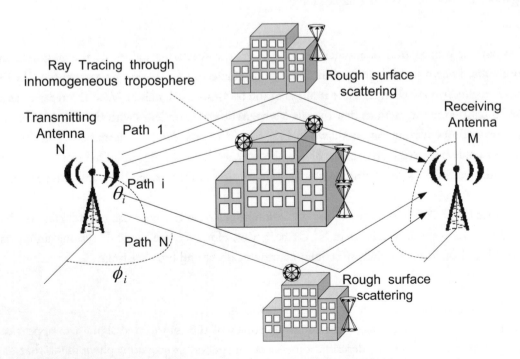

FIGURE 1.4: Three-dimensional model of the wave propagation mechanisms.

- Reflection: the change in direction of a wavefront at an interface between two different media so that the wavefront returns into the medium from which it originated. Fig. 1.3 shows a schematic of the reflection phenomena.

- Diffraction: refers to various phenomena associated with wave propagation, such as the bending, spreading, and interference of waves passing by an object or aperture that disrupts the wave (see Fig. 1.4).

- Refraction: the change in direction of a wave due to a change in its speed. This is most commonly seen when a wave passes from one medium to another, as shown in Fig. 1.3.

- Scattering: a physical process in which radiated waves are forced to deviate from a straight trajectory by one or more electrically small (relative to its wavelength) and localized nonuniformities in the medium through which it passes. In other words, it denotes the dispersal of radiation into a range of directions as a result of physical interactions with electrically small objects. Scattering processes are known in the literature as diffuse reflections (as opposed to specular reflections, or simply reflections, as described above).

The understanding of these electromagnetic phenomena is important to accurately model wireless propagation channels and appropriately design multiantenna systems. An extensive discussion on these phenomena can be found in Refs. [10,11]. In many occasions, wave propagation is explained using geometric optics (also known as ray tracing) and the geometric theory of diffraction (GTD). GTD is an extension of geometric optics that accounts for diffraction. It introduces diffracted rays in addition to the usual (reflected) rays of geometric optics. These rays are produced by incident rays that hit edges, corners, or vertices of boundary surfaces or that graze such surfaces. Then, various laws of diffraction, analogous to the laws of reflection and refraction, can be used to characterize the diffracted rays.

All these phenomena occur in a typical wireless channel as waves propagate and interact with surrounding objects. The nature and origin of the above phenomena can be visualized easily using a three-dimensional model of the wave propagation mechanisms, such as that shown in Fig. 1.4.

1.4   DEFINITION OF A MULTIANTENNA SYSTEM

In this book, we use the term *multiantenna system* in a wide sense because we use it to refer to any multiport antenna (MPA) system in which each port can be associated with distinct and physically separated antennas, with different polarization, with different radiation patterns, or with any combination of the above possibilities. That is, we can distinguish three main categories of MPAs:

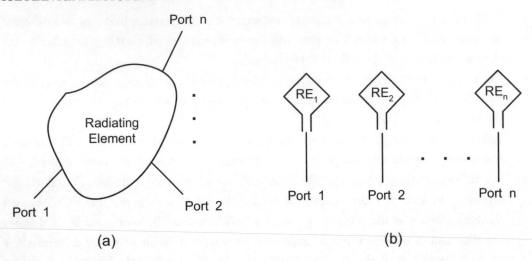

FIGURE 1.5: Various forms of an MPA: (a) MPA consisting of a single radiating element (surface or volume) such as for the MPOA and MMA; (b) MPA consisting of several radiating elements such as in the case of an MEA.

- multielement antenna (MEA)
- multipolarized antenna (MPOA)
- multimode antenna (MMA).

In the case where the ports are associated with distinct antennas, we refer to that MPA as an MEA. If the ports are to be connected to different polarizations, we refer to it as an MPOA. Finally, if the ports are connected to different radiation patterns, we refer to it as an MMA.

Notice that an MPA can adopt different forms. For example, MEA are traditionally composed of multiple single-port radiating elements, as shown in Fig. 1.5a. On the other hand, in MPOA and MMA, the structures are traditionally composed of single radiating elements with multiple ports connected to them, as illustrated in Fig. 1.5b. Despite this, any combination of intermediate forms is, in general, possible.

1.5   CAPACITY IN A MIMO COMMUNICATION SYSTEM

In 1998, Foschini and Gans [1], driven to understand the ultimate limits of bandwidth efficiency, examined the possibility of using MEAs or arrays to improve wireless transmission rates in certain applications dealing with NLOS propagation environments. In information theory, multiantenna systems are known as MIMO communication systems. Such systems operate by using the spatial and/or polarization properties of the multipath channel, thereby offering new dimensions

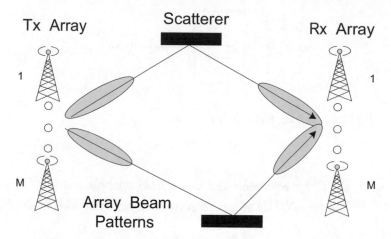

FIGURE 1.6: Simple multipath propagation environment showing two paths between transmitter and receiver. The arrays are capable of resolving the individual multipaths, enabling increased data throughput.

that can be used to enhance communication performance. This working principle is possible because of the capability of MPAs (antenna arrays) in resolving individual multipath components, enabling increased data throughput, as illustrated in Fig. 1.6. In MIMO systems, therefore, the antenna properties as well as the multipath channel characteristics play a key role in determining communication performance.

1.5.1 System Model

Let us assume a MIMO wireless communication system with an MPA at the transmitter having M-accessible ports and with an MPA at the receiver having N-accessible ports. For the remainder of this book, we refer to such a MIMO system as a $M \times N$ system. Such a system is normally described with the following input–output relationship:

$$\mathbf{r} = \mathbf{Hs} + \mathbf{n}, \tag{1.6}$$

where $\mathbf{H} \in \mathcal{C}^{N \times M}$ is the channel matrix where its entries have the magnitude and phase information of the propagation paths between the M transmit and N receive ports. As commented previously, these ports may correspond to different antennas in an array or MEA [2], to different polarizations in an MPA [3,4,12], or to different radiating patterns [5,6] or directional beams [13] in an MMA. Also, these ports may correspond to a combination of the above possibilities. The entries of $\mathbf{H}$ can also be understood as the ratio of received voltage waves at the receive antennas, after propagating through the channel, to the incident voltage waves at the transmit antennas. As shown schematically

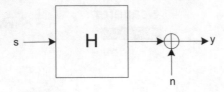

FIGURE 1.7: MIMO input–output relationship.

in Fig. 1.7, the vector $\mathbf{s} \in \mathcal{C}^{M \times 1}$ represents the transmitted symbols, $\mathbf{r} \in \mathcal{C}^{N \times 1}$ are the received symbols, and $\mathbf{n} \in \mathcal{C}^{N \times 1}$ are the AWGN noise components. Notice that $\mathbf{H}$ is of the form

$$\mathbf{H} = \begin{pmatrix} \mathbf{H}_{11} & \cdots & \mathbf{H}_{1M} \\ \vdots & \ddots & \vdots \\ \mathbf{H}_{M1} & \cdots & \mathbf{H}_{NM} \end{pmatrix}, \tag{1.7}$$

where $\mathbf{H}_{ij}$ describes the channel coefficients between the jth transmit and ith receive ports of the MPA, respectively, as shown in Fig. 1.8.

It is important to point out that if the dominant source of noise in the system is from the channel (co–channel interference, channel instability, cosmic radiation, etc.), we may neglect noise additions in the receiver. However, in single-user point-to-point transmission systems (as in our case), the receiver front end is often the major source of noise [14], and thus the noise is added at the receiver, as shown in Fig. 1.7.

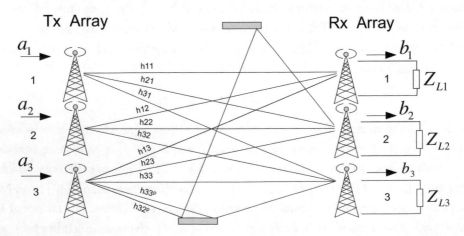

FIGURE 1.8: Representation of the channel matrix coefficients.

Although different channel propagation models are described in Chapter 2, we give here a physical insight on how these channel coefficients are related to typical propagation and antenna parameters. For the sake of simplicity, we consider a narrowband communication scheme, in which a constant frequency response is assumed over the channel. Then, $\mathbf{H}$ is composed of complex scalar elements $\mathbf{H}_{ij}$. Using the path-based model described in Ref. [15], the elements $\mathbf{H}_{ij}$ can be related to the antenna parameters and to the multipath and scattering characteristics of the channel in a very intuitive manner. Using this model, each path p is described by the polarimetric transfer matrix $\mathbf{\Gamma}_{ij}^{p} \in \mathcal{C}^{2\times 2}$, containing its polarization, gain, and phase information (among the jth transmit antenna and ith receive antenna). This matrix is given by

$$\mathbf{\Gamma}_{ij}^{p} = \begin{pmatrix} \Gamma_{\theta\theta_{ij}}^{p} & \Gamma_{\theta,\phi_{ij}}^{p} \\ \Gamma_{\phi\theta_{ij}}^{p} & \Gamma_{\phi\phi_{ij}}^{p} \end{pmatrix}. \tag{1.8}$$

Notice that the entries of $\mathbf{\Gamma}_{ij}^{p}$ contain the information on the free space propagation loss term $\left(\frac{1}{r}\right)$ for each one of the polarizations in addition to also accounting for other losses that result from phenomena such as reflection, diffraction, or scattering with objects present in the channel. The entries $\mathbf{H}_{ij}$ can be expressed as follows:

$$\mathbf{H}_{ij} = \lambda \sqrt{\frac{Z_{j}^{0,T}}{Z_{i}^{0,R}}} \sqrt{\frac{\Re(Z_{i}^{R})}{\Re(Z_{j}^{T})}} \sum_{p=1}^{P} (\mathbf{F}_{i}^{R}((\theta,\phi)_{i}^{p}))^{\dagger} \mathbf{\Gamma}_{ij}^{p} \mathbf{F}_{j}^{T}((\theta,\phi)_{j}^{p}), \tag{1.9}$$

where $Z_{j}^{0,T} = Z_{\alpha_{j}}^{0,T}$ and $Z_{i}^{0,R} = Z_{\alpha_{i}}^{0,R}$ are the characteristic reference impedances of the jth transmit and ith receiver ports, respectively; $Z_{j}^{T} = \mathbf{Z}_{jj}^{T} = \mathbf{Z}_{\alpha_{jj}}^{T}$ and $Z_{i}^{R} = \mathbf{Z}_{ii}^{R} = \mathbf{Z}_{\alpha_{ii}}^{R}$ are the antenna input impedances of the jth transmit and ith receive ports, respectively; $(\theta,\phi)_{j}^{p}$ and $(\theta,\phi)_{i}^{p}$ are the directions of the multipath components associated with the jth transmit and ith receive ports, respectively, and P is the number of relevant paths. With the definition given in Eq. (1.9), $\mathbf{H}$ relates input waves into the transmit antennas to incident waves into the load of the receive antennas. Notice that $\mathbf{F}_{j}^{T}(\theta,\phi) \in \mathcal{C}^{2\times 1}$ and $\mathbf{F}_{i}^{R}(\theta,\phi) \in \mathcal{C}^{2\times 1}$ are the normalized far-field radiation patterns associated with the jth transmit and ith receive ports, respectively. The radiation patterns are represented by vectors because they contain information on the θ and ϕ components. The normalized far-field radiation patterns, defined in Chapter 4, are computed with all the other ports of the MPA terminated with the characteristic reference impedance, and therefore include the coupling effects among adjacent antennas.

1.5.2 General Capacity Expression

According to Foschini and Gans [1], the capacity of a MIMO wireless communication system using M transmit and N receive accessible ports is given by

$$C = \max_{\{R_s : \text{Tr}(R_s) \leq P^{S,\text{av}}\}} \log_2 \frac{\det\left(\mathbf{H}\mathbf{R}_s\mathbf{H}^H + \mathbf{R}_n\right)}{\det\left(\mathbf{R}_n\right)}, \tag{1.10}$$

where $\mathbf{R}_s = E\{\mathbf{s}\mathbf{s}^H\} \in \mathcal{C}^{M \times M}$ is the covariance of the transmitted symbols ($\mathbf{s}$), $\mathbf{R}_n = E\{\mathbf{n}\mathbf{n}^H\} \in \mathcal{C}^{N \times N}$ is the covariance of the noise ($\mathbf{n}$), and $P^{S,\text{av}}$ is the average transmitted power.

If the noise is AWGN, such that the entries of $\mathbf{n}$ are independent and identically distributed (iid) random variables with equal variance $\mathbf{N}_0$, then the previous expression reduces to

$$C = \max_{\{R_s : \text{Tr}(R_s) \leq P^{S,\text{av}}\}} \log_2 \det\left(\mathbf{I} + \frac{\mathbf{H}\mathbf{R}_s\mathbf{H}^H}{N_0}\right), \tag{1.11}$$

where $\mathbf{I} \in \mathcal{C}^{N \times N}$ is the N-dimension identity matrix. Notice that Eq. (1.11) can also be expressed as

$$C = \max_{\{R_s : \text{Tr}(R_s) \leq P^{S,\text{av}}\}} \sum_i^{\min(M,N)} \log_2\left(1 + \frac{\Lambda_{ii}}{N_0}\right), \tag{1.12}$$

where Λ_{ii} is the ith eigenvalue in the eigenvector decomposition (EVD) of $\mathbf{H}\mathbf{R}_s\mathbf{H}^H = \xi \Lambda \xi^H$. ξ is a matrix containing the eigenvectors of $\mathbf{H}\mathbf{R}_s\mathbf{H}^H$. Notice that $\mathbf{H}\mathbf{R}_s\mathbf{H}^H$ is the covariance of the received signal in the absence of noise, and the eigenvalue Λ_{ii} represents the received signal power level in the ith eigenchannel. For the sake of better understanding, an eigenchannel (or eigenmode) can be thought of as an equivalent uncoupled SISO channel between the transmitter and receiver. That is, the EVD of $\mathbf{H}\mathbf{R}_s\mathbf{H}^H$ gives us an insight into the strength and number of significant equivalent uncoupled SISO channels that one could eventually produce and use by means of channel coding techniques.

1.5.3 Uninformed Transmitter Capacity

In the particular case, where the transmitter does not know the channel $\mathbf{H}$, the transmitted power is equally distributed among the transmit antennas, thus $\mathbf{R}_s = \frac{P^{S,\text{av}}}{M}\mathbf{I}$. In this case, the capacity expression reduces to

$$C = \log_2 \det\left(\mathbf{I} + \frac{\rho}{M}\mathbf{H}\mathbf{H}^H\right), \tag{1.13}$$

where $\rho = \frac{P^{S,\text{av}}}{N_0}$ is the average SNR at the transmitter. Notice that Eq. (1.13) does not impose any constraint on linear combination of the antenna outputs—that is, it is a general expression of the capacity, which may or not be attained with current channel coding techniques.

1.5.4   Water-Filling Capacity

In the case that feedback information is available at the transmitter, such that the transmitter and receiver know the channel matrix $\mathbf{H}$, the capacity can be enhanced by using beam-forming techniques and assigning an unequal power distribution among the transmit ports. The optimal power allocation method that one can use is known as water filling or water pouring [16]. To clarify, beam forming refers to the capability to send information among artificially decoupled subchannels, in such a way that the best eigenmodes of the propagation channel—that is, the subchannels with highest gain—are selected. On the other hand, water filling is an optimal method of allocating power on the subchannels created using beam-forming techniques, in which weaker channels are, in general, not used. Let the uncoded transmit vector in the waveform domain be denoted as $\mathbf{s}'$. Assume the singular value decomposition (SVD) of $\mathbf{H}$ given by

$$\mathbf{H} = \mathbf{USV}^H, \tag{1.14}$$

where $\mathbf{U}$ and $\mathbf{V}$ are the left and right singular vectors, respectively, and where $\mathbf{S}$ is a diagonal matrix containing the singular values of $\mathbf{H}$, given by

$$\mathbf{S} = \mathbf{U}^H \mathbf{H} \mathbf{V}. \tag{1.15}$$

We then encode the transmit vector as $\mathbf{s} = \mathbf{V s}'$. Because each element of $\mathbf{s}'$ multiplies the corresponding column of $\mathbf{V}$, this operation suggests that each column of $\mathbf{V}$ represents array weights for each signal stream. The receiver performs the operation $\mathbf{r}' = \mathbf{U}^H \mathbf{r}$, indicating that each row of $\mathbf{U}^H$ represents the receive array weights for each stream. Because $\mathbf{U}$ and $\mathbf{V}$ are unitary, using Eq. (1.6), we obtain

$$\mathbf{r}' = \mathbf{U}^H \mathbf{r} = \mathbf{S s}' + \mathbf{n}', \tag{1.16}$$

where $\mathbf{n}' = \mathbf{U}^H \mathbf{n}$. The noiseless input–output relationship is represented graphically in Fig. 1.9. Because the matrix $\mathbf{S}$ of singular values is diagonal, Eq. (1.16) indicates that $\mathbf{r}'$ is a scaled version of the transmit vector $\mathbf{s}'$ corrupted by additive noise. Therefore, beam forming using the singular

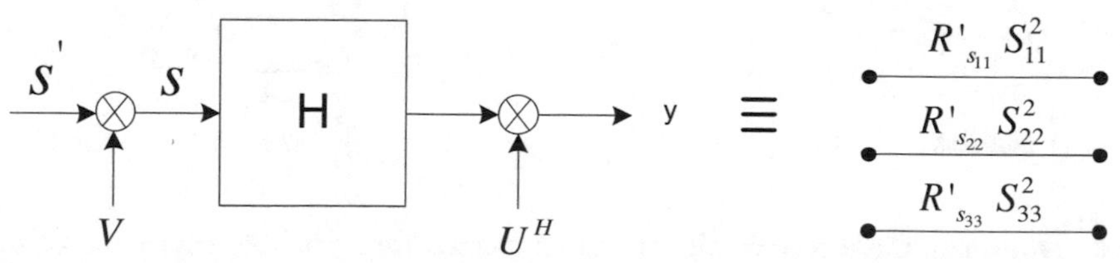

FIGURE 1.9: Noiseless MIMO input–output relationship using water filling.

vectors of $\mathbf{H}$ as array weights has produced a set of radiation patterns (also known in this context as eigenpatterns) that create a set of independent and parallel SISO communication channels in a multipath environment. Using this scheme, the capacity of the system is given by

$$C = \max_{\{\mathbf{R}'_s : \mathrm{Tr}(\mathbf{R}'_s) \leq P^{S,\mathrm{av}}\}} \sum_{i}^{\min(M,N)} \log_2 \left(1 + \frac{\mathbf{R}'_{s_{ii}} S_{ii}^2}{N_0}\right), \qquad (1.17)$$

where $\mathbf{R}'_s = E\{\mathbf{s}'(\mathbf{s}')^H\} = \mathbf{V}^H \mathbf{R}_s \mathbf{V}$, with the optimal solution of $\mathbf{R}'_{s_{ii}}$ given by

$$\mathbf{R}'_{s_{ii}} = \left(v_i - \frac{N_0}{S_{ii}^2}\right)^+. \qquad (1.18)$$

Notice that v_i is such that $\sum_{i}^{\min(M,N)} v_i = P^{S,\mathrm{av}}$, with $(\cdot)^+ = \max(\cdot, 0)$, and where $\mathbf{R}'_{s_{ii}}$ represents the optimal transmitted power in the ith eigenchannel.

As shown in Ref. [2], under certain circumstances, the resultant eigenpatterns can be directly related to the propagation scenario. Consider the example of 10 vertical dipoles along the x-axis (see Fig. 1.10) with a uniform spacing of $\lambda/2$, where λ is the free-space wavelength. Assume that each of the dipoles is aligned in the z-axis direction. Consider now that three sets (also known as clusters)

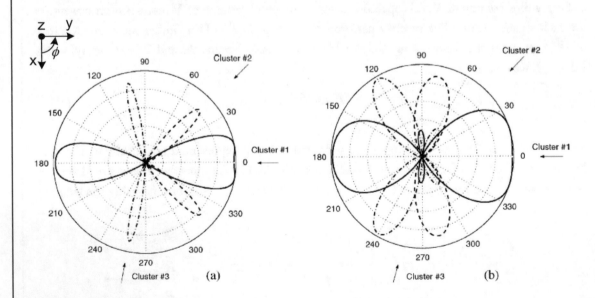

FIGURE 1.10: Dominant three receive eigenpatterns created for a channel consisting of three plane wave clusters centered at the angles indicated by the arrows: (a) 10- and (b) 3-element array. Figure extracted from Ref. [2].

of plane waves with propagation vectors in the horizontal plane (*xy*-plane) arrive at the receiver, with each cluster consisting of a central plane wave with two additional waves, $10°$ on either side of the center, with field strengths 0.7 times the central wave strength. The clusters are centered at $\phi = 0°$ (field strength = 1), $\phi = 45°$ (field strength = 0.7), and $\phi = 260°$ (field strength = 0.85). The receive eigenpatterns corresponding to the three largest singular values, shown in Fig. 1.10a, are focused on these clusters. Notice how the system uses beam forming to focus the radiation patterns toward the desired directions of arrival. Fig. 1.10b shows the eigenpatterns that result when a three-element array is used for the same channel. Here, it becomes clear that if the array cannot resolve the multipath components, then the eigenpatterns achieve a superposition of the waves in an effort to maximize performance.

As an additional example of the previous concepts, let us examine the MIMO performance in a complex environment such as the tunnel shown in Fig. 1.11. The tunnel is a 12 m diameter half-cylinder, and we are interested in assessing the performance of a 4×4 MIMO system at the frequency of 900 MHz. Through electromagnetic simulation, it is possible to compute the field distribution inside the tunnel for the four transmitter antenna positions of the Fig. 1.11. The antennas are spaced 2 m apart and are placed 2 m above the ground. In Fig. 1.12, the field distribution is shown in logarithmic scale for each transmitting antenna. In this case, we consider four receiving antennas placed also 2 m above the ground with a spacing of 2 m and at a distance of 125 m from

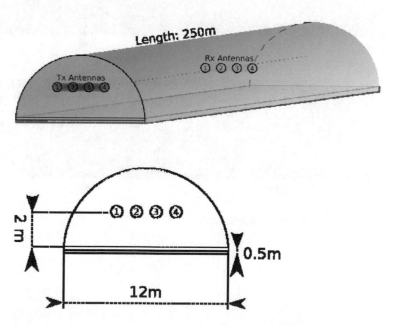

FIGURE 1.11: Example of propagation environment within a tunnel.

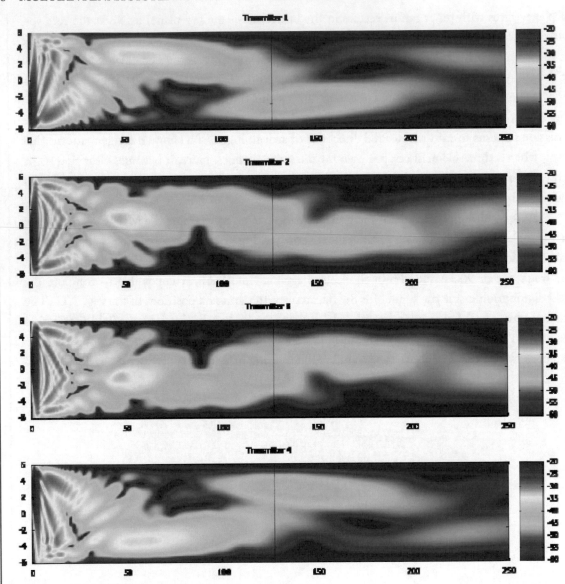

FIGURE 1.12: Field distributions inside the tunnel.

i	1	2	3	4
Λ_{ii}	2.392	1.412	0.136	0.056

TABLE 1.1: Summary of the eigenvalues of $H_0(H_0)^H$

the transmitter. Neglecting mutual coupling between antennas and assuming point sources, the matrix channel can be obtained by relating the input currents at the transmitting antennas with field value at the position of the receiving antennas. For this example, the channel matrix is

$$\mathbf{H}_0 = \mathbf{H}\sqrt{\frac{NM}{\|\mathbf{H}\|_F^2}} = 4 \begin{pmatrix} 0.35\angle_{52.5°} & 0.18\angle_{114.6°} & 0.19\angle_{-175.2°} & 0.14\angle_{-139.5°} \\ 0.18\angle_{114.6°} & 0.26\angle_{148.0°} & 0.40\angle_{-172.6°} & 0.18\angle_{-175.7°} \\ 0.19\angle_{-175.2°} & 0.40\angle_{172.6°} & 0.25\angle_{146.1°} & 0.18\angle_{112.2°} \\ 0.14\angle_{-139.5°} & 0.18\angle_{-175.7°} & 0.18\angle_{112.2°} & 0.35\angle_{51.4°} \end{pmatrix}. \quad (1.19)$$

In our case, $N = M = 4$ and the matrix elements are normalized so the Frobenius norm of the matrix $\mathbf{H}_0$ is 4. In the capacity computation, we are interested in the eigenvalues of $\mathbf{H}_0(\mathbf{H}_0)^H$ that are readily obtained as shown in Table 1.1. Notice that in this case, the trace of $\mathbf{H}_0$, given by the summation of the four eigenvalues, is 4. For the case of uniform power allocation the capacity is given by

$$C = \sum_i^{\min(M,N)} \log_2\left(1 + \frac{\rho_0}{M}\Lambda_{ii}\right), \quad (1.20)$$

where ρ_0 is the average SNR at the receiver. In our case, $M = 4$ and $\rho_0 = 10$, and we can compute not only the total capacity but also the contribution of each eigenvalue to the total capacity, as shown in Table 1.2. This clearly shows that although the overall capacity is 5.6 bps/Hz, the main contribution to the capacity comes from the first two eigenvalues, whereas the other two provide only a small contribution. To gain physical insight on how the four subchannels are formed, we can apply the beam-forming technique from the transmitter side described in this section (water

TABLE 1.2: Summary of the contribution of each eigenvalue to the final capacity for a SNR = 10 dB

i	1	2	3	4
C_i (bps/Hz)	2.8031	2.1808	0.4257	0.1931

filling). That is, we weight each of the transmitters with a column of the matrix $\mathbf{V}$ that results from the SVD decomposition of $\mathbf{H}_0$. From the electromagnetic point of view, this is merely performing the superposition of the field distribution of Fig. 1.12 weighted by the elements of the column vectors of $\mathbf{V}$. In Fig. 1.13, the results for the field distribution inside the tunnel are shown for the four different beams that can be produced. The red line shows the position of the receiver antennas. From the plots, one can easily visualize how the beam forming produces orthogonal subchannels

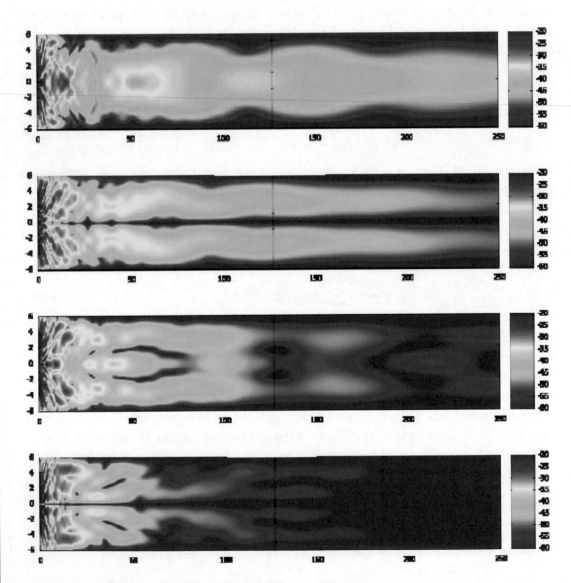

FIGURE 1.13: Field distributions using water filling inside the tunnel.

in the propagation medium. It is also clear that field values at the receiver associated with the third and fourth eigenvalues are much smaller than for the first and second eigenvalues. Our model assumes that equal power is radiated in each subchannel. It is clear that employing the third and fourth subchannels is a waste of the available power and that better capacity would be obtained if the available power is radiated into the first and second subchannels.

1.5.5   Channel Normalization

The above equations express capacity as a function of the transmitted power P^S. The attenuation of the transmission link, influenced by the antennas and other propagation phenomena of the radio channels, is taken into account in $\mathbf{H}$. When performing simulations and analyses using channel matrices obtained from measurements or models (such as the Kronecker model, etc.), it is often useful to be able to properly scale $\mathbf{H}$ to achieve a specified average SNR at the receiver. In fact, to investigate the influence of the correlation properties on the capacity, the channel matrix is often normalized so that it is independent of the attenuation of the channel [2]. When normalizing, the capacity formula becomes a function of the SNR at the receiver. In that case, Eq. (1.13) can be rewritten as

$$C = \log_2 \det\left(\mathbf{I} + \frac{\rho_0}{M}\mathbf{H}_0\mathbf{H}_0^H\right),$$
(1.21)

where ρ_0 is now the SNR at the receiver, averaged over time and among all receive ports. Given a set of Q channel realizations, each one of them denoted by $\mathbf{H}^q$, these matrices should be normalized according to

$$\mathbf{H}_0^q = \mathbf{H}^q \sqrt{\frac{NMQ}{\sum_{q=1}^{Q}||\mathbf{H}^q||_F^2}},$$
(1.22)

where $||\cdot||_F$ represents the Frobenius norm [17]. If $Q = 1$, this implies that each individual channel matrix is normalized to have the specified SNR (ρ_0). If $Q > 1$, the average SNR over the group of Q matrices will be as specified, although the SNR for each individual matrix will fluctuate about this mean value. This allows investigation of the relative variations in received signal strength over an ensemble of measurements or simulations. Notice that ρ_0 is also known in the literature as the average receive SNR of a single antenna system and referred as the SISO SNR [2]. Notice, too, that ρ_0 and ρ have different meanings. In particular, ρ in Eq. (1.13) was defined as the SNR at the transmitter—thus, without including the attenuations of the link—and can be related to ρ_0 by

$$\rho_0 = \rho \frac{||\mathbf{H}||_F^2}{MN}.$$
(1.23)

After normalization, the entries of $\mathbf{H}_0$ are correlated Gaussian random variables. Unless channel normalization is performed, second-order statistics such as channel correlation, which undergo small variations on the magnitude and phase of the channel matrix entries, may appear hidden because of first-order statistics such as path losses.

1.5.6   Ergodic and Outage Capacity

Because $\mathbf{H}$ is a random variable that expresses different realizations of the channel propagation environment, the capacity is also a random variable. There exist two figures of merit that are normally defined for the capacity: the ergodic capacity and the outage capacity. The ergodic capacity, C_E, is the average capacity, and it can be mathematically expressed by

$$C_E = E\left\{ \log_2 \det\left(\mathbf{I} + \frac{\rho}{M}\mathbf{H}\mathbf{H}^H \right) \right\}. \tag{1.24}$$

We also define the $q\%$ outage capacity $C_{\text{out},q}$ of a fading channel as the information rate that is guaranteed for $(100 - q)\%$ of the channel realizations, that is,

$$P(C \le C_{\text{out},q}) = q\%. \tag{1.25}$$

1.6   MIMO CHANNEL MODELS

It is the radio propagation channel that is the crucial determinant of the characteristics of the entire MIMO system. Therefore, accurate modeling of MIMO channels is an important prerequisite for MIMO system design, simulation, and deployment. Analytic MIMO channel models that describe the impulse response (or equivalently, the transfer function) of the channel between the elements of the MPAs at both link ends by providing analytic expressions for the channel matrix are very popular for developing MIMO algorithms, in general.

With the Ricean K factor, K, defined as the ratio of deterministic to scattered power, the channel matrix $\mathbf{H}$ can be expanded into

$$\mathbf{H} = \sqrt{\frac{K}{1+K}}\,\overline{\mathbf{H}} + \sqrt{\frac{1}{1+K}}\,\widetilde{\mathbf{H}}, \tag{1.26}$$

where the entries $\widetilde{\mathbf{H}}$ are, in general, correlated identically distributed unit-variance complex Gaussian random variables with zero mean. These variables are used to describe the scattering nature of the (NLOS) propagation channel. On the other hand, the entries of $\overline{\mathbf{H}}$ are deterministic variables that describe the LOS component of the channel. $\overline{\mathbf{H}}_{mn}$ can be directly computed from the transmit and receive antenna radiation patterns and the propagation loss factor of the LOS component [10,18] as follows

$$\overline{\mathbf{H}}_{mn} = \frac{\lambda}{R} \sqrt{\frac{Z_m^{0,T}}{Z_n^{0,R}}} \sqrt{\frac{\Re(Z_n^R)}{\Re(Z_m^T)}} \cdot (\mathbf{F}_R^n((\theta,\phi)_n^0))^\dagger \mathbf{F}_T^m((\theta,\phi)_m^0), \tag{1.27}$$

where $(\theta,\phi)_m^0$ and $(\theta,\phi)_n^0$ are the angle of departure and the angle of arrival of the LOS path component with respect to the local coordinate systems at the transmitter and receiver, respectively, and R is the distance between the transmitter and the receiver.

For the scattering component of the channel, denoted by $\widetilde{\mathbf{H}}$, the most popular channel models include

- the Kronecker model [18],
- the Weichselberger model [19], and
- the virtual channel representation [19,20].

The Kronecker channel model assumes that the random fading processes at the receiver are uncorrelated to those at the transmitter, which, however, happens in most of the NLOS scenarios with rich scattering. The parameters of this model are the transmit and receive correlation matrices. The Kronecker model became popular because of its simple analytic treatment. However, the main drawback of this model is that it forces both link ends to be separable, irrespective of whether the channel supports this or not. The Kronecker channel model is used in some derivations in this book to provide a simple physical insight. Given a particular channel propagation environment described by $\widetilde{\mathbf{H}}$, its Kronecker channel model can be expressed as

$$\widetilde{\mathbf{H}} = (\mathbf{R}_H^R)^{\frac{1}{2}} \mathbf{W} ((\mathbf{R}_H^T)^{\frac{1}{2}})^\dagger, \tag{1.28}$$

where $\mathbf{W} \in \mathcal{C}^{M \times N}$ is a complex random matrix with all elements being iid Gaussian random variables with zero-mean and unit variance ($N(0,1)$), and $\mathbf{R}_H^T \in \mathcal{C}^{M \times M}$ and $\mathbf{R}_H^R \in \mathcal{C}^{N \times N}$ are the transmit and receive spatial correlation matrices, respectively. These matrices can be computed from measured data as follows:

$$\mathbf{R}_H^R = \frac{E\{\widetilde{\mathbf{H}}(\widetilde{\mathbf{H}})^H\}}{\sqrt{E\{\|\widetilde{\mathbf{H}}\|_F^2\}}}, \tag{1.29}$$

$$\mathbf{R}_H^T = \frac{E\{(\widetilde{\mathbf{H}})^H \widetilde{\mathbf{H}}\}}{\sqrt{E\{\|\widetilde{\mathbf{H}}\|_F^2\}}}. \tag{1.30}$$

Each one of the entries within these matrices can be computed as

$$\mathbf{R}_{H_{ij}}^{R} = \frac{\sum_{m=1}^{M} E\{\widetilde{\mathbf{H}}_{im}(\widetilde{\mathbf{H}}_{jm})^{*}\}}{\sqrt{E\{||\widetilde{\mathbf{H}}||_{F}^{2}\}}}, \tag{1.31}$$

$$\mathbf{R}_{H}^{T} = \frac{\sum_{n=1}^{N} E\{(\widetilde{\mathbf{H}}_{ni})^{*}\widetilde{\mathbf{H}}_{nj}\}}{\sqrt{E\{||\widetilde{\mathbf{H}}||_{F}^{2}\}}}. \tag{1.32}$$

We can define now the channel correlation $\mathcal{R}^{H}$ as the covariance matrix of the vectorization of the channel matrix $\widetilde{\mathbf{H}}$, denoted by $\mathbf{vec}(\widetilde{\mathbf{H}})$. The term $\mathcal{R}^{H}$ can then be computed as

$$\mathcal{R}^{H} = E\{\mathbf{vec}(\widetilde{\mathbf{H}})\,\mathbf{vec}(\widetilde{\mathbf{H}})^{H}\}. \tag{1.33}$$

If the Kronecker channel assumption about the separability of the transmitter and receiver is satisfied by the channel, then

$$\mathcal{R}^{H} = \mathbf{R}_{H}^{T} \otimes \mathbf{R}_{H}^{R}. \tag{1.34}$$

Using the Kronecker channel model for simulations, the first step is to estimate the transmit and receive spatial correlation matrices. Assuming an ideal NLOS propagation scenario, the entries of $\mathbf{R}_{H}^{T}$ and $\mathbf{R}_{H}^{R}$ can be computed, as shown in the next section, using Eq. (1.39). Finally, the channel matrix would be given by Eq. (1.28).

Notice that the Kronecker channel model has been widely used in the literature and several IEEE standards (such as IEEE 802.11n) [21]. When using the Kronecker model to estimate the channel capacity or the probability of error (either when the correlation matrices have been computed from simulations or measured data), each one of the generated channel realizations ($\widetilde{\mathbf{H}}^{p}$), which are given by

$$\widetilde{\mathbf{H}}^{p} = (\mathbf{R}_{H}^{R})^{\frac{1}{2}}\mathbf{W}^{p}((\mathbf{R}_{H}^{T})^{\frac{1}{2}})^{\dagger}, \tag{1.35}$$

must be normalized according to Eq. (1.22). We denote the normalized version of the channel realizations described by Eq. (1.35) as:

$$\widetilde{\mathbf{H}_{0}}^{p} = (\mathbf{R}_{H_{0}}^{R})^{\frac{1}{2}}\mathbf{W}^{p}(\mathbf{R}_{H_{0}}^{T})^{\frac{1}{2}}, \tag{1.36}$$

where $\mathbf{R}_{H_{0}}^{T} \in \mathcal{C}^{M \times M}$ and $\mathbf{R}_{H_{0}}^{R} \in \mathcal{C}^{N \times N}$ are the normalized versions of the transmit and receive correlation matrices, respectively, given by

$$\mathbf{R}_{H_{0}}^{R} = \mathbf{R}_{H}^{R}\sqrt{\frac{NMQ}{\sum_{q=1}^{Q}||\widetilde{\mathbf{H}}^{q}||_{F}^{2}}}, \tag{1.37}$$

$$\mathbf{R}_{H_0}^T = \mathbf{R}_H^T \sqrt{\frac{NMQ}{\Sigma_{q=1}^Q ||\widetilde{\mathbf{H}}^q||_F^2}}.$$ *(1.38)*

In this manner, the entries of $\mathbf{H}_0$ correspond to zero-mean–correlated Gaussian random variables such that $||\mathbf{H}_0||_F^2 = MN$.

On the other hand, the idea of the Weichselberger channel model was to relax the separability restriction of the Kronecker model and allow for any arbitrary coupling between the transmit and receive antennas (i.e., to model the correlation properties at the receiver and transmitter jointly). The Weichselberger model parameters are the eigenbasis of receive and transmit correlation matrices and a coupling matrix.

In contrast to the two prior models, the virtual channel representation models the MIMO channel in the beamspace instead of the eigenspace. In particular, the eigenvectors are replaced by fixed and predefined steering vectors. The model is fully specified by the coupling matrix. This model results in more accurate predictions than those provided by the Kronecker channel model [19] at the expense of less analytic tractability.

Note that for the remainder of the book, unless otherwise specified, we will assume $K \approx 0$, and therefore, $\mathbf{H} \approx \widetilde{\mathbf{H}}$.

1.7   PRELIMINARIES OF ANTENNA DESIGN IN MIMO COMMUNICATION SYSTEMS

The basic principle beyond MIMO antenna design is to reduce the correlation between received signals among the ports of an MPA and to simultaneously maximize the capacity and the transmission power gain of the system. As it will be shown, these three requirements are interrelated and can be satisfied by using various forms of what is known in the literature as antenna diversity or by means of combinations of matching and decoupling networks. In this section, we give a preliminary physical insight on the aforementioned design criteria.

Let us assume, for now, an ideal NLOS-rich scattering environment. In that case, the correlation coefficient among the ith and jth port of an MPA can be expressed as

$$\mathbf{R}_{H_{ij}}^A = \frac{\mathbf{X}_{H_{ij}}^A}{\sqrt{\mathbf{X}_{H_{ii}}^A \mathbf{X}_{H_{jj}}^A}},$$ *(1.39)*

with the covariance terms $\mathbf{X}_{H_{ij}}^A$ and $\mathbf{X}_{H_{ii}}^A$ given by

$$\mathbf{X}_{H_{ij}}^A = \frac{1}{8\pi} \int_{-\pi}^{\pi} \int_0^{\pi} (\mathbf{F}_i^A(\theta,\phi))^H \mathbf{F}_j^A(\theta,\phi) \sin(\theta) \mathrm{d}\theta \mathrm{d}\phi \tag{1.40}$$

$$= \frac{1}{8\pi} \int_{-\pi}^{\pi} \int_0^{\pi} \left((\mathbf{F}_{i,\theta}^A(\theta,\phi))^* \mathbf{F}_{j,\theta}^A(\theta,\phi) + (\mathbf{F}_{i,\phi}^A(\theta,\phi))^* \mathbf{F}_{j,\phi}^A(\theta,\phi) \right) \sin(\theta) \mathrm{d}\theta \mathrm{d}\phi$$

$$\mathbf{X}_{H_{ii}}^A = \frac{1}{8\pi} \int_{-\pi}^{\pi} \int_0^{\pi} \left(|\mathbf{F}_{i,\theta}^A(\theta,\phi)|^2 + |\mathbf{F}_{i,\phi}^A(\theta,\phi)|^2 \right) \sin(\theta) \mathrm{d}\theta \mathrm{d}\phi, \tag{1.41}$$

where $\mathbf{F}_{i,\theta}^A(\theta,\phi)$ and $\mathbf{F}_{i,\phi}^A(\theta,\phi)$ are the θ and ϕ components of the normalized far-field radiation pattern associated with the ith port of the MPA, respectively. Notice that the upper index A can either refer to the transmit (T) or receive (R) MPAs.

In general, it is difficult to show the direct relationship between the capacity distribution and the correlation properties, unless a Kronecker channel model [18] is assumed. A Kronecker channel model [18] can be assumed when the random fading processes at the receiver are uncorrelated with those at the transmitter, as happens in most NLOS scenarios. For the sake of providing a preliminary insight on how the correlation properties may impact capacity, we will assume this model for now. Using normalized channel matrices, $\mathbf{H}_0$ can be expressed as shown in Eq. (1.36). Finally, the capacity expression given in Eq. (1.13) can be rewritten explicitly including the correlation effects at the antennas, as follows [22]:

$$C = \log_2 \det \left(\mathbf{I} + \frac{\rho_0}{M} \mathbf{R}_{H_0}^R \mathbf{W} \mathbf{R}_{H_0}^T \mathbf{W}^H \right) \tag{1.42}$$

where we have used the fact that $\det(\mathbf{I} + \mathbf{AB}) = \det(\mathbf{I} + \mathbf{BA})$.

The ports of an MPA are said to be uncorrelated when $\mathbf{R}_H^T$ and $\mathbf{R}_H^R$ are equal to the identity matrix. Therefore, ideally, one would desire $\mathbf{R}_H^T$ and $\mathbf{R}_H^R$ to be the identity matrix, because in that case, Eq. (1.42) reduces to Eq. (1.21) with $\mathbf{H}_0 = \sqrt{\frac{NM}{\|\mathbf{W}\|_F^2}} \mathbf{W}$. As a result, the capacity of the MIMO communication system is not degraded because of the characteristics of the used MPAs. In other words, the MPAs become transparent to the system from a performance perspective. It is important to remark that the entries of the transmit and receive correlation matrices depend mainly on two factors: the richness of scattering within the channel and the coupling effects among the antennas. Therefore, designing a multiantenna system whose correlation matrix is equal to the identity matrix may not always be possible, especially if the propagation channel is poor in scattering or in keyhole scenarios [23].

The techniques that antenna designers use to optimize the performance of a MIMO communication system from an antenna perspective consist of finding mechanisms to orthogonalize the radiation patterns associated with the ports of the MPAs. For the sake of understanding, we refer to orthogonalization as that antenna design methodology that seeks to achieve $\mathbf{R}_H^T \to \mathbf{I}$ and $\mathbf{R}_H^R \to \mathbf{I}$.

In an ideal NLOS scattered environment, by transmitting and receiving with orthogonal radiation patterns, the propagated signals undergo reflections and diffraction among the scatterers of the channel in such a way that uncorrelated propagation subchannels can be produced. In the case of a more realistic NLOS scenario, this orthogonalization is performed, taking into account the characteristics of the propagation channel by weighting the transmit and receive array radiation patterns, by the intensity of the electromagnetic waves according to its departing or arriving angles. Interestingly, notice that this weighting is the same as the one done using the beam-forming (water-filling) technique described in Section 1.5.4, when the channel is known at the transmitter and receiver.

There exist two main mechanisms to decorrelate antennas: the first one is the use of antenna diversity techniques and the second one is the use of some external network, such as decorrelating networks. In the first case, decorrelation is achieved by engineering the shape of the antennas and their relative location, whereas in the latter case, it is achieved through external circuitry, which linearly combines the received signals to undo the electromagnetic coupling among the antennas.

Despite its lack of convenience, the orthogonality design criteria in the radiation patterns given by

$$\mathbf{R}_{H_{ij}}^A = 1 \qquad i = j \tag{1.43}$$

$$\mathbf{R}_{H_{ij}}^A = 0 \qquad i \neq j \tag{1.44}$$

can easily be transferred to the currents on the antenna by using the following expression given in Ref. [10]:

$$\mathbf{F}_i^A(\theta, \phi) \propto \widehat{r} \times \left[\widehat{r} \times \left[\int_{V'} \mathbf{J}_{V',i}^A(\mathbf{r}') e^{-jk\widehat{\mathbf{k}}\mathbf{r}'} dV' \right] \right], \tag{1.45}$$

in which the radiation patterns are related to the time-varying currents and where $\mathbf{J}_{V',i}^A(\mathbf{r}')$ is the volumetric current density phasor associated with the port i of the MPA, k is the wave number at the operating frequency, $\mathbf{r}'$ is the vector from the coordinate system origin to any point on the source, $\widehat{k}$ is a unit vector in the direction of propagation, and V' is the volume of the antenna containing the volumetric current densities. Notice that if the antenna is a planar antenna occupying an area S', then the current density becomes a surface current density $\mathbf{J}_{S',i}^A(\mathbf{r}')$. Therefore, orthogonality in the radiation patterns implies the following condition on the current:

$$\int_{V'} \left(\mathbf{J}_{V',i}^A(\mathbf{r}') \right)^\dagger \left(\mathbf{J}_{V',j}^A(\mathbf{r}') \right)^* dV' = 0 \qquad i \neq j. \tag{1.46}$$

Of course, in addition to the aforementioned design criteria, there are other important parameters that need to be taken into account when designing an antenna for MIMO systems, such as those typically considered in SISO systems: bandwidth, matching efficiency, transmission rate, channel coding, source coding, and so forth.

1.8   CHANNEL CODING IN MIMO COMMUNICATION SYSTEMS

Unlike with single-antenna systems, using MPAs at either the transmitter and/or the receiver makes it possible to combat detrimental propagation phenomena, such as multipath and others, through appropriate coding of the transmitted signals. There exist three main families of techniques to send information over a MIMO communication system: diversity, multiplexing, and beam forming. All three techniques take advantage of the degrees of freedom offered by using MPAs on a multipath propagation environment. Although they are more fully explained in Chapter 3, we provide here some introductory remarks:

- Diversity techniques, also referred as space–time coding techniques, are used to increase the signal strength of the transmitted symbols and therefore to improve the SNR. They are based on the principle of appropriately sending redundant symbols over the channel, from different antennas.
- Spatial multiplexing (SM) techniques are used to increase the transmission rate of a communication system. They consist of sending different symbols at the same time and using the same frequency but from different antennas.
- Beam-forming techniques are used to either increase the signal strength of the transmitted symbols or the transmission rate, when knowledge of the channel propagation environment is available at both the transmitter and the receiver.

Notice that diversity and multiplexing techniques do not strictly require knowledge of the characteristics of the propagation channel at the transmitter. The channel only needs to be known at the receiver for proper combination of the received signals.

To quantify how good one transmission technique is over another and how close they are to achieving capacity or a certain probability of signal strength, the following figures of merit are normally used: the diversity gain, the multiplexing gain, and the array gain. Diversity gain quantifies the gain on the received SNR as a result of using diversity techniques. Multiplexing gain gives us information on the increase in the transmission rate as a result of using multiplexing techniques, compared with a SISO system. Finally, array gain quantifies the improvement in SNR obtained by coherently combining the signals on multiple transmit or multiple receive antennas.

1.8.1 Diversity Techniques

Channel coding techniques that are used to increase the signal strength of the transmitted symbols in MIMO communication systems are called "diversity techniques." Despite this, the improvement on the SNR that one achieves using these techniques ultimately also has an impact on the capacity of the system. We can classify the existing diversity techniques into three categories:

1. temporal diversity
2. frequency diversity
3. antenna diversity.

Temporal diversity techniques can also be further subdivided into two additional types: time and code diversity. In particular, time diversity uses error-correcting coding and interleaving [24]. Frequency diversity techniques are based on frequency hopping and orthogonal frequency-division multiplexing (OFDM) modulations [25,26]. In both cases, some of these techniques have already been used for a long time in wireless communications, in particular for SISO communication systems, and now these techniques have also been adopted for MIMO systems.

The main idea behind antenna diversity techniques is to produce different replicas of the transmitted signal to the receiver. If these replicas are sent over the propagation channel such that their statistics are independent, when one of them fades, it is less likely that the other copies of the transmitted signal will be in deep fade simultaneously. Thanks to this redundancy, the receiver can decode the transmitted signal even in fading conditions, as long as they all do not

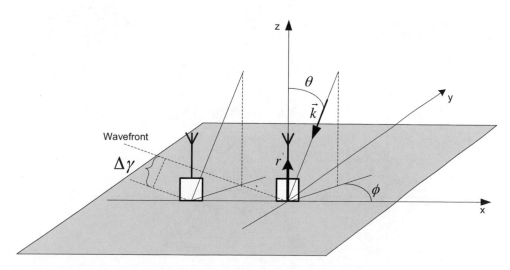

FIGURE 1.14: Graphical representation of the space diversity concept.

fade simultaneously. By inspection of Eq. (1.39), antenna diversity techniques can be classified according to which physical mechanism is used to produce uncorrelated signals at the ports of an MPA. One can identify three kinds of antenna diversity techniques, which, however, are inherently interrelated:

- space diversity
- polarization diversity
- pattern diversity.

Using space diversity techniques [26], equal antennas are spaced apart, so that the magnitude of the radiation pattern at each port is essentially the same, but the phase patterns relative to a common coordinate system are such that the resultant radiation patterns, among different ports of the MPA, become orthogonal. As a result, the received signals are statistically uncorrelated. Let us illustrate this concept with a simple graphical example. Imagine the situation given in Fig. 1.14, where two antennas are placed next to each other at a certain distance. For a plane wave arriving from direction **k**, the received signals at the two antennas will have approximately the same amplitude but different phase, because the wavefront is not parallel to the x-axis and the antennas are not in the same exact position. Such phase difference, $\Delta\gamma$, is in fact the parameter that one needs to optimize

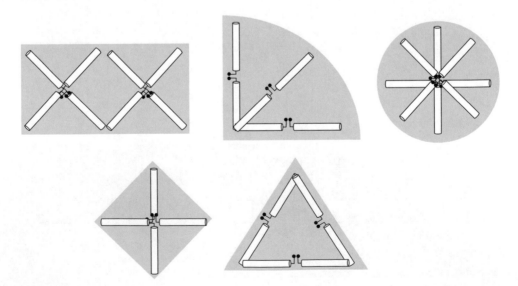

FIGURE 1.15: Different MPOA configuration using polarization diversity techniques.

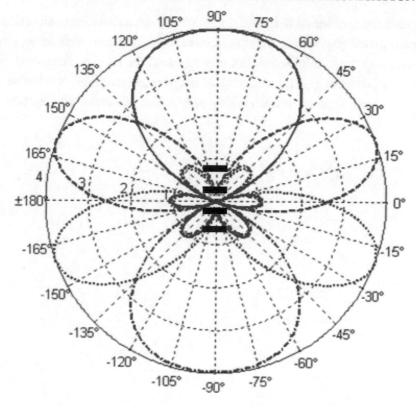

FIGURE 1.16: Possible radiation patterns of an MMA employing pattern diversity techniques. Notice that each beam is spatially (angularly) disjoint from the others.

such that the chances of having a fade in both antennas, in multipath channels, is minimized. The second approach is polarization diversity [5], in which antennas radiate energy using orthogonal polarizations. In this case, the integrand in Eq. (1.39) vanishes because the pattern scalar product is zero, in each direction of the space, due to the vectorial orthogonal characteristic of the fields radiated from the antennas, as a result of using two orthogonal polarizations. Finally, it is also possible to impose orthogonality by producing spatially disjoint radiation patterns. This is done by shaping the radiation patterns associated with the different ports of an MPA, such that the integrand in Eq. (1.39) tends to vanish. This is referred in the literature as pattern diversity [6]. To generate pattern diversity, one can use MMAs, in which case the antennas do not necessarily need to be separated by a physical distance; alternatively, one can use external beam-forming networks, such as a Butler matrix [13]. The concepts of polarization and pattern diversity are graphically represented in Figs. 1.15 and 1.16, respectively.

This classification is rather arbitrary and, for most cases, a combination of all these sources are present on antenna diversity systems. Using the aforementioned diversity techniques, multiple and uncorrelated versions of the transmitted signals may appear at the receiver. Afterward, to improve the performance of a particular MIMO communication system, those signals need to be combined. This combination is normally done through coding or by means of microwave networks.

. . . .

CHAPTER 2

The Radio Channel for MIMO Communication Systems

2.1 INTRODUCTION

The (wireless) radio channel is the medium linking the transmitter and the receiver. We will start with a physical presentation of the channel and follow with its statistical description; finally, we will give a comprehensive characterization of the more significant channel models and parameters.

One of the basic requirements for accurate comprehension of a multiantenna communication system is the proper understanding of the radio channel. The propagation channel normally corresponds to a multipath environment that needs to be appropriately modeled. The way in which the waves interact and propagate through the existing environment between transmit and receive antennas will determine the time, frequency, and space parameters of the radio channel. A proper understanding of the physical mechanisms behind the propagation phenomena will give a clearer insight into the whole system.

2.2 RADIO CHANNEL PROPAGATION: DETERMINISTIC PHYSICAL DESCRIPTION

In general, the radio propagation channel (see Fig. 2.1) can be described as a linear time-variant (LTV) system. This allows a physical wideband description of the channel by the time-variant channel impulse response (CIR), $h(t, \tau)$, in its base band representation [27]. Here, t is used to describe the channel time variation, whereas τ is the delay time parameter. $h(t, \tau)$ describes the relation of the input signal $s(t)$ at the transmit antenna to the output signal $r(t)$ at the receive antenna. These three quantities are related as follow:

$$r(t) = \int_{-\infty}^{\infty} h(t, \tau)s(t - \tau)\mathrm{d}\tau + n(t), \qquad (2.1)$$

where $n(t)$ accounts for the noise and interfering signals.

For the case of a band-pass radio channel (with bandwidth B), the spectral content of the signal and the corresponding channel frequency band are small compared with any frequency in the

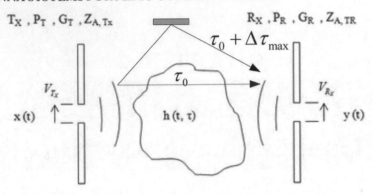

FIGURE 2.1: General geometry of the radio communication channel.

band (e.g., compared with the carrier frequency, f_c). Then $h(t, \tau)$ can be unambiguously described with the radio frequency (RF) band-pass representation $h_{BP}(t, \tau)$ or with its equivalent base band representation $h_{BB}(t, \tau)$, which are related [28] as

$$h_{BB}(t, \tau) = \Re \left\{ h_{BP}(t, \tau) e^{j\omega_c t} \right\}, \tag{2.2}$$

where $\omega_c = 2\pi f_c$. The term $h_{BP}(t, \tau)$ is then the complex envelope of $h_{BB}(t, \tau)$. For the following, the complex envelope $h_{BP}(t, \tau)$ will be used and noted simply as $h(t, \tau)$. A detailed knowledge and characterization of $h(t, \tau)$, which provides the basis for modeling the radio channel, is therefore desirable.

2.2.1   Channel Bandwidth

One of the basic concepts for understanding and modeling a radio channel is the relation between the system bandwidth B (normally limited by the receiver) and the maximum observable delay τ_{max}. The value of B leads to a time resolution of the system on the order of $1/B$. In a multipath environment, the signals will arrive at the receiver after interacting with the intermediate objects lying in the propagation pathway. As a result, the signal $r(t)$ will not be merely a delayed replica of the transmitted signal $s(t)$, but a combination of replicas with different amplitudes, phases, and delays filtered by the receiver, with bandwidth B. Because of the finite system bandwidth, all the signals arriving within a time length of B^{-1} will be time-integrated at the receiver. On the other hand, depending on the structure of the environment, the maximum time extension of the delays ($\Delta\tau_{max}$) that can be physically "observed" by the receiver (e.g., the above noise level) will vary, being in general higher for open areas with few blocking obstacles where the receiver can pick up waves from distant scatterers. By denoting τ^0 the delay time that a wave needs to travel the direct distance d_0 between transmitter and receiver, we can then state that all the pathways have to relay within a volume limited by a traveling time between τ^0 and $\tau^0 + \Delta\tau_{max}$. Converting this time length into distance length by multiplying it with the wave velocity, $(d_{max} = c\Delta\tau_{max})$, $\Delta\tau_{max}$ defines an imaginary maximum volume (in free space, it may be approximated by an ellipse with the transmitter and receiver as the foci and $c\Delta\tau_{max}$ as the width) and only contributions from within this volume need to be accounted for. On the other hand, B^{-1} defines the temporal window for which all contributions arriving inside this temporal period will be integrated by the system into a single time value. If we convert this time length B^{-1} into space length $d_s = cB^{-1}$, we can also visualize this parameter as the dimension of an obstacle zone producing scattered signals that are jointly integrated at the receiver. Thus, all scattering contributions within an elliptical ring of width cB^{-1} at a fixed delay are collected and integrated together (with weighting by the filter function). We can then imagine the electromagnetic acting volume as an ellipse of width $c\Delta\tau_{max}$ divided into elliptical rings of width cB^{-1} defining the different time sampling intervals. The ratio $B\Delta\tau_{max}$ thus plays an important role in determining the channel properties. If $\Delta\tau_{max} \leq B^{-1}$ ($\Delta\tau_{max}B \leq 1$ narrowband system), all scattering contributions are collected in one smeared peak. In this case, the channel is said not to be frequency-selective. In the case of broadband systems, $\Delta\tau_{max}B \gg 1$, and detailed scattering mechanisms are identified as the impulse response leading to frequency selectivity. This ratio is also very significant for digital communications. Because the symbol time length can be approached by B^{-1}, having $\Delta\tau_{max} \leq B^{-1}$ represents that all the replicas of a certain transmitted symbol will arrive at the receiver within the time duration of that pulse, thus not producing contamination of other symbols. Instead, $\Delta\tau_{max} \gg B^{-1}$ will mean that replicas of the original transmitted symbol will be arriving at the receiver where subsequent symbols have already been sent, thus producing intersymbol interference and increasing the bit error rate (BER) of the channel.

2.2.2 Space–Time CIR

To take into account physical mechanisms of the radio channel such as attenuation, vectorial character of the fields, directional behavior of antennas, and the terrain geometry or time variation due to the mobility of antennas or scatterers, the complex propagation channel may be described by its CIR:

$$h(t,\tau) = K_{TR} \int \int (\mathbf{F}^R(t,\Omega_R))^\dagger \cdot \mathbf{h}_{ST}(t,\tau,\Omega_T,\Omega_R) \cdot \mathbf{F}^T(t,\Omega_T)\mathrm{d}\Omega_T\mathrm{d}\Omega_R \qquad (2.3)$$

or, equivalently, by its transfer function,

$$H(t,f) = K_{TR}(f) \int \int (\mathbf{F}^R(t,f,\Omega_R))^\dagger \cdot \mathbf{H}_{ST}(t,f,\Omega_T,\Omega_R) \cdot \mathbf{F}^T(t,f,\Omega_T)\mathrm{d}\Omega_T\mathrm{d}\Omega_R, \qquad (2.4)$$

where K_{TR} is a gain factor based on antenna impedances Z^T and Z^R, $\mathbf{F}^T(t,\Omega_T) \in C^{2\times 1}$ and $\mathbf{F}^R(t,\Omega_R) \in C^{2\times 1}$ are the transmit and receive normalized far-field radiation patterns, respectively, and $\mathbf{h}_{ST} \in C^{2\times 2}$ and $\mathbf{H}_{ST} \in C^{2\times 2}$ are the polarimetric matrix representation of the space–time channel impulse response and channel transfer function, respectively. Eqs. (2.3) and (2.4) describe all the spatial, temporal, directional, and polarimetric characteristics of the wave propagation within the channel, in the time-delay domain and in the time-frequency domain, respectively. Notice that the impulse response $h(t,\tau)$ may be related to its frequency expression, that is, its transfer function, $H(t,f)$, as follows:

$$h(t,\tau) = \int_{-\infty}^{\infty} H(t,f) e^{j2\pi f\tau} df, \qquad (2.5)$$

$$H(t,f) = \int_{-\infty}^{\infty} h(t,\tau) e^{-j2\pi f\tau} d\tau. \qquad (2.6)$$

In the case of multipath propagation scenarios, a good channel modeling approach consists of considering a certain number of discrete multipath components (MPCs). The modeling is similar to the ray-optical approach, where the first interaction point of the radio wave emerging from the transmitter with the closest obstacle determines the direction of departure (DoD) at the transmitter and the last interaction point of the radio wave arriving at the receiver from the closest obstacle determines the direction of arrival (DoA) at the receiver. MPCs can be characterized as single paths or in some cases as groups of paths with similar DoD or DoA and path lengths (usually specified as time delay of arrival τ). These groups of paths with similar characteristics may arise from the so-called clusters of scatterers [29]. Based on this idea, the space–time impulse response, $h(t,\tau)$, may be approached by a combination of a certain number of time-variant components $P(t)$ with specific delays $\tau^p(t)$ (in particular, when cluster-dispersive scatters are considered), DoD $\Omega_T^p(t)$ and DoA $\Omega_R^p(t)$, and distance propagation factors $\Gamma^p(t)$, as follows:

$$\mathbf{h}_{ST}(t,\tau,\Omega_T,\Omega_R) = \sum_{p=0}^{P(t)} \Gamma^p(t) u_\tau(\tau - \tau^p(t)) \mathbf{C}^p \delta(\Omega_T - \Omega_T^p(t)) \delta(\Omega_R - \Omega_R^p(t)), \qquad (2.7)$$

where $u_\tau(\tau)$ is the time dispersion function associated with the MPCs of the channel, and $\mathbf{C}^p$ is a normalized channel polarimetric matrix, describing the interactions of the waves with the object of the channel for the two polarizations, given by Ref. [30]. In the case of non-frequency-selective multipath channels, $u_\tau(\tau)$ becomes the delta function $\delta(\tau)$. Normally, $p = 0$ will denote the free-space propagation path, with propagation factor $\Gamma^0(t)$, which only exists for line-of-sight (LOS). In a pure non-line-of-sight (NLOS) propagation scenario $\Gamma^0(t) = 0$, meaning that there is no LOS MPC. To get a manageable model without losing much generality, it is often accepted that all the path components experience a similar dispersion $u_\tau(\tau)$. In the following the space–time channel, impulse response and transference function will be investigated for different propagation scenarios applicable to multiantenna systems.

2.3   THE IDEAL FREE-SPACE PROPAGATION CASE

To obtain a better understanding of all the electric parameters of a radio channel, the free-space case will be first studied. Later, different scenarios with increased complexity will be introduced, from the basic single-input single-output (SISO) communication link to the different MIMO communication architectures operating on complex wideband multipath scenarios.

2.3.1   Free-Space SISO

Consider the situation of Fig. 2.1, where one transmit and one receive antenna are communicating in a free-space environment, in a direct LOS. Let us express the electrical field radiated by the transmitter as [31]:

$$\mathbf{E}(\mathbf{r}) = -jk\eta_0 \frac{e^{-jkr}}{4\pi r}\mathbf{l}_{\text{ef}}^T(\theta_T^0,\phi_T^0)I_T = -jk\eta_0\frac{e^{-jkr}}{4\pi r}l_{\text{ef}}^T\mathbf{F}^{T,0}(\theta_T^0,\phi_T^0)I_T, \qquad (2.8)$$

where $\mathbf{r}(r,\theta,\phi)$ is the position at which the electric field is calculated, I_T is the electrical current feeding the transmit antenna, $\mathbf{l}_{\text{ef}}^T(\theta,\phi) \in \mathcal{C}^{2\times1}$ is the directional effective transmission length, and $\mathbf{F}^{T,0}(\Omega_T^0)$ is the normalized unit-gain far-field radiation pattern. The output voltage at the receiving antenna can be expressed as

$$V_R = -\frac{1}{I_T}\int_R (\mathbf{J}_{V'}(\mathbf{r}'))^\dagger \mathbf{E}(\mathbf{r}')\mathrm{d}V' = -(\mathbf{l}_{\text{ef}}^R(\theta,\phi))^\dagger\mathbf{E}(\mathbf{r}'), \qquad (2.9)$$

where $\mathbf{l}_{\text{ef}}^R(\theta,\phi) \in \mathcal{C}^{2\times1}$ is the directional effective length of the receive antenna. Substituting Eq. (2.8) in (2.9) and making explicit the time variation:

$$V_R(t) = jk\eta_0\frac{e^{-jkr}}{4\pi r}(\mathbf{l}_{\text{ef}}^R(\Omega_R^0))^\dagger\mathbf{l}_{\text{ef}}^T(\Omega_T^0)I_T(t) = \qquad (2.10)$$

$$= jk\eta_0\frac{e^{-jkr}}{4\pi r}l_{\text{ef}}^T l_{\text{ef}}^R(\mathbf{F}^R(\Omega_R^0))^\dagger\mathbf{F}^{T,0}(\Omega_T^0)\frac{V_T(t)}{Z^T}, \qquad (2.11)$$

where $\Omega_T^0 = (\theta_T^0,\phi_T^0)$ are the elevation and azimuth transmitting angles, $\Omega_R^0 = (\theta_R^0,\phi_R^0)$ are the elevation and azimuth receiving angles, and $r = |\mathbf{r}_R' - \mathbf{r}_T'|$ is the distance between the transmitter and the receiver. Notice that $\mathbf{r}_T'$ and $\mathbf{r}_R'$ are vectors from the origin to the location of the transmit and receive antennas, respectively. We can express the channel transfer function t as the ratio of the voltages at the receive antenna V_R to the voltages at the transmit antenna V_T as

$$H = S_{RT} = \frac{V_R^-}{V_T^+}\Big|_{V_R^+=0} = \frac{V_R}{V_T}\Big|_{a_R=0} = \frac{Z_{RT}^A}{2Z^R} = -\frac{1}{2Z_0 I_T I_R}\int_R(\mathbf{J}_{V'}(\mathbf{r}'))^\dagger\mathbf{E}(\mathbf{r}')\mathrm{d}V', \qquad (2.12)$$

where Z^A_{RT} is the mutual impedance between the transmitter and the receiver and S_{RT} is the transmission coefficient and where we have assumed that the antennas are matched to the source and drain loads.

Expressing the transfer ratio between the power at the receiver, P^D, and the power at the transmitter, P^S, into two alternative ways—in terms of the mutual impedance as in Eq. (2.12) and in terms of the free-space Frii's equation—we find for the case of perfect matching:

$$\frac{P^D}{P^S} = \frac{|Z^A_{RT}|^2}{4R^T R^R} = G^T G^R (\overline{\Gamma}(r))^2 = G^T G^R \left(\frac{\lambda}{4\pi}\right)^2 \left(\frac{1}{r}\right)^{n_{PL}}, \qquad (2.13)$$

where $R^T = \Re(Z^T)$ and $R^R = \Re(Z^R)$ are the real part of Z^T and Z^R, respectively, and $\overline{\Gamma}(r) = \left(\frac{\lambda}{4\pi}\right)\left(\frac{1}{r}\right)^{\frac{n_{PL}}{2}}$ is called the propagation loss (PL) factor, with n_{PL} equal to 2 for free-space. We may rewrite the transfer function given in Eq. (2.12) as

$$H = H(\mathbf{r}'_T, \mathbf{r}'_R) = \sqrt{G^T G^R} \frac{\lambda}{4\pi|\mathbf{r}'_R - \mathbf{r}'_T|} (\mathbf{F}^{R,0}(\Omega^0_R))^\dagger \mathbf{F}^{T,0}(\Omega^0_T) e^{-jk|\mathbf{r}'_R - \mathbf{r}'_T|}, \qquad (2.14)$$

where $\mathbf{F}^{A,0}(\Omega^0_A)$ is the normalized unit-gain far-field radiation pattern (with $A = T, R$). Equivalently,

$$H = H(\mathbf{r}'_T, \mathbf{r}'_R) = \frac{\lambda}{|\mathbf{r}'_R - \mathbf{r}'_T|} (\mathbf{F}^R(\Omega^0_R))^\dagger \mathbf{F}^T(\Omega^0_T) e^{-jk|\mathbf{r}'_R - \mathbf{r}'_T|}, \qquad (2.15)$$

where $\mathbf{F}^A(\Omega^0_A)$ is the normalized far-field radiation pattern of an antenna as defined in Section 4.2.3. Including the effects of the characteristic reference impedances ($Z^{0,T}$ and $Z^{0,R}$) and antenna impedances (Z^T and Z^R), Eq. (2.15) becomes [15]:

$$H = H(\mathbf{r}'_T, \mathbf{r}'_R) = K_{RT}\Gamma(|\mathbf{r}'_R - \mathbf{r}'_T|)(\mathbf{F}^R(\Omega^0_R))^\dagger \mathbf{F}^T(\Omega^0_T) e^{-jkr}, \qquad (2.16)$$

where $\Gamma(r) = \lambda \left(\frac{1}{r}\right)^{\frac{n_{PL}}{2}}$, and the term K_{RT} is given by

$$K_{RT} = \sqrt{\frac{Z^{0,R}}{Z^{0,T}}} \sqrt{\frac{\Re\{Z^T\}}{\Re\{Z^R\}}}. \qquad (2.17)$$

Using the fact that $\mathbf{k}r = kr = k|\mathbf{r}'_R - \mathbf{r}'_T|$, the free-space time delay may be written as $\tau^0 = \frac{(|\mathbf{r}'_T - \mathbf{r}'_R|)}{c} = \frac{r}{c}$, and finally, we can rewrite the transfer function as

$$H = H(\mathbf{r}'_T, \mathbf{r}'_R) = K_{RT}\Gamma(|\mathbf{r}'_R - \mathbf{r}'_T|)(\mathbf{F}^R(\Omega^0_R))^\dagger \mathbf{F}^T(\Omega^0_T) e^{-j2\pi f\tau^0}. \qquad (2.18)$$

The space–time transfer function response can be expressed as

$$\begin{aligned} \mathbf{H}_{ST} &= \mathbf{H}_{ST}(\mathbf{r}'_T, \mathbf{r}'_R, \Omega_T, \Omega_R) \\ &= K_{RT}\Gamma(|\mathbf{r}'_R - \mathbf{r}'_T|)\mathbf{I}_2\delta(\Omega_T - \Omega^0_T)\delta(\Omega_R - \Omega^0_R)e^{-j2\pi f\tau^0}, \end{aligned} \qquad (2.19)$$

where $\mathbf{I}_2 \in \mathcal{C}^{2\times2}$. Notice that H can be computed by substitution of $\mathbf{H}_{ST}$ into the expression given by Eq. (2.4). The CIR is given by

$$h = h(\mathbf{r}'_T, \mathbf{r}'_R) = K_{RT}\Gamma(|\mathbf{r}'_R - \mathbf{r}'_T|)(\mathbf{F}^R(\Omega_R^0))^\dagger\mathbf{F}^T(\Omega_T^0)\delta_\tau(\tau - \tau^0). \tag{2.20}$$

The space–time CIR is given by

$$\begin{aligned}\mathbf{h}_{ST} &= \mathbf{h}_{ST}(\mathbf{r}'_T, \mathbf{r}'_R, \Omega_T, \Omega_R) \\ &= K_{RT}\Gamma(|\mathbf{r}'_R - \mathbf{r}'_T|)\mathbf{I}_2\delta(\Omega_T - \Omega_T^0)\delta(\Omega_R - \Omega_R^0)\delta_\tau(\tau - \tau^0),\end{aligned} \tag{2.21}$$

where $\Omega_T = (\theta_T, \phi_T)$ are the elevation and azimuth transmitting angles, $\Omega_R = (\theta_R, \phi_R)$ are the elevation and azimuth receiving angles and $\mathbf{I}_2 \in \mathcal{C}^{2\times2}$. Similarly, notice that h can be computed by substitution of $\mathbf{h}_{ST}$ into the expression given by Eq. (2.3). Without loss of generality, for the following, τ^0 is assumed to be zero, $\tau^0 = 0$. Then, all the other delays corresponding to additional MPCs, τ^p no longer represent the absolute delay of the MPC, but rather the excess delay of a particular path relative to the free-space LOS delay τ^0. For this reason, h has no dependence in τ in Eq. (2.20).

2.3.2   Free-Space MIMO

Consider now the geometry of Fig. 2.2 with M transmit antennas located at the following positions $\mathbf{r}'_{T_1}, \mathbf{r}'_{T_2}, \cdots, \mathbf{r}'_{T_M}$ and N receive antennas located at positions $\mathbf{r}'_{R_1}, \mathbf{r}'_{R_2}, \cdots, \mathbf{r}'_{R_N}$. Being $\mathbf{S}_i(t,f) = \mathbf{S}_i(\mathbf{r}'_{T_i}, t, f)$, a band-limited signal transmitted by the antenna T_i, the noiseless signal received at the antenna R_j, $\mathbf{R}_j(t,f) = \mathbf{R}_j(\mathbf{r}'_{R_j}, t, f)$, can be expressed as

$$\mathbf{R}_j(\mathbf{r}'_{R_j}, t, f) = \sum_{i=1}^{M}\mathbf{H}_{ji}(\mathbf{r}'_{T_i}, \mathbf{r}'_{R_j})\mathbf{S}_i(\mathbf{r}'_{T_i}, t, f), \tag{2.22}$$

where the transfer function $\mathbf{H}_{ji}(\mathbf{r}'_{T_i}, \mathbf{r}'_{R_j})$ is given by

$$\mathbf{H}_{ji} = \mathbf{H}_{ji}(\mathbf{r}'_{T_i}, \mathbf{r}'_{R_j}) = K_{RT_{ji}}\Gamma(|\mathbf{r}'_{R_j} - \mathbf{r}'_{T_i}|)(\mathbf{F}_j^R(\Omega_{R_{ji}}))^\dagger\mathbf{F}_i^T(\Omega_{T_{ji}})e^{-jk_{ji}(\mathbf{r}'_{R_j} - \mathbf{r}'_{T_i})}, \tag{2.23}$$

where $\Omega_{T_{ji}} = (\theta_{T_{ji}}, \phi_{T_{ji}})$ is the departure angle from the transmit antenna T_i to the receive antenna R_j (relative to the coordinate system of the transmit antenna), $\Omega_{R_{ji}} = (\theta_{R_{ji}}, \phi_{R_{ji}})$ is the arriving angle from the transmit antenna T_i to the receive antenna R_j (relative to the coordinate system of the receive antenna) and where the term $K_{RT_{ji}}$ can be written as

$$K_{RT_{ji}} = \sqrt{\frac{Z_j^{0,R}}{Z_i^{0,T}}}\sqrt{\frac{\Re\{Z_i^T\}}{\Re\{Z_j^R\}}}, \tag{2.24}$$

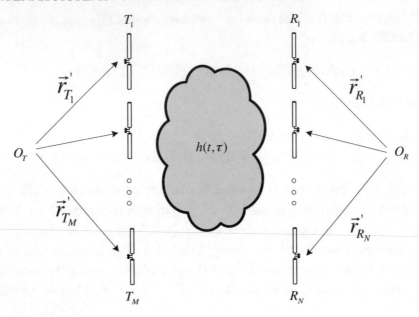

FIGURE 2.2: Geometry of the multiantenna scenario.

with characteristic reference impedance given by $Z_i^{0,T}$ and $Z_j^{0,R}$ and antenna impedances given by Z_i^T and Z_i^T. Notice that $\mathbf{r}_{ji} = (\mathbf{r}'_{R_j} - \mathbf{r}'_{T_i})$ and $\mathbf{k}_{ji}$ are the vector distance and the plane wave vector between transmit antenna T_i and receive antenna R_j, respectively. If we define $r_{ji} = |(\mathbf{r}'_{R_j} - \mathbf{r}'_{T_i})|$, the transfer function can be simply expressed as

$$\mathbf{H}_{ji} = \mathbf{H}_{ji}(\mathbf{r}'_{T_i}, \mathbf{r}'_{R_j}) = K_{RT_{ji}}\Gamma(r_{ji})(\mathbf{F}_j^R(\Omega_{R_{ji}}))^{\dagger}\mathbf{F}_i^T(\Omega_{T_{ji}})e^{-j\mathbf{k}_{ji}\mathbf{r}_{ji}}. \qquad (2.25)$$

Now, if we define $\tau_{ji}^0 = \frac{(|\mathbf{r}'_{T_i} - \mathbf{r}'_{R_j}|)}{c} = \frac{r_{ji}}{c}$ as the free-space time delay from transmit antenna T_i to receive antenna R_j, then the transfer function is equivalent to

$$\mathbf{H}_{ji} = \mathbf{H}_{ji}(\mathbf{r}'_{T_i}, \mathbf{r}'_{R_j}) = K_{RT_{ji}}\Gamma(r_{ji})(\mathbf{F}_j^R(\Omega_{R_{ji}}))^{\dagger}\mathbf{F}_i^T(\Omega_{T_{ji}})e^{-j2\pi f\tau_{ji}^0}. \qquad (2.26)$$

Notice that $\mathbf{H} \in \mathcal{C}^{N \times M}$ is a frequency flat matrix transfer function with the (j,i)th component denoted by $\mathbf{H}_{ji}$. The CIR is given by

$$\mathbf{h}_{ji} = \mathbf{h}_{ji}(\mathbf{r}'_{T_i}, \mathbf{r}'_{R_j}) = K_{RT_{ji}}\Gamma(r_{ji})(\mathbf{F}_j^R(\Omega_{R_{ji}}))^{\dagger}\mathbf{F}_i^T(\Omega_{T_{ji}})\delta_{\tau}(\tau - \tau_{ji}^0). \qquad (2.27)$$

In the time domain, we can express the input–output relationship of the system as

$$\mathbf{r}(t) = \mathbf{h}s(t), \qquad (2.28)$$

where $\mathbf{r}(t) \in \mathcal{C}^{N \times 1}$ is a vector for the voltages received at the N receive antennas, $\mathbf{s}(t) \in \mathcal{C}^{M \times 1}$ is a vector for the voltages applied to the M transmit antennas and $\mathbf{h} \in \mathcal{C}^{N \times M}$ is a channel matrix impulse response with the (j,i)th component denoted by $\mathbf{h}_{ji}$. In the frequency domain, we can write

$$\mathbf{R}(t,f) = \mathbf{H}\mathbf{S}(t,f). \tag{2.29}$$

The above represents a narrowband linear time-invariant (LTI) multiantenna bidirectional polarimetric free space radio channel.

2.4   THE MULTIPATH PROPAGATION CASE

When the transmit and receive antennas are not in free space and when scattering objects lie along the pathway between them, the supposition of a single ray is no longer correct, and additional paths (multipath behavior) have to be considered. That is, in this case, paths different from the direct LOS path between transmitter and receiver will appear with attenuations and delays depending on the particular geometry of the propagation channel. The resulting field will be the coherent combination of the different paths resulting in a PL exponent (n_{PL}) potentially quite different from the free-space exponent $n_{PL} = 2$. One simple case of a multipath scenario is the propagation above a flat surface, as in the case of propagation over a ground plane, which can be approached by a two-path model. In this case, the PL factor for long distances and grazing angles results in $n_{PL} = 4$. Unless specified, for the following, we will assume that there is not an LOS path component.

2.4.1   Multipath SISO

Assume a multipath environment with P significant paths, where each MPC has a departing angle from the transmit antenna $\Omega_T^p = (\theta_T^p, \phi_T^p)$, an arriving angle to the receive antenna $\Omega_R^p = (\theta_R^p, \phi_R^p)$, and a total spatial length $r_T^p + r_R^p$. In the frequency domain, the noiseless system input–output relationship is now given by

$$R(\mathbf{r'}_R, t, f) = H(\mathbf{r'}_T, \mathbf{r'}_R, f) S(\mathbf{r'}_T, t, f), \tag{2.30}$$

where the transfer function $H(\mathbf{r'}_T, \mathbf{r'}_R, f)$ is given by

$$\begin{aligned}
H(f) &= H(\mathbf{r'}_T, \mathbf{r'}_R, f) \\
&= \sum_{p=1}^{P} K_{RT} \Gamma(r_T^p + r_R^p) (\mathbf{F}^R(\Omega_R^p))^\dagger \mathbf{C}_{TR}^p \mathbf{F}^T(\Omega_T^p) e^{-j\left\{\mathbf{k}_T^p \mathbf{r}_T^p\right\}} e^{-j\left\{\mathbf{k}_R^p \mathbf{r}_R^p\right\}}
\end{aligned} \tag{2.31}$$

and $\mathbf{C}^p_{TR}$ is the normalized channel polarimetric matrix, describing the interactions of the waves with the object of the channel for the two polarizations, given by [30]:

$$\mathbf{C}^p_{TR} = \begin{pmatrix} \mathbf{C}^p_{TR_{\theta\theta}} & \mathbf{C}^p_{TR_{\theta\phi}} \\ \mathbf{C}^p_{TR_{\phi\theta}} & \mathbf{C}^p_{TR_{\phi\phi}} \end{pmatrix}. \tag{2.32}$$

On the other hand, $\mathbf{k}^p_T$ and $\mathbf{k}^p_R$ are the plane wave direction vectors, and $\mathbf{r}^p_T$ and $\mathbf{r}^p_R$ are the distance vectors, associated with the pth path from the transmit antenna to the scatter and from the scatter to the receive antenna, respectively. We can rewrite the exponential term from Eq. (2.31) in a more appropriate way [32] to see its frequency dependence:

$$e^{-j\left\{\mathbf{k}^p_T \mathbf{r}^p_T\right\}} e^{-j\left\{\mathbf{k}^p_R \mathbf{r}^p_R\right\}} = e^{-j\left\{\frac{2\pi f}{c}\widehat{\mathbf{k}}^p_T \mathbf{r}^p_T + \frac{2\pi f}{c}\widehat{\mathbf{k}}^p_R \mathbf{r}^p_R\right\}} = e^{-j2\pi f(\tau^p_T + \tau^p_R)} = e^{-j2\pi f\tau^p}, \tag{2.33}$$

and finally we can write the transfer function as

$$H(f) = H(\mathbf{r'}_T, \mathbf{r'}_R, f) = \sum_{p=1}^{P} K_{RT} \Gamma(r^p_T + r^p_R)(\mathbf{F}^R(\Omega^p_R))^\dagger \mathbf{C}^p_{TR} \mathbf{F}^T(\Omega^p_T) e^{-j2\pi f\tau^p}. \tag{2.34}$$

For the sake of better understanding, notice that in the above expressions, we have assumed that the propagating signals only interact with objects in the channel once, and that there is no LOS component. Let us assume now that between two arbitrary paths the frequency changes from $\omega_c - B/2$ to $\omega_c + B/2$ and the delay difference moves from 0 to $\Delta\tau_{\max}$, producing phase differences of $\Delta\tau_{\max}B$ (assuming $\widehat{\mathbf{k}}^p_T\widehat{\mathbf{r}}^p_T \approx 1$). When $\Delta\tau_{\max}B < 1$, those phases are negligible and may be considered frequency-independent (narrowband approach). In the time domain, all the paths arrive at the receiver within a time spread that can be approached by a single delta function and equation Eq. (2.28) still holds. However, when $\Delta\tau_{\max}B \gg 1$ (wideband system), the factor $2\pi f\tau^p = \omega\tau^p$ has to be taken into account, appearing as a frequency dependence on the transfer function (frequency selective channel):

$$R(f,t) = H(f)S(f,t). \tag{2.35}$$

The frequency bandwidth boundary between the two approaches (narrow and wide approaches) is usually stated in terms of the coherence bandwidth of the channel B_c, defined as

$$B_c \approx \frac{1}{\tau_{\max}}. \tag{2.36}$$

The wideband character of the impulse response means that the time delays produced by the different paths are significantly different from one another and much higher than the signal pulse duration. This idea may be expressed with the well-known form of a tapped delay line model:

$$r(t) = \int_{-\infty}^{t} h(\tau)s(t-\tau)d\tau. \tag{2.37}$$

Making explicit the dependence on the location of the transmit and receive antennas, we obtain

$$r(\mathbf{r}'_R, t) = \int_{-\infty}^{t} h(\tau, \mathbf{r}'_T, \mathbf{r}'_R)s(\mathbf{r}'_T, t-\tau)d\tau, \tag{2.38}$$

where the term $h(\tau, \mathbf{r}'_T, \mathbf{r}'_R)$ can be expressed as

$$h(\tau) = h(\tau, \mathbf{r}'_T, \mathbf{r}'_R) = \sum_{p=1}^{P} A^p(\mathbf{r}_T^p, \mathbf{r}_R^p)\delta_\tau(\tau - \tau^p(\mathbf{r}_T^p, \mathbf{r}_R^p)) \tag{2.39}$$

with $A^p(\mathbf{r}_T^p, \mathbf{r}_R^p)$ given by

$$A^p = A^p(\mathbf{r}_T^p, \mathbf{r}_R^p) = K_{RT}\Gamma(r_T^p + r_R^p)(\mathbf{F}^R(\Omega_R^p))^\dagger \mathbf{C}_{TR}^p \mathbf{F}^T(\Omega_T^p) \tag{2.40}$$

and where the time delay $\tau^p(\mathbf{r}_T^p, \mathbf{r}_R^p)$ can be written as

$$\tau^p = \tau^p(\mathbf{r}_T^p, \mathbf{r}_R^p) = \tau_T^p + \tau_R^p = \frac{\mathbf{k}_T^p \mathbf{r}_T^p}{2\pi f} + \frac{\mathbf{k}_R^p \mathbf{r}_R^p}{2\pi f}. \tag{2.41}$$

To calculate the different path losses between transmitter and receiver, the influence of the different obstacles on the propagation has to be considered based on the different propagation mechanisms, such as refraction, diffraction, or scattering (see Fig. 2.3). When a propagation wave impinges on an obstacle with a large size or curvature in terms of the wavelength (electrical size), the wave suffers a

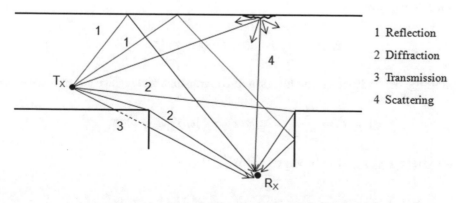

1 Reflection
2 Diffraction
3 Transmission
4 Scattering

FIGURE 2.3: Example of electromagnetic mechanisms in an urban scenario.

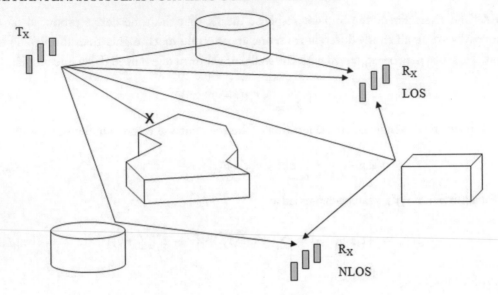

FIGURE 2.4: LOS and NLOS situations for a multipath environment.

significant change in its path that can be expressed in terms of reflection or transmission of the wave and can be modeled by a plane wave incident on a plane interface. When the electrical size is on the order of the wavelength (resonant sizes) or when the object has curvatures and discontinuities, diffraction or scattering play a significant role.

2.4.2   Multipath MIMO

For a MIMO system operating in a multipath environment, as in Fig. 2.4, the voltage at the receive antenna R_i may be expressed as

$$\mathbf{r}_j(t) = \sum_{i=1}^{M} \int_{-\infty}^{t} \mathbf{h}_{ji}(\tau)\mathbf{s}_i(t-\tau)d\tau. \tag{2.42}$$

Making explicit the dependence on the location of the transmit and receive antennas, we obtain:

$$\mathbf{r}_j(\mathbf{r}'_{R_j},t) = \sum_{i=1}^{M} \int_{-\infty}^{t} \mathbf{h}_{ji}(\tau,\mathbf{r}'_{T_i},\mathbf{r}'_{R_j})\mathbf{s}_i(\mathbf{r}'_{T_i},t-\tau)d\tau, \tag{2.43}$$

where the term $\mathbf{h}_{ji}(\tau,\mathbf{r}'_{T_i},\mathbf{r}'_{R_j})$ can be expressed as

$$\mathbf{h}_{ji}(\tau) = \mathbf{h}_{ji}(\tau,\mathbf{r}'_{T_i},\mathbf{r}'_{R_j}) = \sum_{p=1}^{P} A_{ji}^p(\mathbf{r}_{T_i}^p,\mathbf{r}_{R_j}^p)\delta_\tau(\tau-\tau_{ji}^p(\mathbf{r}_{T_i}^p,\mathbf{r}_{R_j}^p)), \tag{2.44}$$

with the term $A_{ji}^{p}(\mathbf{r}_{T_i}^{p}, \mathbf{r}_{R_j}^{p})$ given by

$$A_{ji}^{p} = A_{ji}^{p}(\mathbf{r}_{T_i}^{p}, \mathbf{r}_{R_j}^{p}) = K_{RT_{ji}}\Gamma(r_{T_i}^{p} + r_{R_i}^{p})(\mathbf{F}_{j}^{R}(\Omega_{R_j}^{p}))^{\dagger}\mathbf{C}_{TR_{ji}}^{p}\mathbf{F}_{i}^{T}(\Omega_{T_i}^{p}) \tag{2.45}$$

and where $\Omega_{T_i}^{p} = (\theta_{T_i}^{p}, \phi_{T_i}^{p})$ is the departure angle from the transmit antenna T_i, $\Omega_{R_j}^{p} = (\theta_{R_j}^{p}, \phi_{R_j}^{p})$ is the arriving angle to the receive antenna R_i. The delay associated with the pth MPC from transmit antenna T_i to receive antenna R_j, $\tau_{ji}^{p} = \tau_{ji}^{p}(\mathbf{r}_{T_i}^{p}, \mathbf{r}_{R_j}^{p})$, is given by

$$\tau_{ji}^{p} = \tau_{ji}^{p}(\mathbf{r}_{T_i}^{p}, \mathbf{r}_{R_j}^{p}) = \tau_{T_i}^{p} + \tau_{R_j}^{p} = \frac{\mathbf{k}_{T_i}^{p}\mathbf{r}_{T_i}^{p}}{2\pi f} + \frac{\mathbf{k}_{R_j}^{p}\mathbf{r}_{R_j}^{p}}{2\pi f}. \tag{2.46}$$

Notice that $\mathbf{k}_{T_i}^{p}$ and $\mathbf{k}_{R_j}^{p}$ are the plane wave direction vectors associated with the transmit antenna T_i and receive antenna R_j, respectively, and $\mathbf{r}_{T_i}^{p}$ and $\mathbf{r}_{R_j}^{p}$ are the distance vectors, associated with the pth path from the transmit antenna T_i to the scatter and from the scatter to the receive antenna R_j, respectively. Notice also that $\mathbf{C}_{TR_{ji}}^{p}$ is the normalized channel polarimetric matrix, given by [30]

$$\mathbf{C}_{TR_{ji}}^{p} = \begin{pmatrix} \mathbf{C}_{TR_{ji\theta\theta}}^{p} & \mathbf{C}_{TR_{ji\theta\phi}}^{p} \\ \mathbf{C}_{TR_{ji\phi\theta}}^{p} & \mathbf{C}_{TR_{ji\phi\phi}}^{p} \end{pmatrix}. \tag{2.47}$$

In a more compact matrix form, it can be expressed as

$$\mathbf{r}(t) = \int_{-\infty}^{t} \mathbf{h}(\tau)\mathbf{s}(t - \tau)d\tau, \tag{2.48}$$

where $\mathbf{h}_{ji}(\tau, \mathbf{r}_{T_i}', \mathbf{r}_{R_j}')$ represents the (j, i)th entry of $\mathbf{h}$, and $\mathbf{s}(t)$ and $\mathbf{r}(t)$ are the transmit and receive vectors, respectively. In the frequency domain,

$$\mathbf{R}(t, f) = \mathbf{H}(f)\mathbf{S}(t, f). \tag{2.49}$$

Notice that in the above expressions, we have assumed that the propagating signals only interact with objects in the channel once, and that there is no LOS component. Including any number of interactions with the objects of the channel and including the LOS component, the transfer function is given by

$$\mathbf{H}_{ji}(f) = \mathbf{H}_{ji}(\mathbf{r}_{T_i}', \mathbf{r}_{R_j}', f) = \sum_{p=0}^{P} K_{RT_{ji}}\Gamma(r_{ji}^{p})(\mathbf{F}^{R}(\Omega_{R_j}^{p}))^{\dagger}\mathbf{C}_{TR_{ji}}^{p}\mathbf{F}^{T}(\Omega_{T_i}^{p})e^{-j2\pi f \tau_{ji}^{p}}, \tag{2.50}$$

which may be computed by substitution of space–time response $\mathbf{H}_{ST_{ji}}(f,\Omega_T,\Omega_R)$ into the expression given by Eq. (2.4). Notice that $\mathbf{H}_{ST_{ji}}(f,\Omega_T,\Omega_R)$ is given by

$$
\begin{aligned}
\mathbf{H}_{ST_{ji}}(f,\Omega_T,\Omega_R) &= \mathbf{H}_{ST_{ji}}(\mathbf{r'}_{T_i},\mathbf{r'}_{R_j},f,\Omega_T,\Omega_R) \\
&= \sum_{p=0}^{P} K_{RT_{ji}}\Gamma(r_{ji}^p)\mathbf{C}_{TR_{ji}}^p e^{-j2\pi f\tau_{ji}^p}\delta(\Omega_T-\Omega_{T_i}^p)\delta(\Omega_R-\Omega_{R_j}^p).
\end{aligned}
\tag{2.51}
$$

On the other hand, the CIR is given by

$$
\begin{aligned}
\mathbf{h}_{ji}(\tau) &= \mathbf{h}_{ji}(\tau,\mathbf{r'}_{T_i},\mathbf{r'}_{R_j}) = \sum_{p=0}^{P} A_{ji}^p(r_{ji}^p)\delta_\tau(\tau-\tau_{ji}^p) \\
&= \sum_{p=0}^{P} K_{RT_{ji}}\Gamma(r_{ji}^p)(\mathbf{F}_j^R(\Omega_{R_j}^p))^\dagger \mathbf{C}_{TR_{ji}}^p \mathbf{F}_i^T(\Omega_{T_i}^p)\delta_\tau(\tau-\tau_{ji}^p),
\end{aligned}
\tag{2.52}
$$

which may be computed by substitution of space–time response $\mathbf{h}_{ST_{ji}}(\tau,\Omega_T,\Omega_R)$ into the expression given by Eq. (2.3). Notice that $\mathbf{h}_{ST_{ji}}(\tau,\Omega_T,\Omega_R)$ is given by

$$
\begin{aligned}
\mathbf{h}_{ST_{ji}}(\tau,\Omega_T,\Omega_R) &= \mathbf{h}_{ST_{ji}}(\tau,\mathbf{r'}_{T_i},\mathbf{r'}_{R_j},\Omega_T,\Omega_R) \\
&= \sum_{p=0}^{P} K_{RT_{ji}}\Gamma(r_{ji}^p)\mathbf{C}_{TR_{ji}}^p \delta(\Omega_T-\Omega_{T_i}^p)\delta(\Omega_R-\Omega_{R_j}^p)\delta_\tau(\tau-\tau_{ji}^p),
\end{aligned}
\tag{2.53}
$$

where r_{ji}^p is the total path length from the transmitter to the receiver associated with the pth MPC including multiple reflections, and τ_{ji}^p is the associated time delay including multiple reflections. Finally, notice that the above represents a narrowband linear time-invariant multiantenna bidirectional polarimetric multipath radio channel.

2.4.3 Time-Variant Multipath MIMO

When the transmitter, receivers, or scatterers move at velocity $\mathbf{v}$, their positions change with time, and so does the impulse response of the channel, resulting in a time-variant channel impulse response $\mathbf{h}_{ji}(t,\tau,\mathbf{r'}_{T_i}(t),\mathbf{r'}_{R_j}(t))$, given by

$$
\begin{aligned}
\mathbf{h}_{ST_{ji}}(t,\tau,\Omega_T,\Omega_R) &= \mathbf{h}_{ST_{ji}}(t,\tau,\mathbf{r'}_{T_i}(t),\mathbf{r'}_{R_j}(t),\Omega_T,\Omega_R) \\
&= \sum_{p=0}^{P(t)} K_{RT_{ji}}\Gamma(r_{ji}^p(t))\mathbf{C}_{TR_{ji}}^p(t)\delta(\Omega_T-\Omega_{T_i}^p(t))\delta(\Omega_R-\Omega_{R_j}^p(t))u_\tau(\tau-\tau_{ji}^p(t))
\end{aligned}
$$

$$(2.54)$$

$$\mathbf{h}_{ji}(t,\tau) = \mathbf{h}_{ji}(t,\tau,\mathbf{r'}_{T_i}(t),\mathbf{r'}_{R_j}(t)) = \sum_{p=0}^{P(t)} A_{ji}^p(t,r_{ji}^p(t))u_\tau(\tau - \tau_{ji}^p(t))$$

$$= \sum_{p=0}^{P} K_{RT_{ji}}\Gamma(r_{ji}^p(t))(\mathbf{F}_j^R(t,\Omega_{R_j}^p(t)))^\dagger \mathbf{C}_{TR_{ji}}^p(t)\mathbf{F}_i^T(t,\Omega_{T_i}^p(t))u_\tau(\tau - \tau_{ji}^p(t))$$

$$(2.55)$$

and a time-variant channel transfer function $\mathbf{H}_{ji}(\mathbf{r'}_{T_i}(t),\mathbf{r'}_{R_j}(t),t,f)$ given by

$$\mathbf{H}_{ST_{ji}}(t,f,\Omega_T,\Omega_R) = \mathbf{H}_{ST_{ji}}(\mathbf{r'}_{T_i}(t),\mathbf{r'}_{R_j}(t),t,f,\Omega_T,\Omega_R)$$

$$= \sum_{p=0}^{P(t)} K_{RT_{ji}}\Gamma(r_{ji}^p(t))\mathbf{C}_{TR_{ji}}^p(t)\mathrm{e}^{-j2\pi f\tau_{ji}^p(t)}\delta(\Omega_T - \Omega_{T_i}^p(t))\delta(\Omega_R - \Omega_{R_j}^p(t))$$

$$(2.56)$$

$$\mathbf{H}_{ji}(t,f) = \mathbf{H}_{ji}(\mathbf{r'}_{T_i}(t),\mathbf{r'}_{R_j}(t),t,f)$$

$$= \sum_{p=0}^{P(t)} K_{RT_{ji}}\Gamma(r_{ji}^p(t))(\mathbf{F}^R(t,\Omega_{R_j}^p(t)))^\dagger \mathbf{C}_{TR_{ji}}^p(t)\mathbf{F}^T(t,\Omega_{T_i}^p(t))\mathrm{e}^{-j2\pi f\tau_{ji}^p(t)}, \quad (2.57)$$

where the antenna location vectors depend on the time in the following manner:

$$\mathbf{r'}_{T_i}(t) = \mathbf{r'}_{T_i} + \mathbf{v'}_{T_i}t \tag{2.58}$$

$$\mathbf{r'}_{R_j}(t) = \mathbf{r'}_{R_j} + \mathbf{v'}_{R_j}t. \tag{2.59}$$

Notice that we have assumed that only the transmitter and receiver antenna move and the scatterers remain static. However, the above expression could easily be modified to include the effect of moving scatterers. On the other hand, $\tau_{ji}^p(t)$ is of the form:

$$\tau_{ji}^p(t) = \tau_{ji}^p + \frac{f_d^p(f)}{f}t, \tag{2.60}$$

where $f_d^p(f) = f\frac{v}{c}\cos(\theta_{kv}^p)$ is the Doppler frequency, θ_{kv}^p is the angle formed by the velocity vector and the wave propagation vector. Then, the exponential $\mathrm{e}^{-j2\pi f\tau_{ji}^p(t)}$ can be decomposed into

$$\mathrm{e}^{-j2\pi f\tau_{ji}^p(t)} = \mathrm{e}^{-j2\pi f\tau_{ji}^p}\mathrm{e}^{-j2\pi f_d^p(f)t} = \mathrm{e}^{-j2\pi f\tau_{ji}^p}\mathrm{e}^{-j\omega_d^p t}. \tag{2.61}$$

For the time-variant multipath MIMO channel, the relation between the transmitted signal $\mathbf{s}_i(t)$ and the received signal $\mathbf{r}_j(t)$ can be expressed as

$$\mathbf{r}_j(t) = \int_{-\infty}^t \mathbf{h}_{ji}(t,\tau)\mathbf{s}_i(t-\tau)\mathrm{d}\tau. \tag{2.62}$$

Making explicit the dependence on the location of the transmit and receive antennas, we obtain

$$\mathbf{r}_j(\mathbf{r}'_{R_j}, t) = \int_{-\infty}^{t} \mathbf{h}_{ji}(t, \tau, \mathbf{r}'_{T_i}, \mathbf{r}'_{R_j}) \mathbf{s}_i(\mathbf{r}'_{T_i}, t - \tau) \mathrm{d}\tau, \tag{2.63}$$

whereas the voltage at the R_j receive antenna is given by

$$\mathbf{r}_j(\mathbf{r}'_{R_j}, t) = \sum_{i=1}^{M} \int_{-\infty}^{t} \mathbf{h}_{ji}(t, \tau, \mathbf{r}'_{T_i}, \mathbf{r}'_{R_j}) \mathbf{s}_i(\mathbf{r}'_{T_i}, t - \tau) \mathrm{d}\tau. \tag{2.64}$$

In the frequency domain,

$$\mathbf{R}(t, f) = \mathbf{H}(t, f) \mathbf{S}(t, f). \tag{2.65}$$

As mentioned above, the movement of one of the elements (usually the transmitter, receiver, or the scatterers) produces a frequency shift, called Doppler shift, proportional to $\omega_d = \mathbf{v}\mathbf{k}$ [33]. The range of frequencies over which the Doppler shift extends is called the Doppler spread (B_d) of the channel. Although the Doppler shift is usually low for an individual path, its effect on a multipath environment is significant because the shift is different for every component, creating a frequency

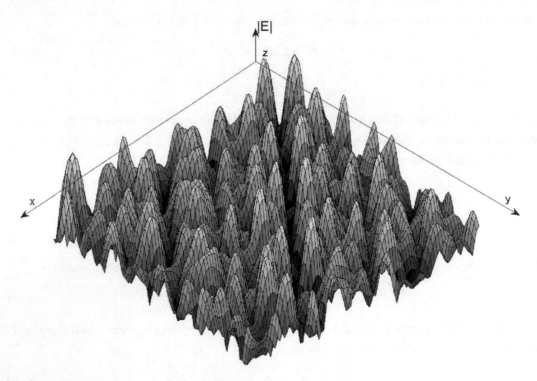

FIGURE 2.5: Amplitude of the irregular wave field representing an interference pattern.

dispersion that finally results in a time variation with fading tips and a low-pass equivalent impulse response of the form

$$\mathbf{h}_{ji}(t,\tau) = \mathbf{h}_{ji}(t,\tau,\mathbf{r}'_{T_i}(t),\mathbf{r}'_{R_j}(t)) = \sum_{p=0}^{P(t)} A_{ji}^p(t,r_{ji}^p(t))e^{j\omega_d^p t}u_\tau(\tau - \tau_{ji}^p), \qquad (2.66)$$

where ω_d^p is the Doppler frequency shift and where we have assumed that $\omega_d^p(f) = \omega_d^p$. The time variation is the result of the relative movement among the transmitter, the receiver, and the environment. Usually, it is formulated in terms of a movement of the receiver antenna through the spatially irregular wave field. The time over which the channel may be considered invariant is usually called the *coherence time of the channel* (T_c) and is related to the rate of change given by the Doppler spread B_d as $T_c \approx \frac{1}{B_d}$. We can interpret this situation as if the receiver were moving through an interference pattern, such as the one shown in Fig. 2.5. Spatially varying fading thus becomes time-varying fading. Because fading dips are approximately half a wavelength apart, this fading is called *small-scale fading*; it is also called *short-term fading* or *fast fading*. The fading rate (number of fading tips per second) depends on the speed of the receiver.

2.4.4   Time-Variant Wideband Multipath MIMO
When wideband systems are considered, the frequency dependence of the complete channel parameters (including the frequency dispersion of antennas) have to be considered, and the channel impulse response becomes

$$\mathbf{h}_{ji}(t,\tau,\mathbf{r}'_{R_j}(t),\mathbf{r}'_{T_i}(t)) = \int_{-\infty}^{\infty} \mathbf{H}_{ji}(\mathbf{r}'_{T_i}(t),\mathbf{r}'_{R_j}(t),t,f)e^{j2\pi f\tau}df, \qquad (2.67)$$

where the transfer function is now given by

$$\begin{aligned}
\mathbf{H}_{ji}(t,f) &= \mathbf{H}_{ji}(\mathbf{r}'_{T_i}(t),\mathbf{r}'_{R_j}(t),t,f) \\
&= \sum_{p=0}^{P(t)} K_{RT_{ji}}(f)\Gamma(r_{ji}^p(t))(\mathbf{F}^R(t,f,\Omega_{R_j}^p(t)))^{\dagger}\mathbf{C}_{TR_{ji}}^p(t)\mathbf{F}^T(t,f,\Omega_{T_i}^p(t))e^{-j2\pi f\tau_{ji}^p(t)},
\end{aligned}$$

$$(2.68)$$

and $K_{RT_{ji}}(f)$ can now be expressed as

$$K_{RT_{ji}}(f) = \sqrt{\frac{Z_j^{0,R}}{Z_i^{0,T}}}\sqrt{\frac{\Re\{Z_i^T(f)\}}{\Re\{Z_j^R(f)\}}}. \qquad (2.69)$$

Diversity, spatial division multiple access, or MIMO systems usually utilize multiantenna elements with antenna separation of less than a few wavelengths. In this case, an alternative simpler model can be used based on the fact that a similar combination of MPCs can be expected at all antenna elements, which allows us to model all the (j, i) channels with respect to a reference function $\mathbf{h}_{ST,\text{REF}}$ (or $\mathbf{H}_{ST,\text{REF}}$) defined between a central transmit element T_{REF} and a central receive element R_{REF}. The channels between all other antenna elements can be referred to $\mathbf{h}_{ST,\text{REF}}$ (or $\mathbf{H}_{ST,\text{REF}}$) by adding the proper delays (or phase shifts) according to their relative locations $\Delta\mathbf{r}'_T$ and $\Delta\mathbf{r}'_R$ to T_{REF} and R_{REF}, respectively, and by taking into account the DoD Ω_T and arrival Ω_R of the MPCs. In this case, the space–time transfer function can be expressed as

$$
\mathbf{H}_{ST_{ji}}(\mathbf{r}'_{T_i}(t), \mathbf{r}'_{R_j}(t), t, f, \Omega_T, \Omega_R) \approx \mathbf{H}_{ST_{\text{REF}}}(\mathbf{r}'_{T_{\text{REF}}}(t), \mathbf{r}'_{R_{\text{REF}}}(t), t, f)
$$
$$
\cdot\, e^{j\frac{2\pi f}{c}\left(\widehat{\mathbf{d}}(\Omega_{T_{\text{REF}}}^p)\Delta\mathbf{r}'_{T_i} + \widehat{\mathbf{d}}(\Omega_{R_{\text{REF}}}^p)\Delta\mathbf{r}'_{R_j}\right)}, \qquad (2.70)
$$

where $\widehat{\mathbf{d}}(\Omega) = (\cos\phi\sin\theta, \sin\phi\sin\theta, \cos\theta)$, and $\mathbf{H}_{ST_{\text{REF}}}(\mathbf{r}'_{T_{\text{REF}}}(t), \mathbf{r}'_{R_{\text{REF}}}(t), t, f, \Omega_T, \Omega_R)$ is given by

$$
\mathbf{H}_{ST_{\text{REF}}}(\mathbf{r}'_{T_{\text{REF}}}(t), \mathbf{r}'_{R_{\text{REF}}}(t), t, f, \Omega_T, \Omega_R) = \sum_{p=0}^{P(t)} K_{RT_{\text{REF}}}(f) \Gamma(r_{\text{REF}}^p(t)) \mathbf{C}_{TR_{\text{REF}}}^p(t)
$$
$$
\cdot\, e^{-j2\pi f\tau_{\text{REF}}^p(t)} \delta(\Omega_T - \Omega_{T_{\text{REF}}}^p(t)) \delta(\Omega_R - \Omega_{R_{\text{REF}}}^p(t)). \qquad (2.71)
$$

On the other hand, let us define $\mathbf{\Gamma}_{ji}^p(t, r_{ji}^p(t)) = \Gamma(r_{ji}^p(t)) \mathbf{C}_{TR_{ji}}^p(t)$ as the transfer matrix between the transmit antenna T_i and the receive antenna R_i associated with the pth MPC. In general, the MPCs are not fully resolved in time but grouped into clusters producing a dispersive behavior effect. Therefore, the transfer matrices $\mathbf{\Gamma}_{ji}(t, r_{ji}^p(t), \tau)$ also have a dependency on the delay time variable τ. For a feasible modeling, the channel description is simplified by assuming that all the components $\mathbf{\Gamma}_{ji}(t, r_{ji}^p(t), \tau)$ of a impulse response experience a similar dispersion that allows to separate the two dependencies as $\mathbf{\Gamma}_{ji}(t, r_{ji}^p(t), \tau) \approx \mathbf{\Gamma}_{ji}^p(t, r_{ji}^p(t))u_\tau(\tau)$ [34] by introducing the dispersion function $u_\tau(\tau)$. In the case of non-frequency-selective MPCs, $u_\tau(\tau)$ becomes $\delta_\tau(\tau)$. If the dispersion of the multipaths cannot be neglected, $u_\tau(\tau)$ can be approximated by an exponential decay with constant τ_c, given by

$$
u(\tau) = \frac{1}{\tau} e^{\frac{\tau}{\tau_\text{c}}} u_\text{step}(\tau), \qquad (2.72)
$$

which corresponds to the cluster power decay phenomena bounded by the unit step function $u_\text{step}(t)$. A common measure of the amount of time dispersion in a channel is the RMS delay spread, σ_τ, defined as

$$
\sigma_\tau = \left(\frac{\sum_{p=1}^{P}(A^p)^2(\tau^p - \overline{\tau})^2}{\sum_{p=1}^{P}(A^p)^2}\right)^{\frac{1}{2}} \qquad (2.73)
$$

$$\overline{\tau} = \frac{\sum_{p=1}^{P}(A^p)^2\tau^p}{\sum_{p=1}^{P}(A^p)^2}. \tag{2.74}$$

It can be shown that the delay spread, expressed in μs, increases with distance approximately as d (km)$^\varepsilon$, where ε=0.5 in urban areas and suburban environments and ε=1 in mountainous regions.

2.5   REPRESENTATIONS OF AN LTV RADIO CHANNEL

In a real environment, because of the high complexity of the physical interactions, including reflections, scattering, refractions, and diffractions, the most appropriate model of the impulse response is from a probabilistic/stochastic point of view, given that $h(t,\tau)$ is a multivariate random process. In most of the cases of interest, however, a parametric approach of the random process may be obtained based on the physically significant parameters previously introduced, such as delays, electromagnetic interaction with the objects of the channel, or Doppler shifts. A general accepted assumption is that $h(t,\tau)$ is a wide-sense stationary processes, where the statistical properties of the channel do not change with time, and with uncorrelated scatterers, where contributions from different delays are uncorrelated. These assumptions are jointly called wide-sense stationary uncorrelated scattering, where time delays τ^p and Doppler shifts f_d^p are henceforth uncorrelated. These conditions are usually fulfilled in mobile communication systems, allowing the channel to be modeled as an LTV system [27].

To deal with the LTV channels, Bello [35] named $h(t,\tau)$ as the input delay spread function, defined as the response of the channel at some time t to a unit impulse function input at some previous time τ seconds earlier. The output of the channel $r(t)$ can then be found by the convolution of the input signal $s(t)$ with $h(t,\tau)$ integrated over the delay variable τ:

$$r(t) = s(t) * h(t,\tau) = \int_{-\infty}^{\infty} h(t,\tau)s(t-\tau)\mathrm{d}\tau. \tag{2.75}$$

As the impulse response of a time-variant system, $h(t,\tau)$, depends on two variables, t and τ, we can perform the Fourier transformations with respect to either (or both) of them. This results in four different, but equivalent representations, as shown in Fig. 2.6. An intuitive interpretation is simpler for slow time-changing channels, where the duration of the impulse response (and the signal) should be much shorter than the time over which the channel changes significantly. The variable t can thus be viewed as an "absolute" time that tells us which impulse response $h(\tau)$ is currently valid. Such a channel is also called *quasi-static*.

Fourier-transforming the impulse response with respect to the variable τ results in the time-variant transfer function $H(t,f)$:

$$H(t,f) = \int_{-\infty}^{\infty} h(t,\tau)\mathrm{e}^{-j2\pi f\tau}\mathrm{d}\tau \tag{2.76}$$

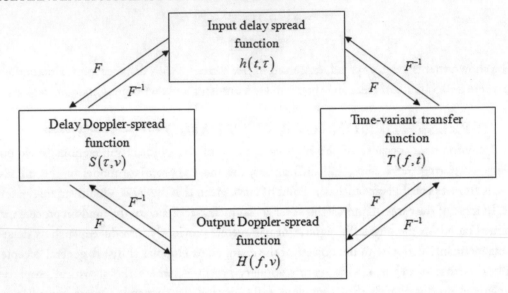

FIGURE 2.6: The Bello functions. Interrelation between different system functions.

$$r(t) = \int_{-\infty}^{\infty} H(t,f)S(f)e^{j2\pi ft}df. \qquad (2.77)$$

The interpretation is again simpler in the case of a quasi-static system, where the input signal is multiplied by the spectrum of the "currently valid" transfer function to give the spectrum of the output signal.

A Fourier transformation with respect to t results in a function called *delay Doppler function*, $S(v, \tau)$, describing the spreading of the impulse signal in the delay and Doppler domains:

$$S(v, \tau) = \int_{-\infty}^{\infty} h(t, \tau)e^{-j2\pi tv}dt. \qquad (2.78)$$

Finally, the spreading function, $S(v, \tau)$, can be transformed with respect to the variable τ, resulting in the Doppler-variant transfer function $B(v, f)$:

$$B(v, f) = \int_{-\infty}^{\infty} S(v, \tau)e^{-j2\pi ft}d\tau. \qquad (2.79)$$

Fig. 2.7 shows the interrelations between the four channel functions for a simulated LTV multipath channel.

From the previous functions, it is possible to obtain the different parameters of the channel, and, in particular, time coherence T_c, the Doppler spread B_d, the bandwidth coherence B_c, and the maximum delay τ_{max}. As we have seen, when the signal bandwidth B is smaller than the coherence

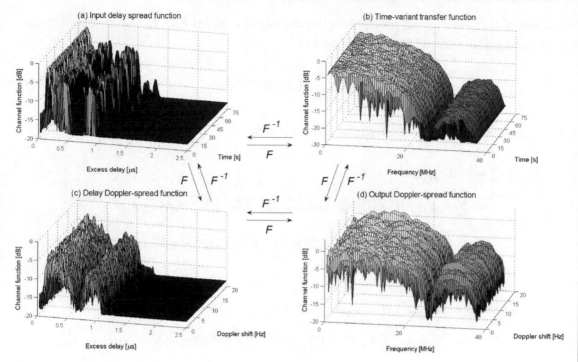

FIGURE 2.7: Simulated Bello functions for a mobile channel. (a) Input delay spread function, (b) delay Doppler spread function, (c) time-variant transfer function, and (d) output Doppler spread function.

bandwidth B_c, the channel is frequency nonselective. Similarly, if we choose a signaling interval t smaller than the time of coherence T_c, we call the channel a slowly fading channel. Finally, when $B \approx \frac{1}{T}$, the conditions for the channel to be frequency nonselective with slow fading imply that the product of τ_{max} and B_d (called *spread factor of the channel*) must satisfy $\tau_{max}B_d < 1$.

2.6   LTV WIDEBAND BIDIRECTIONAL POLARIMETRIC MULTIANTENNA MULTIPATH RADIO CHANNEL MODEL

Consider the previous multiantenna geometry of Fig. 2.2 with M transmit antennas located at positions $\mathbf{r}'_{T_1}, \mathbf{r}'_{T_2}, ..., \mathbf{r}'_{T_M}$ and N receive antennas located at positions $\mathbf{r}'_{R_1}, \mathbf{r}'_{R_2}, ..., \mathbf{r}'_{R_N}$. Being $\mathbf{s}(t)$, the vector of signals transmitted by the M transmitters, the noiseless vector of received signals $\mathbf{r}(t)$ by the N receivers in a time-variant multipath environment may be expressed as

$$\mathbf{r}(t) = \int_{-\infty}^{t} \mathbf{h}(t, \tau)\mathbf{s}(t - \tau)\mathrm{d}\tau, \qquad (2.80)$$

where the $M \times N$ time-variant polarimetric impulse response matrix $\mathbf{H}(t, \tau)$ may be related to the corresponding time-variant frequency response as

$$\mathbf{h}(t, \tau) = \int_{-\infty}^{\infty} \mathbf{H}(t, f) e^{j2\pi f \tau} \mathrm{d}\tau. \tag{2.81}$$

For this general case, the transfer function is now given by

$$
\begin{aligned}
\mathbf{H}_{ji}(t, f) &= \mathbf{H}_{ji}(\mathbf{r}'_{T_i}(t), \mathbf{r}'_{R_j}(t), t, f) \\
&= \sum_{p=0}^{P(t)} K_{RT_{ji}}(f) \Gamma(r_{ji}^p(t)) (\mathbf{F}^R(t, f, \Omega_{R_j}^p(t)))^\dagger \mathbf{C}_{TR_{ji}}^p(t) \mathbf{F}^T(t, f, \Omega_{T_i}^p(t)) e^{-j2\pi f \tau_{ji}^p(t)} \\
&= \int \int (\mathbf{F}^R(t, f, \Omega_T))^\dagger \mathbf{H}_{ST_{ji}}(t, f, \Omega_T, \Omega_R) \mathbf{F}^T(t, f, \Omega_R) \mathrm{d}\Omega_T \mathrm{d}\Omega_R
\end{aligned} \tag{2.82}
$$

$$
\begin{aligned}
\mathbf{H}_{ST_{ji}}(t, f, \Omega_T, \Omega_R) &= \mathbf{H}_{ST_{ji}}(\mathbf{r}'_{T_i}(t), \mathbf{r}'_{R_j}(t), t, f, \Omega_T, \Omega_R) \\
&= \sum_{p=0}^{P(t)} K_{RT_{ji}}(f) \Gamma(r_{ji}^p(t)) \mathbf{C}_{TR_{ji}}^p(t) e^{-j2\pi f \tau_{ji}^p(t)} \delta(\Omega_T - \Omega_{T_i}^p(t)) \delta(\Omega_R - \Omega_{R_j}^p(t)),
\end{aligned} \tag{2.83}
$$

where $K_{RT_{ji}}(f)$ can now be expressed as

$$K_{RT_{ji}}(f) = \sqrt{\frac{Z_j^{0,R}}{Z_i^{0,T}}} \sqrt{\frac{\Re\left\{Z_i^T(f)\right\}}{\Re\left\{Z_j^R(f)\right\}}} \tag{2.84}$$

and where we can define $\mathbf{\Gamma}_{ji}^p(t, r_{ji}^p(t)) = \Gamma(r_{ji}^p(t)) \mathbf{C}_{TR_{ji}}^p(t)$.

Notice that, again, it is also possible to model all the (j, i) channels with respect to a reference function $\mathbf{h}_{ST,REF}$ (or $\mathbf{H}_{ST,REF}$) defined between a central transmit element T_{REF} and a central receive element R_{REF}. The channels between all other antenna elements can be referred to $\mathbf{h}_{ST,REF}$ (or $\mathbf{H}_{ST,REF}$) by adding the proper delays (or phase shifts) according to their relative location $\Delta\mathbf{r}'_T$ and $\Delta\mathbf{r}'_R$, to T_{REF} and R_{REF}, respectively, and by taking into account the DoD Ω_T and arrival Ω_R of the MPCs. In this case, the space–time transfer function can be expressed as

$$
\begin{aligned}
\mathbf{H}_{ST_{ji}}(\mathbf{r}'_{T_i}(t), \mathbf{r}'_{R_j}(t), t, f, \Omega_T, \Omega_R) &\approx \mathbf{H}_{ST_{REF}}(\mathbf{r}'_{T_{REF}}(t), \mathbf{r}'_{R_{REF}}(t), t, f) \\
&\cdot e^{j\frac{2\pi f}{c}\left(\hat{\mathbf{d}}(\Omega_{T_{REF}}^p)\Delta\mathbf{r}'_{T_i} + \hat{\mathbf{d}}(\Omega_{R_{REF}}^p)\Delta\mathbf{r}'_{R_j}\right)},
\end{aligned} \tag{2.85}
$$

where $\widehat{\mathbf{d}}(\Omega) = (\cos\phi\sin\theta, \sin\phi\sin\theta, \cos\theta)$, and $\mathbf{H}_{ST_{\mathrm{REF}}}(\mathbf{r}'_{T_{\mathrm{REF}}}(t), \mathbf{r}'_{R_{\mathrm{REF}}}(t), t, f, \Omega_T, \Omega_R)$ is given by

$$\mathbf{H}_{ST_{\mathrm{REF}}}(\mathbf{r}'_{T_{\mathrm{REF}}}(t), \mathbf{r}'_{R_{\mathrm{REF}}}(t), t, f, \Omega_T, \Omega_R) = \sum_{p=0}^{P(t)} K_{RT_{\mathrm{REF}}}(f)\Gamma(r_{\mathrm{REF}}^{p}(t))\mathbf{C}_{TR_{\mathrm{REF}}}^{p}(t)$$
$$\cdot\, e^{-j2\pi f \tau_{\mathrm{REF}}^{p}(t)} \delta(\Omega_T - \Omega_{T_{\mathrm{REF}}}^{p}(t)) \delta(\Omega_R - \Omega_{R_{\mathrm{REF}}}^{p}(t)).$$

$$(2.86)$$

In the following section, we will give some mathematical descriptions and usual values to generate the modeling of significant parameters (number of MPCs, delay times, and transfer matrices) defining the propagation channel. For each transmitter–receiver matrix element $\mathbf{H}_{ji}$ and for each MPC, the model has to calculate the statistics of the significant parameters completely describing the radio propagation model, which are

- polarimetric transfer matrices ($\Gamma_{ji}^{p}(t, r_{ji}^{p}(t))$), including path loss $\Gamma(r_{ji}^{p}(t))$ and the normalized polarimetric matrices $\mathbf{C}_{TR_{ji}}^{p}(t)$;
- MPCs: number $P(t)$ and delays (excess delay) $\tau_{ji}^{p}(t)$;
- DoD and DoA $\Omega_{T_i}^{p}(t)$ and $\Omega_{R_j}^{p}(t)$ of the transmit and receive antennas.

The total field at the receiver position results from the continuous superposition of different wave components resulting from multiple interactions between the propagation wave and the pathway objects. Measured CIRs typically show a few strong peaks followed by a continuum of lower-intensity additional scattering contributions. The strong peaks can probably be associated with objects with favorable surface reflection or diffraction conditions. The additional contributions probably are produced by much more diffuse scattering mechanisms. The time-frequency behavior of a selected peak reflects characteristics of the wave field resulting from the corresponding scattering element.

Scattering from rough surfaces shows that with relatively small vertical variations in a surface segment, the scattered field becomes diffuse. However, in typical propagation environments, depending on the wavelength, we will find simultaneously that large areas (buildings, streets, lakes, etc.) produce almost specular reflections, whereas other areas (populated surfaces, terrains, forests, vegetation) produce diffuse scattering. A stochastic approach containing both kinds of components is thus necessary.

2.6.1 Polarimetric Transfer Matrices

The first basic parameter in the modeling of a radio channel is the polarimetric transfer matrix. The polarimetric transfer matrix associated with the pth MPC is defined as $\Gamma_{ji}^{p}(t, r_{ji}^{p}(t)) =$

$\Gamma(r_{ji}^{p}(t))\mathbf{C}_{TR_{ji}}^{p}(t)$, with $\mathbf{\Gamma}_{ji}^{p}(t, r_{ji}^{p}(t)) \in \mathcal{C}^{2\times 2}$. This matrix contains information on the propagation factor and the electromagnetic interaction of the waves with the objects of the channel, for each one of the MPCs, including the direct LOS path. A differentiate contribution for the LOS and NLOS components will be considered. When LOS between the transmitter and receiver exists ($p = 0$), the polarimetric transfer matrix depends uniquely on the propagation factor and is modeled as

$$\mathbf{\Gamma}_{ji}^{0}(t, r_{ji}^{0}(t)) = \frac{\lambda}{r_{ji}^{0}(t)}\begin{pmatrix} 1 & 0 \\ 0 & 1 \end{pmatrix}. \tag{2.87}$$

For the rest of the NLOS paths ($p = 1, \cdots, P$) the polarimetric transfer matrix $\mathbf{\Gamma}_{ji}^{p}(t, r_{ji}^{p}(t))$ will also take into account the most significant characteristics of the multipath channel properties, such as

- received power decreasing with distance
- decay of amplitudes with increasing delay time
- slow and fast fading of multipaths
- cross-polarization of the channel.

That is, $\mathbf{\Gamma}_{ji}^{p}(t, r_{ji}^{p}(t))$ is a polarimetric transfer matrix that includes all the losses and depolarization of all scattering processes (reflections, transmissions, diffractions, scattering, etc.) of the pth MPC:

$$\mathbf{\Gamma}_{ji}^{p}(t, r_{ji}^{p}(t), \tau) = \Gamma(r_{ji}^{p}(t))\mathbf{C}_{TR_{ji}}^{p}\frac{1}{\tau}e^{-\frac{\tau}{\tau_{c}}}u_{\text{step}}(\tau), \tag{2.88}$$

where the term $\mathbf{C}_{TR_{ji}}^{p}$ is the so-called normalized transfer matrix, which at the same time can be expanded into [30]:

$$\mathbf{C}_{TR_{ji}}^{p} = \begin{pmatrix} \mathbf{C}_{TR_{ji\theta\theta}}^{p} & \mathbf{C}_{TR_{ji\theta\phi}}^{p} \\ \mathbf{C}_{TR_{ji\phi\theta}}^{p} & \mathbf{C}_{TR_{ji\phi\phi}}^{p} \end{pmatrix} = \begin{pmatrix} C_{TR_{ji\theta\theta}}^{p}e^{j\varphi_{\theta\theta}^{p}} & C_{TR_{ji\theta\theta}}^{p}X_{\theta\phi}e^{j\varphi_{\theta\phi}^{p}} \\ C_{TR_{ji\phi\phi}}^{p}X_{\phi\theta}e^{j\varphi_{\phi\theta}^{p}} & C_{TR_{ji\phi\phi}}^{p}e^{j\varphi_{\phi\phi}^{p}} \end{pmatrix}, \tag{2.89}$$

with $C_{TR_{ji}}^{p} = C_{SF,TR_{ji}}^{p}C_{FF,TR_{ji}}^{p}$. That is,

$$C_{TR_{ji\theta\theta}}^{p} = C_{SF,TR_{ji\theta\theta}}^{p}C_{FF,TR_{ji\theta\theta}}^{p} \tag{2.90}$$

$$C_{TR_{ji\theta\phi}}^{p} = C_{SF,TR_{ji\theta\phi}}^{p}C_{FF,TR_{ji\theta\phi}}^{p} \tag{2.91}$$

$$C_{TR_{ji\phi\theta}}^{p} = C_{SF,TR_{ji\phi\theta}}^{p}C_{FF,TR_{ji\phi\theta}}^{p} \tag{2.92}$$

$$C_{TR_{ji\phi\phi}}^{p} = C_{SF,TR_{ji\phi\phi}}^{p}C_{FF,TR_{ji\phi\phi}}^{p}. \tag{2.93}$$

Depending on the delay time, an exponential decay of the amplitudes can be normally assumed. The average decay time is determined by the variable τ_{c}. The amplitudes of the matrix elements are

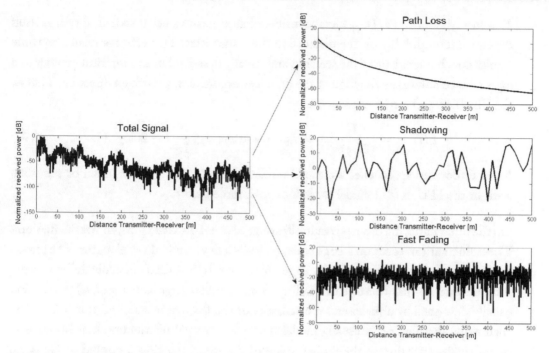

FIGURE 2.8: Radio channel general scenario: representation of the basic properties of the distance propagation factor, for which a steady smooth attenuation is combined with fast fluctuations, called *slow and fast fading*.

scaled by $\Gamma(r_{ji}^p(t))$ in a way that the total received power will follow the power distance law with the power exponent n_{PL}. The channel polarimetric cross-correlation is determined by $X_{\theta\phi}$ and $X_{\phi\theta}$, which normally have a value around 0.1. The amplitudes are usually log-normal distributed for the slow-fading terms $C_{SF,TR_{ji\theta\theta}}^p$ and Nakagami distributed for the fast-fading terms $C_{FF,TR_{ji\theta\theta}}^p$. The phases φ^p are usually supposed to be uniformly distributed.

2.6.2   Distance Propagation Factor

In this section, we describe the basic properties of the distance propagation factor, $\Gamma(r)$, which depends on the distance among the transmitter and the receiver and for which a steady smooth attenuation is combined with fast fluctuations, called *slow and fast fading* (see Fig. 2.8). The slow- and fast-fading factors are referred to by $C_{SF,TR_{ji}}^p$ and $C_{FF,TR_{ji}}^p$, respectively. These fluctuations eventually have a significant impact on the BER, and they may be explained as the combination of the three basic effects [28]:

- The steady smooth PL (in general considered as a mean value in statistical terms) itself depends monotonically on the distance to the transmitter. This effect is related to some significant characteristic of the scenario and usually described in an empirical manner as a log-distance power law function of the path loss exponent n_{PL}, with an offset L_0. That is, the PLs ($L(r)|_{\text{dB}}$) can be expressed as

$$L(r)|_{\text{dB}} = \frac{1}{(\Gamma(r))^2}\Big|_{\text{dB}} = L_0|_{\text{dB}} + 10 n_{\text{PL}} \log(\frac{r}{1m}) - 20\log(\lambda). \qquad (2.94)$$

More elaborated approaches are the Okumara–Hata model [36] for urban or suburban areas or the ITU-R [37] model for indoor environments.

- On the other hand, the mean power itself, averaged over 10 wavelengths, shows fluctuations. These fluctuations occur on a large scale, typically a few hundred wavelengths. The reason for these variations is shadowing by large objects and is thus fundamentally different from the interference that causes small fading. This is called *large-scale* (or *slow*) *fading* and usually described by a mean and the statistics of the fluctuations around this mean. The shadowing distribution is usually modeled as a log normal distribution that in our case describes the variation of the decibel value of the mean signal as a normal or Gaussian distribution. If the mean value and standard deviation of the envelope $C_{SF,TR_{ji}}^p|_{\text{dB}}$ are $\mu_{C_{SF,TR_{ji}}^p|_{\text{dB}}}$ and $\sigma_{C_{SF,TR_{ji}}^p|_{\text{dB}}}$, respectively, its probability density function/distribution is then given by

$$p_{C_{SF,TR_{ji}}^p|_{\text{dB}}}(C_{SF,TR_{ji}}^p|_{\text{dB}}) = \frac{1}{\sqrt{2\pi\sigma_{C_{SF,TR_{ji}}^p|_{\text{dB}}}^2}} e^{\left(-\frac{(C_{SF,TR_{ji}}^p|_{\text{dB}} - \mu_{C_{SF,TR_{ji}}^p|_{\text{dB}}})^2}{2\sigma_{C_{SF,TR_{ji}}^p|_{\text{dB}}}^2}\right)}, \qquad (2.95)$$

where $\sigma_{C_{SF,TR_{ji}}^p|_{\text{dB}}}$ can vary from 5 to 12 dB depending on the environment. A value of 8 dB is typically used. The mean value $\mu_{C_{SF,TR_{ji}}^p|_{\text{dB}}}$ can be taken as the expected median signal level found using the path loss previously predicted.

- On a very-short-distance scale, power fluctuates around a mean value. These fluctuations happen on a scale that is comparable with one wavelength (see Fig. 2.5) and therefore are called *small-scale* (or *fast-*) *fading* fluctuations. Usual mathematical approaches consist of the Rice distribution for LOS situations (for situations with a dominant component with power A^2 plus weaker random components with power σ^2) that transforms on a σ^2 power Rayleigh distribution for NLOS). A practical parametric-fit function is the Nakagami distribution [29] characterized by two parameters: $m = \dfrac{E\left\{\left(C^{p}_{FF,TR_{ji}}\right)^{2}\right\}^{2}}{Var\left\{\left(C^{p}_{FF,TR_{ji}}\right)^{2}\right\}}$, controlling the

basic shape of the distribution, and $\Omega = E\left\{\left(C^{p}_{FF,TR_{ji}}\right)^{2}\right\}$, the mean square value. The probability density function/distribution of the amplitude $C^{p}_{FF,TR_{ji}}$ is then given by

$$p_{C^{p}_{FF,TR_{ji}}}\left(C^{p}_{FF,TR_{ji}}\right) = \frac{2}{\Gamma_{g}(m)}\left(\frac{m}{\Omega}\right)^{m} r^{2m-1} e^{\left(\frac{-m}{\Omega^{2}}\left(C^{p}_{FF,TR_{ji}}\right)^{2}\right)} \qquad (2.96)$$

for $C^{p}_{FF,TR_{ji}} \geq 0, m \geq 0.5$, with $\Gamma_{g}(x)$ defined as

$$\Gamma_{g}(x) = \int_{0}^{\infty} t^{n-1} e^{-t} dt \text{ for } x > 0. \qquad (2.97)$$

For $m = 1$, the Nakagami distribution becomes a Rayleigh distribution, with $\Omega = \sigma^{2}$. The parameter m depends on the type of environment, whereas Ω depends on the total power of the CIR realization.

2.6.3 Nature of the MPCs

To approach a real time-variant scenario, the channel model will consist of the combination of a certain number of paths, generated with a rate λ_{G}, existing over a certain period δ_{L}, eventually recombining at a rate λ_{R} and finally disappearing, whereas other new paths will appear. A suitable description for such a path behavior is a generation–recombination scheme based on Poisson processes according to the following expectations [38]:

$$E\{P\} = \frac{\lambda_{G}}{\lambda_{R}}, \qquad (2.98)$$

where $E\{P\}$ is the expectation of a certain number P of MPCs, with a probability for an MPC to remain from one realization to the next one, given by

$$P_{\text{remain}}(\delta_{P}) = e^{-\lambda_{R}\delta_{P}}, \qquad (2.99)$$

where δ_{P} represents the movement through the scenario for a certain time interval Δt calculated from the velocities of the transmitter $\mathbf{v'}_{T}(t)$, receiver $\mathbf{v'}_{R}(t)$, as well as from the fluctuations of scatterers $P_{F}v_{S}$. That is, $\delta_{P} = \int_{t}^{t+\Delta t}(|\mathbf{v'}_{T}(t)| + |\mathbf{v'}_{R}(t)| + P_{F}v_{S})dt$, where the idea is that a certain percentage of P_F MPCs contains moving scatterers with a mean velocity v_S. Finally, the expectation of new paths based on a Poisson distribution is given by

$$E\{P_{\text{new}}\} = \frac{\lambda_G}{\lambda_R}\left(1 - e^{-\lambda_R\delta_P}\right). \qquad (2.100)$$

In certain cases, the properties of new paths may be slightly different from that of paths already existing in the previous realization, called *inherited paths*. The delay times (τ^p) are usually seen as a stochastic term assumed to be uniformly distributed in the interval $(0, \Delta\tau_{\text{max}})$

$$P_{\Delta\tau^p}(\Delta\tau^p) = \frac{1}{\Delta\tau_{\text{max}}} \text{ for } 0 \le \Delta\tau^p \le \Delta\tau_{\text{max}}. \qquad (2.101)$$

Notice that Eq. (2.101) describes the delay times for new paths. Those inherited paths are determined by a shift according to the movement of the transmitter and the receiver, taking into account the path angles.

2.6.4   DoD and DoA

When modeling the DoD and DoA at the transmit and receive antennas, it is found that they strongly depend on the actual positions of the transmitter and receiver. One way to reduce this dependence is referring the angles to the direct line between the transmitter and the receiver. A second step is to link the angular directions with the delay time of a certain path. Early paths are expected to be close to the direct transmitter–receiver path with similar DoDs and DoAs. On the other hand, later ones may have very different angular statistics and arrive, for example, from backward directions. Because of the reciprocal channel properties, the modeling approaches of DoD and DoA are identical.

In line with our previous discussion, to define a direction in spherical coordinates referenced to the transmitter–receiver direct line, two angles ϑ' and ψ' are needed. ϑ' (colatitude) and ψ' (longitude) may be seen as the local usual spherical coordinates angles θ and ϕ, referenced to a z-axis coincident with the transmitter–receiver interconnection line. The distribution of both angular dependencies is normally based on a combination of experimental measurement and geometric properties of the environment. Normally, combinations of distributions are used. As a example for an indoor environment, for ϑ', a mix distribution consisting of uniform and Nakagami distributions may be used. For ψ' when transmitter and receiver are almost at the same height, horizontal and vertical planes (corresponding to the specular reflections in walls) are emphasized, whereas for very different height, an almost uniform distribution may be used.

2.6.4.1   Modeling of Different Channel Types.   To model real environments, a variety of methods have been applied. Some are approaches based on simple modeling of canonical obstacles such

as reflections by surfaces, diffraction by edges, transmission by obstacles, and eventually scattering by rough surfaces. A number of methods are parametric (sometimes also called *empirical*) in the sense that a few parameters describe the path loss based on experimental results. More recently, deterministic predictions (sometimes called *physical models'*), such as ray tracing and the GTD, or even more advanced methods such as finite difference time domain techniques, have been applied to the deterministic prediction of indoor or microcell propagation. These models may yield wideband information and the statistics of multipath propagation directly. These data can be used for site-specific predictions, provided that sufficient detail of the scenario geometry and materials are available, but, in general, they are complicated to set up. The kind of model chosen depends on the level of desired estimation and the availability of geographical data of the coverage area. Additional details are explained below, and some parameters are presented in Table 2.1.

When dealing with particular scenarios, some generic channel properties, based on the environment structure, may be observed [39]. Based on this, real channels are often roughly classified into four main classes, although the borders between them are not quite precise because some scattering environments are not separable in a simple manner:

1. Indoor environments: a basic reflection scattering mechanism produces a high number of MPCs with low time dispersion. The impulse response results from multiple reflections at interior and exterior walls and transmission through walls. Because of the short distances, the delays are usually very small (tenths of a nanosecond).

TABLE 2.1: Measured parameters (around 2 GHz) for wireless channel modeling

Environment	Path Loss Power Exponent n_{PL}	Power Standard Deviation σ(dB)	Typical RMS Delay Spread Values (μs)
Free space	2	0	0
Indoor building (LOS)	1.6–1.8	4–7	0.01–0.1
Indoor building (NLOS)	4–6	5–12	
Tunnels	1.8–2.2	5–9	0.02
Urban area (shadowed)	3–5	11–17	3–5
Suburban area	2.5–3.5	10–14	< 1
Hilly terrain	3–6	5–12	3–10

2. Town centers: strongly irregular, structured environments characterized by high-density buildings; the buildings usually are very large and may have smooth surfaces because their surface roughness is much smaller than the dimension of a wavelength. The impulse response results from a diffuse contribution (no LOS components are usually observed) from the diffraction and scattering of structures in the immediate neighborhood of the transmitter and receiver, created by a sequence of reflections from buildings walls with small delays (a few tens or hundredths of nanoseconds) that are typically concentrated in a delay window of about 1–2 μs. For a narrowband system, the impulse response may be modeled by a broad peak, the time-selective behavior being Rayleigh distributed with a broad angular distribution of the incoming or outgoing waves.

3. Rural areas: open areas formed by quite regular environments with a low density of strong scatterers such as buildings. Because of the low shadowing, a broad peak (as a result of the superposition of the waves coming from the direct LOS path with those coming from the scattering interactions with neighbor objects) and additional narrow peaks with longer delays may easily be found. Depending on the terrain, 2–10 peaks with typical delays of up to 15 μs are found. Because of smaller amplitude fluctuations as a result of the coherent contributions of the direct and specular reflections, a high value for the parameter m in the Nakagami distribution is produced. That is, m may rise up to $m \approx 15$. The typical characteristics of this channel type are a set of strong peaks quantified by their number, delays, average power, and the relation among these quantities. Strong peaks can be assumed to result from pointlike scatterers for which two sequential free-space propagations (transmitter to scatterer and scatterer to receiver) results in a r^{-4} dependence (similar to the usual radar equation) in the propagation factor. Fluctuations of the average power around the average distance law are large because of the fluctuations of the reflection coefficients (10-dB log-normal distribution may be used). For the time-selective behavior, a Nakagami m distribution may be used with a narrow/concentrated angular distribution.

4. Hilly and mountainous regions: reflections produced by mountain slopes, which have large scattering coefficients and time dispersion because of their size, produce strong contributions up to very distant points (some tenths of a kilometer), resulting in larger delays (tenths of a microsecond) clustered within relative delays of about 10 μs (3 km).

2.7   QUALITY OF THE MIMO CHANNEL MODEL

Using multiple transmit and receive antennas, it is possible to achieve very high channel spectral efficiency in highly scattering environments. These high spectral efficiencies are enabled by the fact that a scattering environment makes the signal from every individual transmitter appear highly uncorrelated at each of the receive antennas. In a sense, the scattering environment acts like a very large aperture that makes it possible for the receiver to resolve the individual incoming signals from the transmitter, or, equivalently, to increase the number of the channel spatial models [32]. This high spectral efficiency, and consequently, the channel capacity, increases if the signals arriving at the receivers are uncorrelated. Maximum diversity gain (DG) and capacity are therefore achieved when $E\{\mathbf{H}_{ji}\mathbf{H}_{kl}^{*}\} = 0$ for $i \neq k$ and $l \neq j$. The effect of fading correlation between antenna elements results in the generally occurring correlation of signals received by two antennas [40], with the correlation coefficient depending both on the antenna separation (in wavelengths) and the angular spectrum of the incoming radio wave. For a more general case of an spatial wave described by the power azimuth and elevation profile $\boldsymbol{\Psi}_{\theta}(\Omega)$ and $\boldsymbol{\Psi}_{\phi}(\Omega)$ for both polarizations θ and ϕ, the correlation coefficient may be expressed as

$$\mathbf{R}_{H_{ji}} = \frac{\mathbf{X}_{H_{ji}}}{\sqrt{\mathbf{X}_{H_{ii}}^{R}\mathbf{X}_{H_{jj}}^{R}}}, \qquad (2.102)$$

where the term $\mathbf{X}_{H_{ji}}$ may be expressed as

$$\begin{aligned}
\mathbf{X}_{H_{ji}} = \int_{-\pi}^{\pi}\int_{0}^{\pi} \Big(& \frac{XPR}{1+XPR}\mathbf{F}_{i_{\theta}}(\Omega)(\mathbf{F}_{j_{\theta}}(\Omega))^{*}\boldsymbol{\Psi}_{\theta}'(\Omega) \\
& + \frac{1}{1+XPR}\mathbf{F}_{i_{\phi}}(\Omega)(\mathbf{F}_{j_{\phi}}(\Omega))^{*}\boldsymbol{\Psi}_{\phi}'(\Omega)\Big)\sin{(\theta)}\theta\mathrm{d}\phi \qquad (2.103)
\end{aligned}$$

and where XPR (-5 to -10 dB) is the cross polarization ratio of the power between the θ and ϕ component at the receiver. Experimental results show that the power azimuth spectrum may be modeled for both polarizations by a combination of Laplacian functions depending on the number of clusters of scatterers, and the power elevation profile by a Gaussian or sinusoidal function. The total power angular spectrum may then be expressed by the product of both distributions, normalized so that $\int \boldsymbol{\Psi}_{\phi}'(\Omega)\mathrm{d}\Omega = 1$ and $\int \boldsymbol{\Psi}_{\theta}'(\Omega)\mathrm{d}\Omega = 1$.

For simplicity, assume now that the rays arrive mostly on the azimuth plane. Then, for the case of an array of omnidirectional antennas, the correlation coefficient for any pair of these antennas separated by a distance d_{ji}, in which their radiation patterns are related by $\mathbf{F}_{j}(\Omega) = \mathbf{F}_{i}(\Omega)\mathrm{e}^{-jkd_{ji}}$, becomes

$$\mathbf{X}_{H_{ji}} = \int_{0}^{2\pi} \mathrm{e}^{jkd\cos{(\phi-\psi)}}\boldsymbol{\Psi}_{\phi}(\phi)\mathrm{d}\phi, \qquad (2.104)$$

where ϕ is the azimuth angle of incidence of each plane wave and ψ is the angle of orientation of the two element axis, set to $\pi/2$ for the broadside case considered here. In the case of uniform illumination $\boldsymbol{\Psi}_{\phi}(\phi) = \frac{1}{2\pi}$ and $\mathbf{R}_{H_{ji}} = J_{0}(kd_{ji})$, where J_{0} is the 0th-order Bessel function. To achieve near-complete decorrelation in this case, the antenna elements should be spaced about $\lambda/2$

apart. For more concentrated incident illuminations, the minimum antenna separation to produce decorrelation tends to increase.

There exist other parameters that describe the quality of the propagation channel, such as the mean effective gain (MEG), the (DG), capacity, and other, which will be described later in this book.

2.7.1 Frequency-Selective Channels

Although the capacity expressions given in this book assume a frequency-flat channel, a generalized expression for a frequency-dependent channel must also take into account the frequency dependence (adding additional diversity to the channel) as follows:

$$C = \frac{1}{B} \int_B C(f) \mathrm{d}f. \qquad (2.105)$$

2.7.2 Channel Time-Variant Predictions

Effectiveness of the MIMO technology is highly dependent on the type and quality of channel state information (CSI) at the receiver, and in the case of rank-deficient channels, also at the transmitter. In practice, CSI is obtained by dedicating a small fraction of the transmission capacity to train the receiver. If the channel varies slowly relative to the symbol period, the training consumption is acceptable; otherwise, the available transmission bandwidth may be quickly exhausted, leading to low effective channel capacities. Consequently, good metrics need to be used for a good temporal variation prediction [41].

2.7.3 Channel Scattering Dimension

On the other hand, even when the correlation among transmitter or receiver antennas is small, the quality of the system may be reduced because of a low dimensionality of the channel (low rank), produced by a lack of scattering quality. This may happen, for example, because of a low number of scatterers (limited number of interacting objects) or because of the existence of a kind of degenerated path known as a pinhole or keyhole. In the first case, capacity increases with the number of antennas up to a certain point when the reduced number of potential paths produces a saturation effect. In the keyhole case, on the other hand, we can have a rich scattering environment at the proximities of the transmitter and receiver. However, these advantages are not usable because in between the transmitter and the receiver local environments, there is only one or a reduced number of propagation paths (long stretch patch, single diffraction edge, etc.) resulting in a low-rank channel matrix.

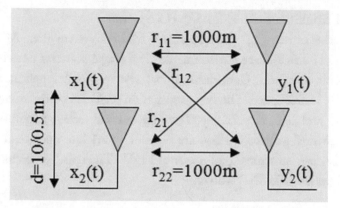

FIGURE 2.9: Geometry of the problem: 2×2 MIMO system without scatterers.

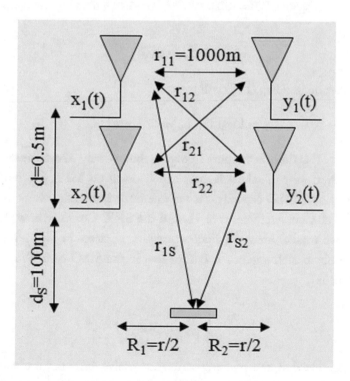

FIGURE 2.10: Geometry of the problem: 2×2 MIMO system with one scatterer.

2.7.4   Example: Capacity of a 2×2 MIMO System

As a final example, the capacity of a simple 2×2 MIMO system (i.e., $M = 2$ and $N = 2$) will be calculated for (1) a free-space environment, (2) a single scatterer deterministic case, and (3) a multipath statistic approach. Consider a global system mobile application using a 2×2 multiantenna geometry in free space. The two propagation scenarios (without and with scatterers) are shown in Figs. 2.9 and 2.10, respectively. The transmit and receive antennas are polarization matched, have an individual gain of 0 dBm, are separated by 1 km, operate at 1.9 GHz with a bandwidth of 200 kHz, and transmit a total power of 1 mW. The noise temperature of the receiver is 300 K. For a SISO system, the SNR would be

$$P^D = P^S G^T G^R \left(\frac{\lambda}{4\pi r} \right)^2 = P^S G^T G^R (\overline{\Gamma}(r))^2 = 1.58 \cdot 10^{-13} \, W(-98 \text{ dB}m) \qquad (2.106)$$

$$N_0 = kT_e B = 8.28 \cdot 10^{-16} \, W(-120.8 \text{ dbm}) \qquad (2.107)$$

$$\rho_0 = 190.8(22.8 \text{ dB}) \qquad (2.108)$$

The capacity for the system (see Eq. (1.2)) will be

$$C_{1 \times 1} = \log_2(1 + \rho_0) = 7.6 \text{bps/Hz}. \qquad (2.109)$$

If we now suppose that there is not pure free space, but we have a certain amount of scattering in the environment, then we can take advantage of it using an MPA (an MEA in this case) and improve the (DG) or MIMO capacity. As we mentioned, the improvement depends on the correlation among the channel coefficients of **H**, and the SNR. Consider now the 2×2 MIMO system consisting of two parallel identical dipoles placed as depicted in Fig. 2.9. Notice that now, the transmitted power from each antenna is half of that in the SISO case. We can now calculate the propagation matrix as

$$\mathbf{H} = \overline{\Gamma}(r) \begin{pmatrix} e^{-jkr_{11}} & e^{-jkr_{12}} \\ e^{-jkr_{21}} & e^{-jkr_{22}} \end{pmatrix}. \qquad (2.110)$$

Let us consider first that the two transmit and receive antennas are very close together ($d = 0.5$ m $= 3.5\lambda$). Using $r = 1000$ m, $f = 1.9$ GHz, for $d = 0.5$ m the eigenvalues of the channel matrix are $\lambda_1 = 2.0$ and $\lambda_1 = 0$, and the channel capacity is given by $C_{2 \times 2} = \sum \log_2 \left(1 + \frac{\rho_0}{2} \lambda_i \right) = 7.6$ bps/Hz. For $d = 10$ m, the eigenvalues of the channel matrix are $\lambda_1 = 1.41$ and $\lambda_1 = 0.60$, and the channel capacity is given by $C_{2 \times 2} = \sum \log_2 \left(1 + \frac{\rho_0}{2} \lambda_i \right) = 12.95$ bps/Hz, which gives a high

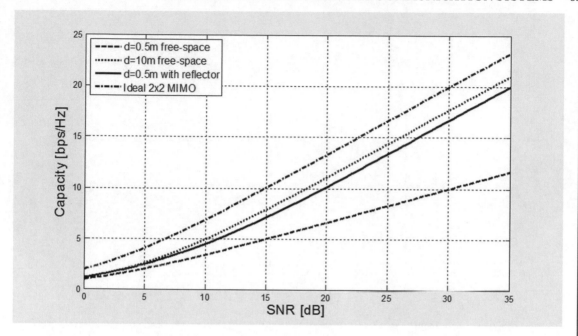

FIGURE 2.11: Capacities of the systems for the different cases.

increase in capacity. In the case of having a single deterministic scatter, the channel matrix is given by

$$\mathbf{H} = \overline{\Gamma}(r) \begin{pmatrix} e^{-jkr_{11}} + 0.6\left(e^{-jkr_{1S}} + e^{-jkr_{S1}}\right) e^{-jkr_{12}} + 0.6\left(e^{-jkr_{1S}} + e^{-jkr_{S2}}\right) \\ e^{-jkr_{21}} + 0.6\left(e^{-jkr_{2S}} + e^{-jkr_{S1}}\right) e^{-jkr_{22}} + 0.6\left(e^{-jkr_{2S}} + e^{-jkr_{S2}}\right) \end{pmatrix}. \quad (2.111)$$

Then, with $d = 0.5$m, the eigenvalues of the channel matrix are $\lambda_1 = 2.14$ and $\lambda_1 = 0.2$, and the channel capacity is given b $C_{2\times2} = \sum \log_2\left(1 + \frac{\rho_0}{2}\lambda_i\right) = 12.01$ bps/Hz.

When the scattering content increases, using the Kronecker model, we may calculate the capacity in a probabilistic manner and obtain the correlation between the different paths. For the ideal case of perfect uncorrelated paths, $\lambda_1 = 2.0$ and $\lambda_1 = 2.0$, and the channel capacity is given by $C_{2\times2} = \sum \log_2\left(1 + \frac{\rho_0}{2}\lambda_i\right) = 15.17$ bps/Hz. Finally, in Fig. 2.11 is shown a capacity plot of the systems for the different cases versus SNR.

. . . .

CHAPTER 3

Coding Theory for MIMO Communication Systems

3.1 OVERVIEW OF MIMO CODING TECHNIQUES

In recent years, much research has been done in the field of channel coding for MIMO communication systems. An extensive review can be found in Refs. [2, 7, 25, 26, 42, 43], where the latest results and advances are presented. Channel coding techniques are used to improve the performance of MIMO communication systems in multipath NLOS propagation environments.

Depending on its maximization goal and the knowledge of the channel characteristics at the receiver and/or the transmitter, the MIMO channel coding techniques can be classified into three main categories:

1. space–time coding or diversity
2. SM
3. beam forming.

In the following sections, we will review these coding techniques and their design criteria. To do this, we consider a wireless communication system where the transmitter has M-accessible ports and the receiver (decoder) has N-accessible ports. We refer to such a MIMO system as a $M \times N$ system. In general, the goal of MIMO coding techniques is to take advantage of the spatial, polarization, time, and frequency domains offered by the channel propagation environment to improve the reliability of the transmitted signals, the transmission data rate, or both. For the remainder of this chapter, we assume the normalized version of the channel matrix given by $\mathbf{H}_0$ in Eq. (1.22) and the following dimensions $\mathbf{H}_0 \in \mathcal{C}^{M \times N}$. Unless specified, in general, we will assume that $\mathbf{H}_0 = \mathbf{W}$ where $\mathbf{W} \in \mathcal{C}^{M \times N}$ is a complex random matrix with all elements being iid Gaussian random variables with distribution $N(0, 1)$.

In addition to the above techniques, the received signals at the ports of an MPA need to be combined to improve the performance of MIMO systems. The most relevant combining methods will also be reviewed. Finally, notice that the knowledge of the channel characteristics or the channel matrix itself at the receiver and/or transmitter is also known in the literature as CSI.

3.2 PERFORMANCE MEASURES IN CODING THEORY

Let us begin by defining the main figures of merit that are used to quantify the performance level of the different MIMO channel coding techniques. These performance measures are the diversity, multiplexing, and array gains. In the literature, the channel correlation is also often considered as a relevant figure of merit.

3.2.1 Definition of Diversity Gain

The diversity gain (DG) is a figure of merit used to quantify the performance level of diversity techniques. The main idea behind diversity techniques is to produce different replicas of the transmitted signal to the receiver. If these replicas are sent over the propagation channel such that their statistics are independent, when one of them fades, it is less probable that all the other copies of the transmitted signal will fade simultaneously. Because of this redundancy, the receiver can still decode the transmitted signal even in fading conditions. A formal definition for (DG), G_d, is given in Ref. [26]:

$$G_d = - \lim_{\gamma \to \infty} \frac{\log P_e}{\log \gamma},\qquad(3.1)$$

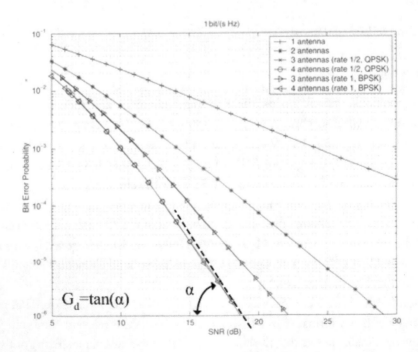

FIGURE 3.1: BER versus SNR curves for different MIMO configurations (figure extracted from [26]).

where P_e is the error probability at a SNR level equal to γ. Therefore, (DG), also known as diversity order, is the slope of the error probability curve in terms of the received SNR in a log–log scale. Fig. 3.1 shows the BER versus SNR curves of different MIMO configurations compared with an uncoded SISO (1×1) system. Notice that the curves for the 2×1 MIMO systems decay twice as rapidly, as the SNR increases, than those of the SISO system, revealing that the diversity order of MIMO systems is equal to 2. The (DG) in Eq. (3.1) describes the increment on the average SNR because of the use of diversity techniques. However, the (DG) can also be defined as the increment of the SNR at a given probability, normally 1% or 10%. Such (DG) can easily be calculated by looking at the cumulative distribution function (CDF) curves of the SNR, and comparing the combined SNR using some specific diversity technique with the SNR of an uncoded SISO communication system.

Finally, notice that for a binary phase-shift keying (BPSK) modulation and a quadrature phase-shift keying (QPSK) modulation using the Gray mapping of symbols [26], the P_e is equivalent to the BER. Unless specified, for the rest of this chapter, we assume either of these two constellations.

3.2.2 Definition of Multiplexing Gain

When multiple antennas are used to maximize the transmission rate, SM is the desired technique. The basic principle behind these schemes is to send distinct symbols from different transmit ports and use appropriate decoding techniques to separate these symbols at the receiver. Following the approach of Ref. [26], the SM gain (SMG) can be defined as

$$G_{sm} = \lim_{\gamma \to \infty} \frac{R}{\log \gamma} \qquad (3.2)$$

where R is the rate of the code at the transmitter and is a function of the SNR. Notice that the relationship of the SMG to the transmission rate is similar to that of the (DG) to the probability of error. This is because, in fact, the SMG measures how far the rate R is from the capacity, C. The range of the SMG is from 0 to $\min(N, M)$.

3.2.2.1 Tradeoffs Between Diversity and Multiplexing Gain.

A MIMO system can simultaneously obtain both (DG) and multiplexing gain, but there is a fundamental tradeoff between how much of each type of gain any signaling scheme can extract. There exist a theorem that quantifies such a tradeoff [44], which says that if we assume M transmitting antennas and N receive antennas, for a given SMG $G_{sm} = k$, where $k = 0, 1, ..., \min(N, M)$ is an integer, then the maximum (DG) is given by $G_d(k) = (M - k)(N - k)$, if the block length of the code is greater than or equal to $N + M - 1$. The optimal tradeoff curve is achieved by connecting the points $(k, G_d(k))$ by lines, as shown in Fig. 3.2. For example, for $M = 4$ and $N = 3$, a SM gain of $G_{sm} = 2$ can only be obtained

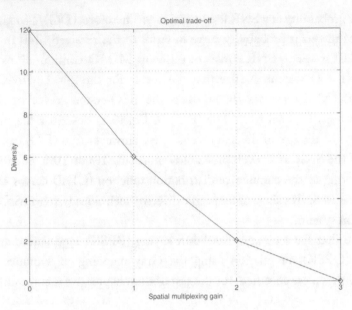

FIGURE 3.2: Optimal tradeoff curve between diversity and multiplexing gain. (Figure extracted from [26])

with a maximum (DG) of $G_d = 2$. On the other hand, if no SMG exists, $G_{sm} = 0$, then a maximum (DG) of $G_d = MN = 12$ can be achieved.

3.2.3   Definition of Array Gain

Beam-forming gain, also known as array gain, is the improvement in SNR obtained by coherently combining the signals on multiple transmit or multiple receive antennas—that is, it is the increase in average received SNR obtained by coherently combining the incoming and outgoing signals. It requires channel knowledge both at the transmitter and at the receiver. The beam-forming gain is easily characterized as a shift of the BER versus SNR curve due to the gain in SNR. Finally, notice that *beam forming* is a term traditionally associated with phase-array processing or smart antennas in wireless communications.

3.2.4   Definition of Channel Correlation

Correlation among the entries of $\mathbf{H}_0$ has been observed to influence the capacity [2] and the (DG) [45] of a MIMO communication system. Let us define the channel correlation matrix, $\mathcal{R}_{H_0} \in \mathcal{C}^{MN \times MN}$, as

$$\mathcal{R}_{H_0} = E\{\mathbf{vec}(\mathbf{H}_0)\mathbf{vec}(\mathbf{H}_0)^H\}, \tag{3.3}$$

where $\mathbf{vec}(\mathbf{H}_0) \in \mathcal{C}^{MN \times 1}$. Observe that the number of correlation coefficients between all elements $\mathbf{H}_{0ij}$ in $\mathbf{H}_0$ is $N^2 M^2$. Thus, it is difficult to assess the correlation properties. A simple way to assess whether the correlation is high or low is to consider only the transmit and receive correlation. As shown in Section 1.6, in many NLOS scenarios with rich scattering, the random fading processes at the receiver are uncorrelated with those at the transmitter, and a Kronecker channel propagation model [46] can be assumed. In that case, a good approximation for $\mathcal{R}_{H_0}$ is given by

$$\mathcal{R}_{H_0} = \mathbf{R}_{H_0}^R \otimes \mathbf{R}_{H_0}^T \qquad (3.4)$$

where $\mathbf{R}_{H_0}^T$ and $\mathbf{R}_{H_0}^T$ are the normalized transmit and receive correlation matrices, respectively, as defined in Section 1.6. Let us now recall that using the Kronecker model, $\mathbf{H}_0$ can be written as

$$\mathbf{H}_0 = (\mathbf{R}_{H_0}^T)^{\frac{1}{2}} \mathbf{W}((\mathbf{R}_{H_0}^R)^{\frac{1}{2}})^{\dagger}, \qquad (3.5)$$

where $\mathbf{W} \in \mathcal{C}^{M \times N}$ is a complex random matrix with all elements being iid Gaussian random variables with distribution $N(0,1)$. The entries of $\mathbf{H}_0$ correspond to zero-mean correlated Gaussian random variables such that $||\mathbf{H}_0||_F^2 = MN$.

To understand how correlation affects the capacity of a MIMO system, we must relate these two quantities. Because in this chapter, $\mathbf{H}_0$ has dimensions $\mathbf{H}_0 \in \mathcal{C}^{M \times N}$, Eq. (1.13) must be rewritten as

$$C = \log_2 \det\left(\mathbf{I} + \frac{\rho_0}{M}\mathbf{H}_0{}^H\mathbf{H}_0\right) = \log_2 \det\left(\mathbf{I} + \frac{\rho_0}{M}\mathbf{R}_{H_0}^R\mathbf{W}^H\mathbf{R}_{H_0}^T\mathbf{W}\right). \qquad (3.6)$$

As shown in Ref. [25], the number of equivalent uncoupled subchannels that contribute to the total capacity in a MIMO system is not degraded when $\mathcal{R}_{H_0}$ is a full rank matrix, in which case $\mathbf{R}_{H_0}^T$ and $\mathbf{R}_{H_0}^R$ must also be full rank.

On the other hand, for most space–time codes (STCs), having a full-rank correlation matrix $\mathcal{R}_{H_0}$ assures that maximum (DG) can be achieved. As with the capacity, it is important to relate the diversity order of a MIMO system with the correlation properties of the antennas. In this way, we can predict and evaluate the performance of any MPA from a diversity perspective in a MIMO wireless communication system. Different space–time architectures have distinct expression for the diversity order; therefore, to illustrate, we give here the expression for a system using STCs. We assume again a Kronecker channel model [18] and an STC designed such that it can achieve full diversity in an ideal NLOS situation. Such code is called a *full-rank code*. Using Ref. [45], the probability of having an error, called *pairwise error probability* (PEP), $P(\mathbf{C}^1 \rightarrow \mathbf{C}^2)$, in a MIMO communication system using STCs can be written as

$$P(\mathbf{C}^1 \rightarrow \mathbf{C}^2) \leq \left(\frac{\rho_0}{4}\right)^{r(\mathbf{R}_{H_0}^T) r(\mathbf{R}_{H_0}^R)} \prod_{l=1}^{r(\mathbf{R}_{H_0}^R)-1} (\lambda_l(\mathbf{R}_{H_0}^R))^{-r(\mathbf{R}_{H_0}^T)} \qquad (3.7)$$

$$\cdot \prod_{i=1}^{r(\mathbf{R}_{H_0}^T)-1} (\lambda_i((\mathbf{C}^1 - \mathbf{C}^2)^\dagger \mathbf{R}_{H_0}^T (\mathbf{C}^1 - \mathbf{C}^2)^*))^{-r(\mathbf{R}_{H_0}^R)},$$

where $r(\mathbf{A})$ denotes the rank of $\mathbf{A}$, and $\lambda_i(\mathbf{A})$ denotes the ith eigenvalue of the matrix $\mathbf{A}$. Notice that the above expression gives us an upper bound on the PEP for such a system using a particular MPA such that the transmit and receive correlation matrices are given by $\mathbf{R}_{H_0}^T$ and $\mathbf{R}_{H_0}^R$, respectively. As shown also in Ref. [45], the diversity order of this system is $G_d = r(\mathbf{R}_{H_0}^T) r(\mathbf{R}_{H_0}^R)$.

The effect of antenna correlation on the diversity order of a MIMO system using space–time block codes (STBCs) can be observed in Fig. 3.3, where the BER versus SNR curves of a 2×2 MIMO system are shown over a correlated propagation channel with correlation ξ among its entries. Notice that as the correlation value ξ increases from 0 to 1, the slope of the BER versus SNR curve decreases, thereby revealing the detrimental effect of correlation on the (DG) of a MIMO communication system. On the other hand, having $\mathbf{R}_H^T = \mathbf{I}$ and $\mathbf{R}_H^R = \mathbf{I}$ ensures not only maximum (DG) but also maximum coding gain [25, 26].

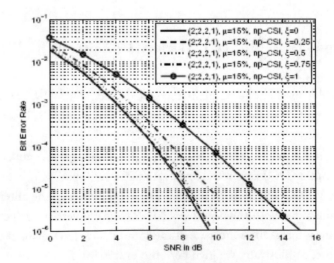

FIGURE 3.3: BER versus SNR curves for a 2×2 MIMO system over a correlated propagation channel with correlation ξ among its entries.

In Chapter 4, it will be shown that the channel correlation is determined by two factors: the mutual coupling characteristics of the MPAs and the scattering characteristics of the channel propagation environment. That is, correlation may come about because of propagation or antenna characteristics.

3.3   COMBINING METHODS

The multiple versions of the signals accessible at the N receiving ports of an MPA need to be combined to improve the performance of a MIMO communication system. We assume that CSI is known at the receiver, and in most cases, it can be assumed that this information is perfect (good assumption for slow-fading channels). In reality, CSI is obtained through channel estimation techniques involving training sequences [47] or other methods [48].

Because, in this chapter, $\mathbf{H}_0$ has dimensions $\mathbf{H}_0 \in \mathcal{C}^{M \times N}$, we must rewrite the input–output relationship given in Eq. (1.6), into

$$\mathbf{r} = \mathbf{s}\mathbf{H}_0 + \mathbf{n}, \tag{3.8}$$

where $\mathbf{r} \in \mathcal{C}^{1 \times N}$ includes all received signals during one time slots at the N receiving antennas, $\mathbf{s} \in \mathcal{C}^{1 \times M}$ represents the transmitted signals, and $\mathbf{n}^{1 \times N}$ contains the noise components. For simplicity, and without loss of generality, assume one transmit antenna, $M = 1$. In this case, Eq. (3.8) can be further simplified into

$$\mathbf{r} = s\mathbf{h}_0 + \mathbf{n}, \tag{3.9}$$

where s is a scalar and represents the transmitted symbol and $\mathbf{h}_0 \in \mathcal{C}^{1 \times N}$. We can distinguish among three different flavors of combining methods:

1. Maximal ratio combining (MRC): This method coherently combines the N signals at the receiver [49]. The signals are previously weighted according to their SNR. MRC performs the following operation:

$$y = \mathbf{r}\mathbf{h}_0{}^H = \mathbf{c}||\mathbf{h}_0||_F^2 + \mathbf{n}\mathbf{h}_0{}^H, \tag{3.10}$$

where y is the received symbol after combining. It can be shown that the average receive SNR of a system employing MRC is given by $\rho_{0_{MRC}} = N\rho_0$. In the case of an LOS propagation scenario, this method becomes equivalent to performing phase-array beam forming. Using MRC, N RF chains are needed at the output ports of the receive MPA. A variant of MRC exists when the weight is fixed to one. In such a case, the method is called *equal gain combining* (EGC). EGC performs the following operation:

$$y = \mathbf{r}\Phi^{\mathbf{h}_0} = \mathbf{c}\mathbf{h}_0\Phi^{\mathbf{h}_0} + \mathbf{n}\Phi^{\mathbf{h}_0}, \tag{3.11}$$

where $\Phi = e^{j\angle \mathbf{h}_0}$ is a matrix containing only the phase information of $\mathbf{h}_0$. Each entry is of the form $\Phi_{ij} = e^{j\angle \mathbf{h}_{0\,ij}}$. It can be shown that the average receive SNR of a system employing EGC is given by $\rho_{0\,EGC} = (1 + \frac{\pi}{4}(N-1))\rho_0$ [25, 26]. MRC outperforms EGC, although EGC is, in general, easier and cheaper to implement than MRC.

2. Selection combining (SC): This method picks the signal with the highest SNR. This requires that the receiver is monitoring all the branches, and hence, N RF chains are needed, after the MPA. SC performs the following operation:

$$i_{\text{opt}} = \arg \max_i (\text{abs})(\mathbf{r})) \tag{3.12}$$

$$y = \mathbf{r}_{i_{\text{opt}}}, \tag{3.13}$$

where $\max(\cdot)$ returns the maximum entry of a vector, with $\mathbf{r}_i$ being the ith entry of $\mathbf{r}$. It can be shown that the average receive SNR is in this case given by $\rho_{0\,SC} = \sum_{m=1}^{N} \frac{1}{m}\rho_0$ [26].

3. Switching combining (SWC): This method picks an antenna and uses it until the SNR falls below a certain threshold. Then, it switches to the next antenna. The system does not monitor all the branches, and hence, only one RF chain is needed. SWC is the only method that does not need to have CSI at the receiver.

3.4   SPACE–TIME CODING

Qualitatively speaking, the goal of STCs [42, 43] is to achieve the maximum (DG) of MN, the maximum coding gain, and therefore to provide the lowest error probability. To do so, an STC encoder sends redundant symbols from different antennas in different time slots. On the other hand, in a typical wireless communication system, the mobile transceiver has limited available power through a battery and should be a small physical device. To improve the battery life, low complexity encoding and decoding is crucial; thus, these systems normally use very simple decoding techniques, such as maximum likelihood (ML) linear decoding [25, 26] or ML decoding by pairs of symbols.

The drawback of space–time coding compared with multiplexing techniques is its maximum possible transmission rate. To achieve full diversity, one cannot transmit more than one symbol per time slot. That is, it can only be transmitted on a single data stream. Therefore, the use of STCs over MPAs can result in a smaller probability of error for the same throughput or capacity as that of a SISO channel. The maximum throughput of STCs is one symbol per channel used, for any number of transmit ports. Notice that in other words, the goal of STCs is to increase the SNR of the system through diversity. The linear improvement of SNR with the number of antennas will

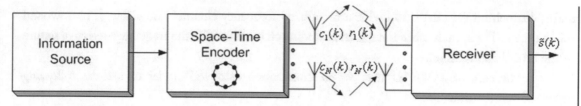

FIGURE 3.4: Schematic of a MIMO communication systems using STCs.

result in a logarithmic increase of the transmission rate. This is because the goal of STCs is not to maximize the throughput.

Fig. 3.4 shows an schematic of a MIMO communication systems using STCs. As seen on the figure, the generated symbols from the source are passed through an STC encoder before being sent to the antennas. As shown in Fig. 3.5, at the receiver, the received signals from each port are decoded thanks to the CSI and the inherent structure of the used codes.

Despite the fact that the above techniques are typically used in narrowband flat channels, they can also be used on wideband channels by explicitly exploiting the frequency domain. This extended architecture is known as STC-OFDM and uses space–time frequency coding in wideband frequency-selective channels. STC-OFDM is a space–time coding technique that in addition uses the frequency dimension, introducing more degrees of freedom to the system. Theoretically, it can achieve a diversity of up to MNJ, where J is the number of frequency subcarriers. The

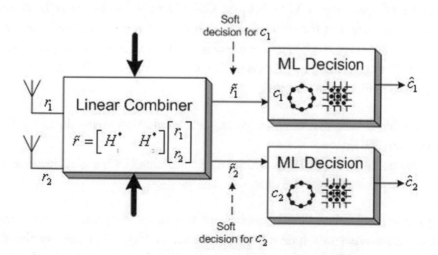

FIGURE 3.5: Receiver structure on a MIMO communication systems using the Alamouti STC.

main idea behind OFDM is to divide a broadband frequency channel into a few, J, narrowband subchannels. Then, each subchannel is a flat-fading channel despite the frequency-selective nature of the broadband channel.

In the case of a MIMO wireless communication using STCs, let us use the following input–output relationship:

$$\mathbf{R} = \mathbf{CH}_0 + \mathbf{N}, \tag{3.14}$$

where $\mathbf{R} \in \mathcal{C}^{T \times N}$ is the received matrix that includes all received signals during t time slots at the N receiving antennas, $\mathbf{C} \in \mathcal{C}^{T \times M}$ is the matrix collecting all the signals transmitted during t time slots from the M transmitting antennas, also called a *codeword*, and $\mathbf{N} \in \mathcal{C}^{T \times N}$ is the noise matrix. Notice that the transmitted codeword $\mathbf{C}$ is constructed from the set of symbols that one desires to transmit, given by $\mathbf{s} \in \mathcal{C}^{1 \times M}$. The channel $\mathbf{H}_0$ is assumed to be quasi-static where the path gains are constant over a frame length $T_c > T$ and is allowed to change from frame to frame. That is, the coherence time T_c of the channel is larger than t.

3.4.1   Classification of Space–Time Coding Techniques

Space–time coding techniques can be classified according to different points of view. By focusing on the availability of CSI at the transmitter, we can distinguish among open- and close-loop STC techniques [25].

For the case of open-loop STC techniques, CSI is assumed to be perfectly known at the receiver, and the performance of these techniques is normally described, as in our case, assuming that MRC is being used at the receiver as a combining technique. However, they can also be used using SC at the receiver. The main two flavors of STCs that do not use CSI at the transmitter can be classified into two additional subcategories:

1. space–time block code (STBCs), which target the reduction of the decoding complexity and maximum (DG);
2. space–time trellis codes (STTCs), which target maximum (DG) and an increase in coding gain, at the expense of larger decoding complexity.

On the other hand, when CSI is available at the transmitter, linear beam forming can be applied at the transmitter to improve the performance of STCs. We refer to this family of techniques as close-loop STCs. In what follows, we give a brief introduction of each one of these techniques.

3.4.2   Space–Time Block Codes

In an STBC, the data stream to be transmitted is encoded in blocks, or codewords **C**, which are distributed among M ports of the MPA and across t time slots. Despite the name, an STBC can be considered as a modulation scheme for MPAs that provides full diversity and very low complexity encoding and decoding but, in general, cannot provide coding gains. The simpler block codes were envisioned by Alamouti [50] and later by Tarokh et al. [42, 51].

Typically, STBCs are obtained from orthogonal designs [51] because they provide simple decoding, based on ML decoding [26], resulting in each symbol being decoded separately using only linear processing and, second, achieving maximum diversity. Such codes are called *orthogonal STBCs* (OSTBC). OSTBCs are obtained based upon the design of a generator matrix **G**. The generator matrix of STCs has similarities and differences with the conventional concept of generating matrix for linear codes in the classical SISO error coding theory. Both of them represent the redundancy of the corresponding codes, although the redundancy in linear block codes is used to correct errors and corresponds to coding gain, whereas the redundancy in STBCs is utilized to achieve (DG). A generalized complex orthogonal design of size M is an $T \times M$ orthogonal matrix **G** with complex entries $x_1, -x_1, ..., x_K, -x_K$, their conjugates $x_1^*, -x_1^*, ..., x_K^*, -x_K^*$, and multiples of these indeterminate variables by $j = \sqrt{-1}$ or $-j$, such that

$$a\mathbf{G}^H\mathbf{G} = (|x_1|^2 + ... + |x_K|^2)\mathbf{I}, \tag{3.15}$$

where $x_1, ..., x_K$ are indeterminate variables that are replaced by symbols from a specific constellation during encoding, and t represents the number of time slots for transmitting one block of symbols. The rate R of the code is given by $R = \frac{K}{T}$.

For real constellations of symbols, such as in BPSK, it is possible to design full rate and full diversity codes in addition to simple ML decoding, for any number of antennas as shown in Ref. [26] as long as the block size is $T = \min(2^{4c+d})$, where the minimization is over all possible integer values c and d in the set $\{c \geq 0, d \geq 0 | 8c + 2^d \geq M\}$. In general, to achieve full rate, $T = K > M$. If we desire minimum delay codes, such that $T = K = M$, then if the symbol constellation is real, real orthogonal designs of rate, one that simultaneously achieve maximum diversity, exist if and only if $M = 2, 4, 8$. If the constellation is complex, such as in QPSK, this only happens if $M = 2$, and this is the case of the well-known Alamouti code [50]. In those cases, the generator matrix is a $M \times M$ orthogonal matrix. The Alamouti code [50] is given by the following generating matrix:

$$\mathbf{G} = \begin{pmatrix} x_1 & x_2 \\ -x_2^* & x_1^* \end{pmatrix}. \tag{3.16}$$

Finally, notice that half-rate, $R = \frac{1}{2}$, complex OSTBCs exist for any number of transmit antennas. Notice also that because the probability of error, P_e, of such codes depends on the diversity order, the larger the SNR, the larger the improvement on the P_e as compared with an uncoded transmission.

It is important to see which is the operation performed by OSTBCs on the received vector $\mathbf{R}$ to decode the transmitted symbols. For the sake of simplicity and without loss of generality, we assume one receiving antenna, in particular, the nth receiving antenna. Therefore, we can set $N = 1$, such that $\mathbf{r}_n \in \mathcal{C}^{T \times 1}$. For more than one receive antenna, we can use MRC to come up with similar formulas. Assume a real constellation. We begin by applying the transpose operator to $\mathbf{r}_n = \mathbf{CH}_{0*n} + \mathbf{n}_n$, and we obtain $\mathbf{r}_n^\dagger = \mathbf{h}_{0n}^\dagger \mathbf{C}^\dagger + \mathbf{n}_n^\dagger$. Notice that $\mathbf{H}_{0*n}$ corresponds to the nth column of $\mathbf{H}_0 \in \mathcal{C}^{M \times N}$. In Ref. [26], it is shown that $\mathbf{r}_n^\dagger$ can always be rewritten in the form

$$\mathbf{r}_n^\dagger = \mathbf{s}\Omega_n + \mathbf{n}_n^\dagger, \tag{3.17}$$

where $\mathbf{s} \in \mathcal{C}^{1 \times M}$ is the vector containing the transmitted symbols, and $\Omega_n \in \mathcal{C}^{M \times T}$ is a matrix containing the channel information and the characteristics of the code. Using the orthogonality property for OSTBCs together with real constellations, we have

$$\Omega_n \Omega_n^\dagger = \sum_{m=1}^{M} |\mathbf{H}_{0\,mn}|^2 \mathbf{I}. \tag{3.18}$$

Then, at the receiver, we perform the following operation:

$$\mathbf{y}_n^\dagger = \mathbf{r}_n^\dagger \Omega_n^\dagger = \left(\sum_{m=1}^{M} |\mathbf{H}_{0\,mn}|^2 \right) \mathbf{s} + \mathbf{n}_n^\dagger \Omega_n^\dagger. \tag{3.19}$$

In the case of having a receive MPA, one can use MRC, which performs the following operation to combine the signal from all the receiving ports:

$$\mathbf{y} = \sum_{n=1}^{N} \mathbf{y}_n^\dagger = \left(\sum_{n=1}^{N} \sum_{n=1}^{M} |\mathbf{H}_{0\,mn}|^2 \right) \mathbf{s} + \sum_{n=1}^{N} \mathbf{n}^\dagger \Omega^\dagger. \tag{3.20}$$

Using complex constellations, a similar approach can be found in Ref. [26]. One can easily observe the benefits of using OSTBCs on MIMO systems because the instantaneous receive power in these systems is given by

$$P^{\mathrm{D,OSTBC+MRC}} = \left(\sum_{n=1}^{N} \sum_{n=1}^{M} |\mathbf{H}_{0\,mn}|^2 \right) E[|s|^2], \tag{3.21}$$

where $E[|s|^2]$ represents the energy per symbol. Finally, the average receive SNR is given by $\rho_{0\,\mathrm{OSTBC+MRC}} = MN\rho_0$.

As noted above, full-rate orthogonal designs with complex elements in their transmission matrix and that can achieve full diversity and separate decoding are impossible for more than two antennas. With the introduction of quasi-orthogonal STBC (QOSTBC) [26], one can achieve pairwise decoding, rate one, and full diversity for more than two antennas and using a complex constellation. To achieve such a rate, QOSTBCs relax the requirement of decoding complexity by decoding pairs of symbols separately. In addition, QOSTBCs use rotated constellations of the original constellation for some of the transmitted symbols to maximize the coding gain. With OSTBCs, this degree of freedom was not available. Both OSTBCs and QOSTBCs have exactly the same encoding complexity although they have different decoding complexity. This decoding complexity is, however, very small in both cases.

3.4.3   Space–Time Trellis Codes

If we further relax the requirement of decoding complexity, STTCs offer a larger coding gain that can be used to improve the BER performance of a multiantenna system with respect to STBCs. STTCs combine modulation and trellis coding to transmit information over MPAs and MIMO channels. Trellis coding is a well-known technique from SISO error correcting theory to increase the coding gain of the codes. The goal is to construct trellis codes satisfying the design criteria of STCs. That is, they must be designed to achieve full diversity. In addition, it is desirable to have a rate one code and of course among all possible full-rate, full-diversity codes, the one that provides us with the highest coding gain. The encoding and decoding of STTCs is based upon the design of a trellis or finite-state machine composed of several states. It is, however, possible to define as well a generating matrix $\mathbf{G}$ as for STBCs. A complete overview of the design and properties of STTCs can be found in Refs. [25, 26]. STTCs are distinct from STBCs; because STTCs have memory and the encoder's output is a function of the current and previous blocks of data, STTCs fit in the category of convolutional codes. Finally, notice that an STTC intended to be used with a M-port MPA is designed by assigning M constellation symbols to every transmission in the trellis under some particular rules. For an STTC with a rate b bits/s/Hz and a diversity of G_{d}, at least $2^{b(G_d-1)}$ states are needed in the finite state machine of the trellis. The ML decoding is done using the Viterbi algorithm [25, 26], and it is possible to improve the performance of STTCs by increasing the number of trellis states and hence the complexity of the decoding.

STTCs, like STBCs, are designed based on criteria that seek to minimize the PEP, that is, the probability of mistaking two codewords. In the case of STTCs, a codeword corresponds to a data frame. Therefore, STTCs provide lower frame error rate (FER) probability than OSTBCs. On the other hand, the errors in STTCs are bursty—that is, when a frame is in error, many of its symbols are in error because of the trellis architecture. Therefore, in general, the BER of STTCs may not be as good as other competitive structures, such as OSTBCs. This is because the smaller block size of OSTBCs reduces the chance of bursty errors. The choice, however, between FER and BER as performance measures depends heavily on the application. It is possible to improve the BER of an STTC by using super-orthogonal STTC (SOSTTC) [26, 52].

SOSTTCs bring together the benefits of STBCs and STTCs to boost the performance of space–time coding architectures. The main idea behind those codes is to consider the STBC as a modulation scheme for MPAs. Here, it is important to know that multiplying an STBC-generating matrix $\mathbf{G}$ by a unitary matrix $\mathbf{U}$ gives us a different STBC generating matrix or code but with the same properties as the original code. In this way, each of the possible orthogonal matrices generated by a particular STBC can be viewed as points in a high-dimensional space constellation. Then, the trellis code task is to select one of these high-dimensional signal points to be transmitted based on the current state and input bits. For a SOSTTC using $T \times M$ STBCs, picking a trellis branch emanating from a state is equivalent to transmitting MT symbols from M transmit antennas in t time intervals. By doing so, the diversity of the corresponding STBC is guaranteed. Notice that before, in STTCs, choosing a trellis branch was equivalent to transmitting M symbols from M

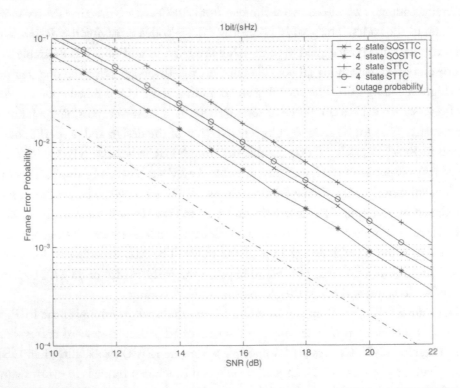

FIGURE 3.6: BER versus SNR curves comparing the performance of STTCs and SOSTTCs. Figure extracted from Ref. [26].

transmit antennas in one time slot. As in the case of STTCs, the design is composed of two steps: the design of the trellis and the set partitioning that assigns a certain symbol (or generating matrix in SOSTTCs) to a certain state in the trellis. The encoding of SOSTTCs is similar to STTCs, and the ML decoding can be done using the Viterbi algorithm as well. The decoding complexity of SOSTTCs is smaller than that of an STTC with the same number of states at a given rate. This is because the orthogonality of the STBC building blocks can be utilized to reduce the decoding complexity. Finally, to summarize, SOSTTCs provide full-diversity, full-rate, and huge coding gains, and they clearly outperform STTCs in terms of coding gain and performance, as shown in Fig. 3.6. Notice, however, that although they achieve great BER and FER performance, the decoding complexity of STTCs and SOSTTCs are still much larger than that of STBCs.

All of the abovementioned space–time coding techniques are, in general, more appropriate for low SNR regimes. In large SNR regimes, SM techniques are a better choice.

3.4.4   Beam Forming

When CSI is available at the transmitter, beam forming can be applied at the transmitter to improve the performance of STCs. Such system is also called *close-loop STC*, and Fig. 3.7 shows a schematic of its architecture. Using beam forming, the transmitting strategy depends on the quality of the CSI. That is, the number of independent beams that can be created depends on the quality of the CSI and, in practical, scenarios depends also on the channel scattering response and the MPA characteristics. At the same time, the CSI depends on the speed of variation of the channel (channel

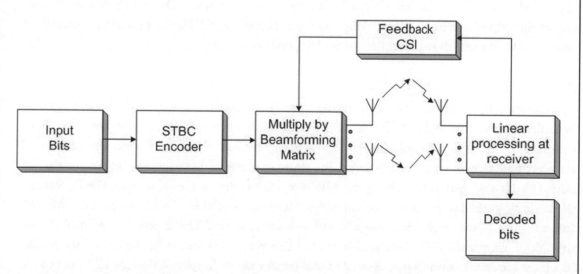

FIGURE 3.7: System architecture of a MIMO system using STBCs and beam forming. Notice the presence of a CSI feedback channel from the receiver to the transmitter.

fading), that is, its coherence time. beam-forming techniques normally have as a goal to maximize the array gain or to establish some tradeoff between diversity and array gain.

Using beam forming, before the codewords are sent out from the M transmit antennas, they are multiplied by a beam-forming matrix $\mathbf{P}$. The actual transmitted codeword can be expressed as $\overline{\mathbf{C}} = \mathbf{CP}$, where $\mathbf{C}$ is the predetermined un-beam-forming STC codeword. $\mathbf{C}$ assures that the scheme will exploit the diversity of the system, whereas $\mathbf{P}$ focuses the energy in the desired directions. The coefficients of $\mathbf{P}$ can be found from the SVD of the channel matrix $\mathbf{H}_0$, and $\mathbf{P}$ is in fact constructed from the eigenvectors of this decomposition. At the receiver, the codeword $\overline{\mathbf{C}}$ is decoded as described in the open-loop approach, in Section 3.4.2.

With perfect CSI, beam forming uses multiple beams, up to M. Those beams can be used to achieve uncorrelated paths, and the system can be thought as applying space–time coding to M directive antennas. That is, the symbols that were originally assigned to a specific port are now assigned to an eigenmode of propagation or, in other words, to a specific beam; hence, each transmitted symbol is weighted with different coefficients. Because each symbol is beam formed, all the symbols are assigned to all the antennas at the same time. In this case, water-filling is used as a power allocation criteria [26], as shown in Section 1.5.4.

On the other hand, practical schemes normally provide some limited feedback, and therefore, in general, the CSI is not perfect. With nonperfect CSI, normally, a fewer number of beams or just the predominant one is used. Let us assume that one beam is used. The antenna system in this case is also called an *adaptive array* or *phase array*. Now, the M symbols are assigned to the same eigenmode, and hence, they are weighed with the same coefficients. This simple but useful case can be thought of as a mechanism to improve the performance of STCs by tuning the transmitted codeword to fit the channel conditions as well as possible.

3.5 SM TECHNIQUES

The use of multiple antennas for increasing the capacity or spectral efficiency of MIMO channels has been demonstrated in Refs. [25, 26, 42, 43]. Therefore, one may transmit at a higher throughput compared with SISO channels. In fact, the capacity analysis shows that when the number of transmit and receive ports are the same, the capacity grows at least linearly by the number of ports [1]. Hence, instead of utilizing the MPAs to achieve the maximum possible (DG), one can use multiple antennas to increase the throughput (transmission rate). This is the goal of SM. SM transmits up to one symbol per transmit port and per time slot. This is equal to M symbols per time slot for a system with M-accessible transmitting ports. Consequently, the use of SM results in a huge increase in throughput, about M times larger than in diversity techniques. The drawback of SM is that most decoding algorithms with reasonable complexity only work when the number of receive ports is at least as many as the number of transmit ports. Another drawback is that they can only achieve a diversity of N, and hence, they cannot achieve the maximum diversity of MN, increasing the probability of error. Summarizing, this extra throughput is achieved in a tradeoff with lower diversity and much higher decoding complexity in addition to the need for a higher number of receive ports.

If the CSI is not available at the transmitter, the maximum achievable capacity that one could achieve with the best SM technique is given by Eq. (1.13), which corresponds to the uninformed transmitter capacity. Notice that for the capacity to increase linearly as a function of the antennas, $N \geq M$. In such a case, the (DG) for each substream is given by $N - M + 1$ [25, 26]. On the other hand, if perfect CSI is available at the transmitter, then beam forming can be used in combination with SM, and together with an optimal power allocation criteria, the capacity is given by that in Eq. (1.17), which corresponds to the water-filling capacity expression.

3.5.1 Bell Labs Layered Space–Time

To exploit the potential of SM, Gerard J. Foschini proposed a layered space–time architecture that was christened Bell Labs Layered Space–Time (BLAST) architecture [53]. In BLAST, multiple data streams are transmitted simultaneously and on the same frequency using a transmit MPA. These different streams can be separated and successfully decoded at the receiver using another MPA. The total transmit power is preserved irrespective of the number of transmit ports. Hence, there is no increase in the amount of interference caused to other users. The transmitter needs no information about the channel, which eliminates the need for fast feedback links. There exist two different versions of BLAST, the D-BLAST and the V-BLAST. The Shannon limit can be approached with D-BLAST, and even attained, but with significant complexity. On the other hand, V-BLAST is much simpler and still attains a hefty portion of the Shannon spectral efficiency but with smaller decoding complexity. The encoder of the D-BLAST is very similar to that of V-BLAST. However, the main difference is in the way that the signals are transmitted from different antennas. In V-BLAST, all signals from each layer are transmitted from the same antenna, whereas in D-BLAST, they are shifted before transmission. This shifting increases the decoding complexity but boosts its performance.

In SM, ML decoding is possible although its complexity is more critical than with STBCs, because it requires a full search of the ML codeword. Other more efficient methods to perform decoding exist, such as sphere decoding and equalization. Equalization techniques are used to separate different symbols. Among the equalization techniques, zero-forcing and minimum mean-squared error equalizer are the most popular. In particular, BLAST uses some of the above equalization techniques. They seek to find the best signal that represents each of the symbols and then decode the symbol using the detected signal. In detecting the best representation of each symbol, the effects of the other symbols are considered as interference. Therefore, all the findings on equalization to remove intersymbol interference can be used in this new context. The final objective is to separate the symbols with a minimum enhancement of the noise. For more detailed information, there is a complete description of those architectures and the possible methods for designing their decoders in Refs. [25, 26].

3.5.2 Combined Diversity Multiplexing Techniques

It is possible to simultaneously combine the use of SM with STCs [25, 26]. The main idea is to divide the M transmit ports into several groups, J. Then, STCs are used to achieve maximum diversity among each group while J independent substreams are transmitted simultaneously. This is a way to design a system that trades off between those two techniques. It can also be used with beam forming when CSI is available at the transmitter, offering a wide variety of possible schemes.

3.6 DESIGN CRITERIA FOR MIMO CODING TECHNIQUES

In this section, we present the main design criteria that are used in the design of MIMO wireless communication systems using STCs and SM.

3.6.1 Symbol Decision Criteria

Let us first review the symbol decision criteria. Assuming the input–output relationship described in Eq. (3.14), to build a decoder for such techniques, one has to consider the distribution of the received signals for a known codeword and channel matrix, $p(\mathbf{R}|\mathbf{C}, \mathbf{H}_0)$, given by

$$p(\mathbf{R}|\mathbf{C}, \mathbf{H}_0) = \left(\frac{\rho_0}{\pi}\right)^{\frac{N \times N}{2}} e^{\left\{\frac{-Tr((\mathbf{R}-\mathbf{CH}_0)^H(\mathbf{R}-\mathbf{CH}_0))}{N_0}\right\}} = \left(\frac{\rho_0}{\pi}\right)^{\frac{N \times N}{2}} e^{\left\{-\rho_0 ||\mathbf{R}-\mathbf{CH}_0||_F^2\right\}}. \quad (3.22)$$

In STCs, ML decoding is used as a decision criterion. The name of ML decoding arises from the fact that this method of decoding decides in favor of a codeword that maximizes $p(\mathbf{R}|\mathbf{C}, \mathbf{H}_0)$. Because the noise is Gaussian distributed, for a fixed $\mathbf{C}$ and $\mathbf{H}_0$, the receiver vector $\mathbf{R}$ is also a multivariate, multidimensional Gaussian random variable, as shown in Ref. [26], and hence the ML decoding is equivalent to minimizing the Frobenius norm of $||\mathbf{R} - \mathbf{CH}_0||_F^2$, such that the decoded codeword is given by

$$\widehat{C} = \arg \min_{C}\{\mathrm{Tr}((\mathbf{R} - \mathbf{CH}_0)^H(\mathbf{R} - \mathbf{CH}_0))\}$$

$$= \arg \min_{C}\{\sum_{n=1}^{N}(\mathbf{H}_{0*n}^H\mathbf{C}^H\mathbf{CH}_{0*n} - 2Re(\mathbf{H}_{0*n}^H\mathbf{C}^H\mathbf{r}_n))\}, \quad (3.23)$$

where $\mathbf{H}_{0*n}$ is the nth column of $\mathbf{H}_0$ and $\mathbf{r}_n$ is the nth column of $\mathbf{R}$. From Eq. (3.23) it is possible to see that, using MRC, the decoding complexity increases linearly with the number of receiving antennas. This is the main reason to use ML decoding. If OSTBCs are used, the decision rule becomes even simpler because in that case the symbols can be decoded separately [25, 26].

3.6.2 Diversity and Coding Gain Design Criteria

In the analysis of the design criteria of STCs, we need to remember that their objective is to reduce the probability of error. For this reason, the design criteria are based upon the reduction of the PEP [26, 51]. There exist many design criteria for STCs and for different propagation conditions. Some of the most relevant are

1. rank and determinant criterion and
2. trace criterion.

Here we will discuss the rank and determinant criterion that concerns the (DG) and also the coding gain. To qualitatively illustrate the above criteria, we assume for now an STC with only two codewords, $\mathbf{C}^1$ and $\mathbf{C}^2$. The analysis for the case with larger number of codewords is outside the framework of this book, and a good reference on it can be found in Refs. [25, 26, 51]. To come up with a design criterion for STCs, first, it is necessary to quantify the effects of mistaking two codewords, $\mathbf{C}^1$ and $\mathbf{C}^2$, with one another. As it is shown in Ref. [26], the PEP, $P(\mathbf{C}^1 \rightarrow \mathbf{C}^2)$, can be computed from the PEP when a fixed known channel is assumed, $P(\mathbf{C}^1 \rightarrow \mathbf{C}^2|\mathbf{H}_0)$, and then the average error can be calculated by computing the expected value over the distribution of $\mathbf{H}_0$. In general, the term $P(\mathbf{C}^1 \rightarrow \mathbf{C}^2|\mathbf{H}_0)$ depends on maximizing $||(\mathbf{C}^1 - \mathbf{C}^2)\mathbf{H}_0||_F^2$ in the following way:

$$P(\mathbf{C}^1 \rightarrow \mathbf{C}^2|\mathbf{H}_0) = Q\left(\sqrt{\frac{\rho_0}{2}}||(\mathbf{C}^1 - \mathbf{C}^2)\mathbf{H}_0||_F^2\right) \leq \frac{1}{2}e^{-\frac{\rho_0}{4}||(\mathbf{C}^1-\mathbf{C}^2)\mathbf{H}_0||_F}, \qquad (3.24)$$

where ρ_0 is the average receive SISO SNR, and Q is the Q function [26]. The term $||(\mathbf{C}^1 - \mathbf{C}^2)\mathbf{H}_0||_F^2$ can be expanded into

$$||(\mathbf{C}^1 - \mathbf{C}^2)\mathbf{H}_0||_F^2 = \text{Tr}(\mathbf{H}_0{}^H A(\mathbf{C}^1, \mathbf{C}^2)\mathbf{H}_0) = \text{Tr}(\mathbf{H}_0{}^H \mathbf{V}^H \Lambda \mathbf{V} \mathbf{H}_0) = \sum_{n=1}^{N} \sum_{m=1}^{M} \lambda_m |\beta_{m,n}|^2.$$
$$(3.25)$$

Notice that $\mathbf{A}(\mathbf{C}^1, \mathbf{C}^2) = \mathbf{D}(\mathbf{C}^1, \mathbf{C}^2)^H \mathbf{D}(\mathbf{C}^1, \mathbf{C}^2)$, with $\mathbf{D}(\mathbf{C}^1, \mathbf{C}^2) = \mathbf{C}^2 - \mathbf{C}^1$, and the SVD have been applied to $\mathbf{A}(\mathbf{C}^1, \mathbf{C}^2) = \mathbf{V}^H \Lambda \mathbf{V}$. Λ is a diagonal matrix with eigenvalues $\{\lambda_1, ... \lambda_M\}$ and M and N being the number of accessible ports from the transmitting and receiving antenna, respectively. $\beta_{m,n}$ denote the (m, n)th element of $\mathbf{V}\mathbf{H}_0$. From Eq. (3.25), we can see that the PEP depends on the rank of the $\mathbf{A}(\mathbf{C}^1, \mathbf{C}^2)$ matrix, as well as on the distribution of $|\beta_{m,n}|$. The larger the number of nonzero singular values λ_m and the larger its value, the lower the error probability. Notice that the number of nonzero values can be related to the (DG) G_d, whereas their value can be related to the coding gain G_c.

Assume a Rayleigh propagation environment, where $\mathbf{H}_0$, as defined in Section 1.5.5, has a probability density function given by $p_{\mathbf{H}_0}(\mathbf{H}_0)$ and $|\beta_{n,m}|$ is Rayleigh distributed. In that case, the PEP can be expressed as

$$P(\mathbf{C}^1 \to \mathbf{C}^2) \leq \frac{4^{rN}}{(\prod_{m=1}^{r} \lambda_m)^N \rho_0^{\,rN}} = (G_c \rho_0)^{-G_d}, \qquad (3.26)$$

where the (DG) is given by $G_d = rN$, the coding gain by $G_c = (\prod_{m=1}^{r} \lambda_m)^{\frac{1}{r}}$, and $r \leq M$ is the number of nonzero eigenvalues. Notice that the (DG) is equal to the rank of the $\mathbf{A}(\mathbf{C}^1, \mathbf{C}^2)$ matrix multiplied by the number of receiver antennas. The coding gain relates to the product of the nonzero eigenvalues of the matrix $\mathbf{A}(\mathbf{C}^1, \mathbf{C}^2)$ or, equivalently, to the determinant of the $\mathbf{A}(\mathbf{C}^1, \mathbf{C}^2)$. The larger this determinant, the larger the coding gain (determinant criterion for the coding gain). The coding gain distance is defined as $\mathrm{CGD}(\mathbf{C}^1, \mathbf{C}^2) = \det(\mathbf{A}(\mathbf{C}^1, \mathbf{C}^2))$. In the general case where there exists a correlation among the entries in the channel matrix, the (DG) as well as the coding gain may be reduced. Notice also that, in Eq. (3.26), it has been assumed that MRC is being used at the receiver as a combining technique.

Therefore, a good design criterion to guarantee full diversity is to make sure that for all possible codewords $\mathbf{C}^i$ and $\mathbf{C}^j$, $i \neq j$ obtained from a given generating matrix $\mathbf{G}$, the matrix $\mathbf{A}(\mathbf{C}^i, \mathbf{C}^j)$ is full rank (rank criterion). Then, to increase the coding gain for a full diversity code, an additional good design criterion is to maximize the minimum determinant of the matrices $\mathbf{A}(\mathbf{C}^i, \mathbf{C}^j)$ for all $i \neq j$ (determinant criterion). It is important to know that the rank criterion described above, as well as the other criteria, are valid for both STTCs and STBCs.

3.6.3   Array Versus DG Design Criteria

In Refs. [25, 26], it is shown that the optimal beam-forming matrix, depending on the quality of the CSI available at the transmitter, can be obtained by minimizing the maximum value of the PEP, which is the one that dominates the performance. This provides a criterion to choose the optimal beam-forming matrix $\mathbf{P}$. Following the approach in Ref. [26], assume that partial CSI is known at the transmitter, given by the conditional mean of the path gains and the covariance between different path gains. Assume also that the true channel matrix is represented by $\mathbf{H}_0$, whereas the estimated channel matrix is represented by $\widehat{\mathbf{H}}_0$. To qualitatively illustrate the criteria, we assume once again an STC with only two codewords, $\mathbf{C}^1$ and $\mathbf{C}^2$.

Because CSI is available at the transmitter, we are not interested in computing the PEP itself, $P(\mathbf{C}^1 \to \mathbf{C}^2)$, but the PEP, conditional to the fact that we have some estimation of the channel matrix at the transmitter, $P(\mathbf{C}^1 \to \mathbf{C}^2 | \widehat{\mathbf{H}}_0)$. The maximum PEP for a given channel estimation

$P(\mathbf{C}^1 \rightarrow \mathbf{C}^2 | \widehat{\mathbf{H}}_0)$ can be computed from the PEP when a fixed and an estimated known channels are assumed, $P(\mathbf{C}^1 \rightarrow \mathbf{C}^2 | \mathbf{H}_0, \widehat{\mathbf{H}}_0)$, and then, the average error can be calculated by computing the expected value over the distribution $p_{\mathbf{H}_0 | \widehat{\mathbf{H}}_0}(\mathbf{H}_0 | \widehat{\mathbf{H}}_0)$. The conditional PEP is $P(\mathbf{C}^1 \rightarrow \mathbf{C}^2 | \mathbf{H}_0, \widehat{\mathbf{H}}_0)$ and also depends on maximizing $||(\mathbf{C}^1 - \mathbf{C}^2)\mathbf{H}_0||_F^2$ in the following way:

$$P(\mathbf{C}^1 \rightarrow \mathbf{C}^2 | \mathbf{H}_0, \widehat{\mathbf{H}}_0) = Q(\sqrt{\frac{\rho_0}{2}} ||(\mathbf{C}^1 - \mathbf{C}^2)\mathbf{H}_0||_F^2) \leq \frac{1}{2} e^{-\frac{\rho_0}{4}||(\mathbf{C}^1 - \mathbf{C}^2)\mathbf{H}_0||_F}. \qquad (3.27)$$

However, notice that now we will integrate using the distribution $p_{\mathbf{H}_0 | \widehat{\mathbf{H}}_0}(\mathbf{H}_0 | \widehat{\mathbf{H}}_0)$ given in Ref. [26], instead of $p_{\mathbf{H}_0}(\mathbf{H}_0)$. The upper bound $V(\mathbf{C}^1, \mathbf{C}^2)$ on $P(\mathbf{C}^1 \rightarrow \mathbf{C}^2 | \widehat{\mathbf{H}}_0) \leq V(\mathbf{C}^1, \mathbf{C}^2)$ is the design criterion. The design criterion can be proven to result in the minimization of this upper bound $V(\mathbf{C}^1, \mathbf{C}^2)$ that at the same time is the result of the sum of two terms $V(\mathbf{C}^1, \mathbf{C}^2) = \min(V_1 + V_2)$. The first term, V_1, is a function of the channel mean and becomes the dominating factor when perfect CSI is available. The second term, V_2, is basically the same as the determinant criterion for STCs. Therefore, V_2 is a good indication of the degree of spatial diversity of the system. When the quality of the CSI is low, the second term becomes dominant, and most of the performance gains come from diversity. Optimizing the beam-forming matrix $\mathbf{P}$ based on the above design criterion results in a system that performs well in the case of partial CSI while providing optimal performance in the other cases. When there is no CSI, the transmit energy is evenly distributed on all the beams, and the system behaves like a system using only STBCs. When the CSI at the transmitter is perfect, all the energy is allocated on the strongest beam, becoming a one-directional beam forming.

3.6.3.1 Power Allocation Criteria.
Two power allocation criteria exist for STCs and SM: uniform power loading and water-filling (also called *water-pouring*). These two method were explained in detail in Sections 1.5.3 and 1.5.4.

3.7   PERFORMANCE ANALYSIS

In this section, we qualitatively review the performance of different space–time coding and SM architectures on different propagation scenarios. The spectral efficiency or the (DG) advantage of MIMO is clearly very large in all cases with respect to the baseline single-antenna system. When using MEAs with sufficient scattering, the advantage is particularly strong.

3.7.1   Performance of Space–Time Coding Techniques

Fixing the channel to be that of iid Gaussian entries, STTCs architectures, in general, outperform STBCs architectures in terms of FER. In terms of BER, some STBCs may outperform STTCs because of the bursty errors in STTCs. However, as noted previously, SOSTTCs boost the performance of STTCs far above any other technique. In terms of complexity, any STBC architecture is much simpler than any STTCs. Curiously, the complexity of SOSTTCs is smaller than that of STTCs. The reason why STTCs outperform STBCs is because of larger decoding complexity and higher coding gain. If CSI is available at the transmitter, STTCs and STBCs with beam forming outperform their counterparts without beam forming. As a function of the number of ports, as those increase, in both STBCs and STTCs, the BER improves, while its outage capacity increases logarithmically because only one data stream of information is sent and no multiplexing exists.

3.7.2   Performance of SM Techniques

In the analysis of the performance of SM techniques on different propagation scenarios, we need to remember that their objective is to maximize the transmission rate by simultaneously sending multiple substreams of information. Let us start by looking at how the various SM techniques behave as a function of the scattering richness, represented by the K factor, K. The K factor is defined as the ratio of deterministic to scattered power on a given channel. In terms of spectral

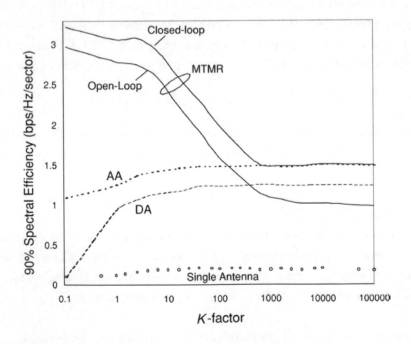

FIGURE 3.8: Performance of SM techniques (BLAST) over channel propagation environments with different K factors. Figure extracted from Ref. [54]

efficiency, BLAST with beam forming (known as close-loop BLAST or BLAST with CSI at the transmitter) outperforms all the other techniques in highly scattered environments, as shown in Fig. 3.8. Nonetheless, its robustness is remarkable because its spectral efficiency does not drop significantly with Ricean K factors as large as $K = 10$, which covers most cases of practical interest [54]. On the other hand, phase-array (adaptive array) beam forming using one single beam improves capacity monotonically with K. In the limit nonscattered scenario, when $K \rightarrow \infty$, the performance of phase arrays using one data stream becomes identical to that of the close-loop BLAST, clearly indicating that phase array is the most adequate architecture in such conditions.

Focusing now on the number of ports, in iid Gaussian channels, it can be observed that with phase array, the efficiency growth is roughly logarithmic, whereas with BLAST, even in open loop, the growth is basically linear [55]. Finally, open-loop BLAST, if sufficient scattering is present, yields very close spectral efficiency to the close-loop BLAST, and hence, it becomes a very attractive option because it eliminates the need for fast feedback links while still providing extremely large average and peak efficiencies. In terms of minimizing the complexity of the system, in rich scattering environments, the best option is to use open-loop BLAST. As scattering richness decreases ($K > 100$), the best option becomes a phase-array architecture.

•   •   •   •

CHAPTER 4

Antenna Modeling for MIMO Communication Systems

4.1 INTRODUCTION

With the introduction of MIMO wireless communications systems [1], it has become necessary to integrate system level and channel characteristics parameters into the design of multiantenna systems. In addition, the modeling of any MPA must lead to its simple interconnection with the other parts of a MIMO communication system. To do so, and to be able to correctly assess the performance of such systems, a network theory approach needs to be used. In particular, each one of the components that composes the RF chain of a MIMO communication system, such as the antennas, propagation channel, and so forth, has been modeled as a network and described by a scattering matrix [56].

A scattering matrix describes the relationship among the incident and reflected waves in a particular microwave network. Let us consider now the Bth component of a MIMO system. Its scattering matrix representations is of the form

$$\begin{pmatrix} \mathbf{b}_1^B \\ \mathbf{b}_2^B \end{pmatrix} = \begin{pmatrix} \mathbf{S}_{11}^B & \mathbf{S}_{12}^B \\ \mathbf{S}_{21}^B & \mathbf{S}_{22}^B \end{pmatrix} \begin{pmatrix} \mathbf{a}_1^B \\ \mathbf{a}_2^B \end{pmatrix}, \tag{4.1}$$

where $\mathbf{a}_1^B \in \mathcal{C}^{X \times 1}$ and $\mathbf{b}_1^B \in \mathcal{C}^{X \times 1}$ are the inward and outward propagating wave vectors (with units $\sqrt{W}$), respectively, of port group 1 with X ports, and $\mathbf{a}_2^B \in \mathcal{C}^{Y \times 1}$ and $\mathbf{b}_1^B \in \mathcal{C}^{Y \times 2}$ are the inward and outward propagating wave vectors, respectively, of port group 2 with Y ports, all of them referred to the Bth component of the system. Notice that an antenna group is defined as a set of ports that are at the same physical interface of the RF component and numbered using the same subindex. The submatrices $\mathbf{S}_{ii}^B$ represent the reflection coefficient of the wave vector $\mathbf{a}_i^B$ to $\mathbf{b}_i^B$, and $\mathbf{S}_{ij}^B$, with $i \neq j$, represents the transmission coefficient from $\mathbf{a}_j^B$ to $\mathbf{b}_i^B$.

If the kth port group corresponds to the input ports of the Bth component, then the instantaneous real power delivered to the port group k is given by

$$P_k^B = (\mathbf{a}_k^B)^H \mathbf{a}_k^B - (\mathbf{b}_k^B)^H \mathbf{b}_k^B, \tag{4.2}$$

and the average real power delivered to the port group k, $P_k^{B,\mathrm{av}}$, is henceforth given by $E\{P_k^B\}$. Notice also that the instantaneous real power delivered to the ith port of group k is then given by

$$P_{k_i}^B = (\mathbf{a}_{k_i}^B)^H \mathbf{a}_{k_i}^B - (\mathbf{b}_{k_i}^B)^H \mathbf{b}_{k_i}^B. \tag{4.3}$$

On the other hand, if the kth port group corresponds to the output ports of the Bth component, then the instantaneous real power delivered to the port group k is given by

$$P_k^B = (\mathbf{b}_k^B)^H \mathbf{b}_k^B - (\mathbf{a}_k^B)^H \mathbf{a}_k^B. \tag{4.4}$$

We will use $\mathbf{Z}_k^0$ to refer to the characteristic reference impedance matrix associated with the kth group of the system. $\mathbf{Z}_k^0$ is always a diagonal matrix, and unless specified, we will assume the characteristic reference impedances to be the same for all the ports within a group but different among groups. That is, $\mathbf{Z}_k^0 = Z_k^0 \mathbf{I}$.

The voltages at the port group k can be computed using

$$\mathbf{U}_k^B = Z_k^{0\frac{1}{2}} (\mathbf{a}_k^B + \mathbf{b}_k^B). \tag{4.5}$$

The relationship given in Eq. (4.5) is called the *Heavyside transformation* [15]. On the other hand, the currents at the port group k are given by

$$\mathbf{I}_k^B = Z_k^{0-\frac{1}{2}} (\mathbf{a}_k^B - \mathbf{b}_k^B). \tag{4.6}$$

Notice that we have assumed that $\mathbf{a}_k$ and $\mathbf{b}_k$ correspond to root mean square (RMS) values, instead of peak values. Otherwise, in the previous expressions, one would need to additionally multiply by a factor of $\frac{1}{2}$. Finally, for the purposes of this book, and unless otherwise specified, we will assume the characteristic reference impedances of the all groups to be the same, that is, $Z_k^0 = Z^0$ for all k.

4.2   GENERAL MODEL FOR AN MPA

It is well-known that when the antennas of an MEA are placed at a close distance, they will couple and interact with each other [10]. As result, their electrical parameters, such as the radiation pattern and the input impedance, will change. Therefore, each one of the antennas of the MEA cannot be considered as independent/isolated elements. A similar thing happens in any MPA structure. Such effects have to be taken into account for an MPA model to be accurate. In

Refs. [15, 57, 58], a scattering parameters description is used to model MPAs. The model includes all the aforementioned effects, and it is the one followed in this book.

4.2.1   Scattering Matrix Representation

As shown in Refs. [57, 59, 60], it is convenient to expand the radiated fields of an MPA in terms of spherical waves, whose angular dependence is simply related to the various spherical modes (harmonics). By doing this, any electromagnetic field can be fully characterized by the amplitudes of these various spherical waves, with respect to an arbitrary coordinate system origin. This representation in terms of spherical waves is equivalent to the introduction of terminals with incident and reflected wave amplitudes for each of the spherical waves, and allows us to describe an MPA by means of a scattering matrix.

Thus, we model an MPA with n-accessible ports, as a multiport network. Using this network representation, an MPA is normally assumed to be enclosed within a virtual sphere of radius a, as shown in Fig. 4.1a. a is the minimum radius that encloses the MPA. Outside this sphere, the radiated fields are described by a finite number of spherical modes that depend on the electrical size of the MPA, in terms of the wavelength. The MPA ports are divided into two categories: the wave-guided ports (α ports) and the radiated ports (β ports) (see Fig. 4.1b). The α ports correspond to the n physically accessible guided wave ports of the MPA, whereas the β-ports are associated with the set of spherical modes, the superposition of which describes the radiation characteristics of the MPA [57, 59].

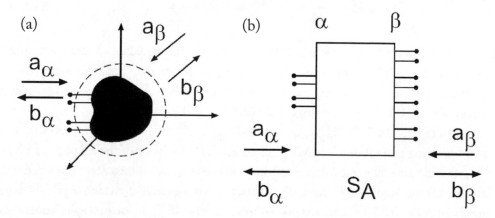

FIGURE 4.1: (a) Modeling of an MPA in terms of incoming and outgoing spherical modes. Notice that MPA is enclosed within a virtual sphere of radius a. (b) Distribution of the α ports and β ports within the network representation of an MPA.

Let us define $\mathbf{S}^A$ as the scattering block matrix of an MPA. The expanded scattering model of an MPA is then given by

$$\mathbf{b}^A = \begin{pmatrix} \mathbf{b}^A_\alpha \\ \mathbf{b}^A_\beta \end{pmatrix} = \begin{pmatrix} \mathbf{S}^A_{\alpha\alpha} & \mathbf{S}^A_{\alpha\beta} \\ \mathbf{S}^A_{\beta\alpha} & \mathbf{S}^A_{\beta\beta} \end{pmatrix} \begin{pmatrix} \mathbf{a}^A_\alpha \\ \mathbf{a}^A_\beta \end{pmatrix} = \mathbf{S}^A \mathbf{a}^A, \tag{4.7}$$

where $\mathbf{a}^A_\alpha \in \mathcal{C}^{n\times 1}$ and $\mathbf{b}^A_\alpha \in \mathcal{C}^{n\times 1}$ are vectors describing the conventional incident and reflected scattering variables at the α ports of the MPA and $\mathbf{a}^A_\beta \in \mathcal{C}^{\infty\times 1}$ and $\mathbf{b}^A_\beta \in \mathcal{C}^{\infty\times 1}$ are infinite-dimensional column vectors representing the complex amplitudes of the incoming and outgoing spherical waves, respectively. Despite this, in practice, $\mathbf{a}^A_\beta$ and $\mathbf{b}^A_\beta$ are finite dimensional vectors because, in general, the number of significant β ports is finite. The upper-index A can either refer to the transmit (T) or receive (R) antennas.

On the other hand, $\mathbf{S}^A_{\alpha\alpha} \in \mathcal{C}^{n\times n}$ is a scattering matrix describing the coupling and reflection coefficients among the α ports of the MPA. These coefficients can also be expressed in terms of the impedance matrix, $\mathbf{Z}^A \in \mathcal{C}^{n\times n}$, given by

$$\mathbf{Z}^A = Z^0(\mathbf{I} + \mathbf{S}^A_{\alpha\alpha})(\mathbf{I} - \mathbf{S}^A_{\alpha\alpha})^{-1}. \tag{4.8}$$

The impedance matrix $\mathbf{Z}^A$, as well as the scattering matrix $\mathbf{S}^A_{\alpha\alpha}$, of a microwave network or MPA can be related to the voltages and the currents. The element Z^A_{ij} can be found by driving the port j with the current $\mathbf{I}^A_{\alpha_j}$, open-circuiting all the other ports, and measuring the open-circuit voltage at port i. That is,

$$Z^A_{ij} = \frac{\mathbf{U}^A_{\alpha_i}}{\mathbf{I}^A_{\alpha_j}} \text{ when } \mathbf{I}^A_{\alpha_k} = 0 \text{ for } k \neq j. \tag{4.9}$$

It is worthwhile to mention that the scattering matrix $\mathbf{S}^A$, the impedance matrix $\mathbf{Z}^A$, and the admittance matrix $\mathbf{Y}^A = \mathbf{Z}^{A-1}$ have the same set of eigenvectors and that their eigenvalues are simply algebraically related. Therefore, the approach followed here, based on a scattering matrix representation of an MPA, does not change if $\mathbf{Z}^A$ or $\mathbf{Y}^A$ is used instead.

The matrices $\mathbf{S}^A_{\alpha\beta} \in \mathcal{C}^{n\times\infty}$, $\mathbf{S}^A_{\beta\alpha} \in \mathcal{C}^{\infty\times n}$, and $\mathbf{S}^A_{\beta\beta} \in \mathcal{C}^{\infty\times\infty}$, describe the receiving, transmitting, and scattering properties of the MPA, respectively. In each one of the columns of $\mathbf{S}^A_{\beta\alpha}$ and rows of $\mathbf{S}^A_{\alpha\beta}$, we will name them modal transmitting and receiving radiation patterns, respectively, to distinguish them from conventional far-field radiation pattern quantities, which explicitly depend on the angular coordinates. That is, the ith column of $\mathbf{S}_{\beta\alpha}$ ($\mathbf{S}^A_{\beta\alpha\cdot_i}$), for example, contains the spherical harmonics coefficients of the modal radiation pattern of the MPA when the ith port is excited, and all the other ports are terminated with its characteristic reference impedance Z^0. Notice that $\mathbf{S}^A_{\beta\alpha_i\cdot}$ represents the ith row of $\mathbf{S}^A_{\beta\alpha}$.

Finally, the (i,j)th elements of $\mathbf{S}^A_{\alpha\alpha}$, $\mathbf{S}^A_{\beta\alpha}$, $\mathbf{S}^A_{\alpha\beta}$, and $\mathbf{S}^A_{\beta\beta}$ are defined as

$$\mathbf{S}^A_{\alpha\alpha_{ij}} = \frac{\mathbf{b}^A_{\alpha_i}}{\mathbf{a}^A_{\alpha_j}} \text{ when } \mathbf{a}^A_{\alpha_k} = 0 \text{ for } k \neq j; \tag{4.10}$$

$$\mathbf{S}^A_{\beta\alpha_{ij}} = \frac{\mathbf{b}^A_{\beta_i}}{\mathbf{a}^A_{\alpha_j}} \text{ when } \mathbf{a}^A_{\alpha_k} = 0 \text{ for } k \neq j; \tag{4.11}$$

$$\mathbf{S}^A_{\alpha\beta_{ij}} = \frac{\mathbf{b}^A_{\alpha_i}}{\mathbf{a}^A_{\beta_j}} \text{ when } \mathbf{a}^A_{\beta_k} = 0 \text{ for } k \neq j; \tag{4.12}$$

$$\mathbf{S}^A_{\beta\beta_{ij}} = \frac{\mathbf{b}^A_{\beta_i}}{\mathbf{a}^A_{\beta_j}} \text{ when } \mathbf{a}^A_{\beta_k} = 0 \text{ for } k \neq j. \tag{4.13}$$

4.2.2 Fundamental Scattering Properties

We review now some of the fundamental scattering properties of the MPAs. Assume a lossless MPA. In that case, power must be conserved, which implies $(\mathbf{a}^A)^H\mathbf{a}^A = (\mathbf{b}^A)^H\mathbf{b}^A$, and consequently, the scattering block matrix $\mathbf{S}^A$ must be unitary [56]. That is, $(\mathbf{S}^A)^H\mathbf{S}^A = \mathbf{I}$, and therefore, it must hold

$$(\mathbf{S}^A_{\beta\alpha})^H\mathbf{S}^A_{\beta\alpha} = \mathbf{I} - (\mathbf{S}^A_{\alpha\alpha})^H\mathbf{S}^A_{\alpha\alpha} \tag{4.14}$$

$$(\mathbf{S}^A_{\alpha\beta})^H\mathbf{S}^A_{\alpha\beta} = \mathbf{I} - (\mathbf{S}^A_{\beta\beta})^H\mathbf{S}^A_{\beta\beta} \tag{4.15}$$

$$(\mathbf{S}^A_{\alpha\alpha})^H\mathbf{S}^A_{\alpha\beta} = -(\mathbf{S}^A_{\beta\alpha})^H\mathbf{S}^A_{\beta\beta} \tag{4.16}$$

$$(\mathbf{S}^A_{\alpha\beta})^H\mathbf{S}^A_{\alpha\alpha} = -(\mathbf{S}^A_{\beta\beta})^H\mathbf{S}^A_{\beta\alpha}. \tag{4.17}$$

In general, however, for a lossy antenna, $(\mathbf{S}^A_{\beta\alpha})^H\mathbf{S}^A_{\beta\alpha} \leq \mathbf{I} - (\mathbf{S}^A_{\alpha\alpha})^H\mathbf{S}^A_{\alpha\alpha}$. Notice that a lossless antenna also satisfies $\mathbf{S}^A(\mathbf{S}^A)^H = \mathbf{I}$, and therefore, it must hold $\mathbf{I} - (\mathbf{S}^A_{\alpha\alpha})^H\mathbf{S}^A_{\alpha\alpha} = (\mathbf{I} - \mathbf{S}^A_{\alpha\alpha}(\mathbf{S}^A_{\alpha\alpha})^H)^*$ and $(\mathbf{S}^A_{\beta\alpha})^\dagger(\mathbf{S}^A_{\beta\alpha})^* = \mathbf{I} - \mathbf{S}^A_{\alpha\alpha}(\mathbf{S}^A_{\alpha\alpha})^H$. On the other hand, for a reciprocal MPA, it must be satisfied $\mathbf{S}^A_{\alpha\beta} = (\mathbf{S}^A_{\beta\alpha})^\dagger$ [56, 58], and the expanded scattering model of a reciprocal MPA is then given by

$$\mathbf{b}^A = \begin{pmatrix} \mathbf{b}^A_\alpha \\ \mathbf{b}^A_\beta \end{pmatrix} = \begin{pmatrix} \mathbf{S}^A_{\alpha\alpha} & (\mathbf{S}^A_{\beta\alpha})^\dagger \\ \mathbf{S}^A_{\beta\alpha} & \mathbf{S}^A_{\beta\beta} \end{pmatrix} \begin{pmatrix} \mathbf{a}^A_\alpha \\ \mathbf{a}^A_\beta \end{pmatrix} = \mathbf{S}^A\mathbf{a}^A. \tag{4.18}$$

Therefore, reciprocal MPAs have equal transmit and receive radiation properties. For an MPA in reception, one may assume for now that $\mathbf{a}_\alpha^A = \mathbf{0}$ and that the incoming modal amplitude vector $\mathbf{a}_\beta^A$ generates the output

$$\mathbf{b}_\alpha^A = \mathbf{S}_{\alpha\beta}^A \mathbf{a}_\beta^A, \tag{4.19}$$

where the elements of $\mathbf{a}_\beta^A$ must be normalized such that $|\mathbf{b}_\alpha^A|$ equals the received power. $\mathbf{b}_\alpha^A$ is then the absorbed wave by the antenna. For an MPA in transmission, $\mathbf{a}_\beta^A = \mathbf{0}$, and the outgoing modal amplitudes are

$$\mathbf{b}_\beta^A = \mathbf{S}_{\beta\alpha}^A \mathbf{a}_\alpha^A, \tag{4.20}$$

whereas the reflection back into the antenna ports is

$$\mathbf{b}_\alpha^A = \mathbf{S}_{\alpha\alpha}^A \mathbf{a}_\alpha^A. \tag{4.21}$$

Using the above notation, we can also define other important quantities of an MPA. For an MPA in reception, it is sometimes interesting to know the scattered fields, $\mathbf{b}_{\text{scat}}^A$. The scattered fields are defined as the difference between the reflected fields when an antenna is present and those when no antenna is present. The former field is given by $\mathbf{S}_{\beta\beta}^A \mathbf{a}_\beta^A$ and the latter field by $\mathbf{I}\mathbf{a}_\beta^A$. Thus, the modal amplitudes corresponding to the fields scattered by an MPA—that is, the waves reradiated back into the surrounding space—can be expressed as

$$\mathbf{b}_{\text{scat}}^A = \mathbf{b}_\beta^A - \mathbf{a}_\beta^A = (\mathbf{S}_{\beta\beta}^A - \mathbf{I})\mathbf{a}_\beta^A. \tag{4.22}$$

To understand the meaning of the term *scattered field*, one must know that the physical mechanism through which an antenna absorbs power from a wave $\mathbf{a}_\beta^A$ incident from the surrounding space is destructive interference, that is, the antenna scatters so as to cancel some of the incident fields.

4.2.3 Normalized Far-Field Radiation Pattern

For a reciprocal MPAs, as assumed in this book, we define the normalized far-field radiation pattern associated with the α port i of an MPA, $\mathbf{F}_i^A(\theta, \phi) \in \mathcal{C}^{2\times 1}$, with the following modal expansion:

$$\mathbf{F}_i^A(\theta, \phi) = \sum_{j=1}^{\infty} \mathbf{S}_{\beta\alpha_{ji}}^A \mathbf{\Upsilon}_j(\theta, \phi), \tag{4.23}$$

where θ and ϕ are the zenith and azimuth angles, respectively, in the spherical coordinate system, and $\mathbf{\Upsilon}_j(\theta, \phi) \in \mathcal{C}^{2\times 1}$ forms a complete set of orthonormal transverse spherical harmonic functions as defined in Refs. [61, 57, 58]. The upper-index A can either refer to the transmit (T) or receive (R) antennas. Notice that the normalized far-field radiation pattern associated with the ith port is computed, whereas all the other ports are terminated with the characteristic reference impedance,

Z^0. For a receiving MPA, the normalized far-field radiation pattern is equivalently defined using the coefficients of $\mathbf{S}^A_{\alpha\beta_{i^*}}$. For simplicity, a shorthand notation based on a single index j has been used here to designate the double index of the associated vector spherical harmonic as well as the spherical mode type (E or H) [57, 58]. The mode orthogonality can then be expressed as

$$\int_{-\pi}^{\pi}\int_0^{\pi}(\mathbf{\Upsilon}_j(\theta,\phi))^H\mathbf{\Upsilon}_i(\theta,\phi)\sin(\theta)\mathrm{d}\theta\mathrm{d}\phi = \delta_{ji}, \qquad (4.24)$$

where δ_{ji} is a function defined through the following expression:

$$\delta_{ji} = \begin{cases} 1 \text{ if } i=j \\ 0 \text{ if } i \neq j. \end{cases} \qquad (4.25)$$

Notice that the radiation pattern $\mathbf{F}^A_i(\theta,\phi)$ is a two-dimensional vector composed of θ-($\mathbf{F}^A_{i_\theta}(\theta,\phi)$) and ϕ-polarized components ($\mathbf{F}^A_{i_\phi}(\theta,\phi)$), as shown

$$\mathbf{F}^A_i(\theta,\phi) = \mathbf{F}^A_{i_\theta}(\theta,\phi)\widehat{\theta} + \mathbf{F}^A_{i_\phi}(\theta,\phi)\widehat{\phi} = \begin{pmatrix} F^A_{i_\theta}(\theta,\phi) \\ F^A_{i_\phi}(\theta,\phi) \end{pmatrix}, \qquad (4.26)$$

where $\widehat{\theta}$ and $\widehat{\phi}$ are the zenith and azimuth unitary vectors of the spherical coordinate system, respectively. Finally, notice also from Eq. (4.23) that the coefficients of $\mathbf{S}^A_{\beta\alpha_{*i}}$ are the coefficients of the spherical mode expansion [58] of $\mathbf{F}^A_i(\theta,\phi)$ and that $\mathbf{F}^A_i(\theta,\phi)$ is the radiation pattern including the coupling effects from the surrounding antennas, which by definition are taken into account in the $\mathbf{S}^A_{\beta\alpha}$ term. Eq. (4.23) also shows that the radiation pattern associated with a particular port within an MPA will be the superposition of the driven port pattern and the pattern of the coupled ports, weighted by the induced currents. In general, the shape of the resultant radiation pattern will depend on both the mutual coupling characteristics and the load attached to the coupled elements.

For any given port, the coefficients of the modal far-field radiation pattern must be normalized, such that they satisfy [57]:

$$\int_{-\pi}^{\pi}\int_0^{\pi}(\mathbf{F}^A_i(\theta,\phi))^H\mathbf{F}^A_i(\theta,\phi)\sin(\theta)\mathrm{d}\theta\mathrm{d}\phi = (\mathbf{S}^A_{\beta\alpha_{*i}})^H\mathbf{S}^A_{\beta\alpha_{*i}} \leq 1 - (\mathbf{S}^A_{\alpha\alpha_{*i}})^H\mathbf{S}^A_{\alpha\alpha_{*i}}, \quad (4.27)$$

where the equality applies to a lossless antenna and where $\mathbf{S}^A_{\alpha\alpha_{*i}}$ represents the ith column of $\mathbf{S}^A_{\alpha\alpha}$.

For an MPA in transmission, using the above normalization, the radiated electric field ($\mathbf{E}^T_i(\mathbf{r},\theta,\phi)$) and radiated magnetic field ($\mathbf{H}^T_i(\mathbf{r},\theta,\phi)$), at a given point at coordinates $(\mathbf{r},\theta,\phi)$, associated with the ith port of that MPA, can be expressed as follows:

$$\mathbf{E}^T_i(\mathbf{r},\theta,\phi) = a^A_{\alpha_{Ti}}\sqrt{\eta_0}\frac{\mathrm{e}^{-jkr}}{r}\mathbf{F}^A_i(\theta,\phi) \qquad \text{V/m} \qquad (4.28)$$

$$\mathbf{H}_i^T(\mathbf{r}, \theta, \phi) = \mathbf{a}_{\alpha_{Ti}}^A \frac{1}{\sqrt{\eta_0}} \frac{e^{-jkr}}{r} \mathbf{F}_i^A(\theta, \phi) \qquad \text{A/m}, \qquad (4.29)$$

where $\eta_0 = \sqrt{\frac{\mu_0}{\varepsilon_0}} = 377\ \Omega$ is the free-space characteristic impedance, and k is the wave number at the operating frequency.

In the literature, the radiation pattern is sometimes normalized to have unit gain. We refer to the latest by $\mathbf{F}_i^{A,0}(\theta, \phi)$. This last one relates to $\mathbf{F}_i^A(\theta, \phi)$ by making explicit the directivity parameters as follows, $\mathbf{F}_i^A(\theta, \phi) = \sqrt{\frac{G_i}{4\pi}} \mathbf{F}_i^{A,0}(\theta, \phi)$. Therefore, notice that the directivity parameters are implicitly included in $\mathbf{F}_i^A(\theta, \phi)$.

4.2.4 Fundamental Radiation Parameters

Let us define now some of the fundamental radiation parameters associated with MPAs. We begin by defining the average power density (average Poynting vector), associated with the ith port of a transmitting MPA, as

$$\mathbf{W}_i^T(\mathbf{r}, \theta, \phi) = \text{Re}(\mathbf{E}_i^T(\mathbf{r}, \theta, \phi) \times \mathbf{H}_i^{T*}(\mathbf{r}, \theta, \phi)) \qquad \text{W/m}^2. \qquad (4.30)$$

Notice that a $\frac{1}{2}$ term should be added to multiply the above expression if $\mathbf{E}_i^T(\mathbf{r}, \theta, \phi)$ and $\mathbf{H}_i^T(\mathbf{r}, \theta, \phi)$ represent field peak values instead of RMS values. In the previous expression, we have assumed a harmonic time variation of the electric and magnetic fields, and the average has been calculated over one period. The total radiated power from the ith α_T port, $P_{\alpha_{Ti}}^{av}$, is then computed using

$$P_{\alpha_{Ti}}^{T,\text{av}} = \int_{-\pi}^{\pi} \int_0^{\pi} \mathbf{W}_i^T(\mathbf{r}, \theta, \phi) d\mathbf{S} \qquad \text{W}, \qquad (4.31)$$

where $d\mathbf{S} = r^2 \sin(\theta) d\theta d\phi$. Finally, the radiation intensity is defined as $U_i^T(\theta, \phi) = r^2 |\mathbf{W}_i^T(\mathbf{r}, \theta, \phi)|$, and the total radiated power can also be computed as

$$P_{\alpha_{Ti}}^{T,\text{av}} = \int_{-\pi}^{\pi} \int_0^{\pi} U_i^T(\theta, \phi) \sin(\theta) d\theta d\phi. \qquad (4.32)$$

If we use now the fact that $|\mathbf{W}_i^T(\mathbf{r}, \theta, \phi)| = \frac{1}{\eta_0}(\mathbf{E}_i^T(\mathbf{r}, \theta, \phi))^H \mathbf{E}_i^T(\mathbf{r}, \theta, \phi) = \frac{1}{\eta_0}|\mathbf{E}_i^T(\mathbf{r}, \theta, \phi)|^2$, then the total radiated power can finally be expanded into

$$\begin{aligned}
P_{\alpha_{Ti}}^{T,\text{av}} &= \frac{1}{\eta_0} \int_{-\pi}^{\pi} \int_0^{\pi} |\mathbf{E}_i^T(\mathbf{r}, \theta, \phi)|^2 r^2 \sin(\theta) d\theta d\phi \\
&= (\mathbf{a}_{\alpha_{Ti}}^T)^* \left(\int_{-\pi}^{\pi} \int_0^{\pi} (\mathbf{F}_i^T(\theta, \phi))^H \mathbf{F}_i^T(\theta, \phi) \sin(\theta) d\theta d\phi \right) \mathbf{a}_{\alpha_{Ti}}^T,
\end{aligned} \qquad (4.33)$$

and finally, using Eq. (4.27), it reduces to

$$P_{\alpha_{Ti}}^{T,\text{av}} = (\mathbf{a}_{\alpha_{Ti}}^T)^* (\mathbf{S}_{\beta_T \alpha_{T^*i}}^T)^H \mathbf{S}_{\beta_T \alpha_{T^*i}}^T \mathbf{a}_{\alpha_{Ti}} \leq (\mathbf{a}_{\alpha_{Ti}}^T)^* \left(1 - (\mathbf{S}_{\alpha_T \alpha_{T^*i}}^T)^H \mathbf{S}_{\alpha_T \alpha_{T^*i}}^T\right) \mathbf{a}_{\alpha_{Ti}}$$

$$= |\mathbf{a}_{\alpha_{Ti}}^T|^2 \left(1 - (\mathbf{S}_{\alpha_T \alpha_{T^*i}}^T)^H \mathbf{S}_{\alpha_T \alpha_{T^*i}}^T\right). \tag{4.34}$$

We now define $\mathbf{F}^T(\theta, \phi) \in \mathcal{C}^{2 \times n}$ as the vector containing the radiation patterns associated with all the ports of a particular MPA, referred to a common coordinate system, and given by

$$\mathbf{F}^T(\theta, \phi) = (\mathbf{F}_1^T(\theta, \phi), \cdots, \mathbf{F}_n^T(\theta, \phi)). \tag{4.35}$$

Similarly as before, $\mathbf{F}^T(\theta, \phi)$ must be normalized as well, such that

$$\int_{-\pi}^{\pi} \int_0^{\pi} (\mathbf{F}^T(\theta, \phi))^H \mathbf{F}^T(\theta, \phi) \sin(\theta) d\theta d\phi = (\mathbf{S}_{\beta_T \alpha_T}^T)^H \mathbf{S}_{\beta_T \alpha_T}^T \leq \mathbf{I} - (\mathbf{S}_{\alpha_T \alpha_T}^T)^H \mathbf{S}_{\alpha_T \alpha_T}^T. \tag{4.36}$$

In this case, the radiated electric field from all n antennas, $\mathbf{E}(\mathbf{r}, \theta, \phi) \in \mathcal{C}^{2 \times n}$, is given by

$$\mathbf{E}^T(\mathbf{r}, \theta, \phi) = \sum_i^n \mathbf{E}_i^T(\mathbf{r}, \theta, \phi) = \sqrt{\eta_0} \frac{e^{-j\beta_T r}}{r} \mathbf{F}^T(\theta, \phi) \mathbf{a}_{\alpha_T}^T. \tag{4.37}$$

Taking into account all the ports of the MPA, the total radiated power by the MPA, $P_{\alpha_T}^{T,\text{av}}$, can be expressed as

$$P_{\alpha_T}^{T,\text{av}} = \frac{1}{\eta_0} \int_{-\pi}^{\pi} \int_0^{\pi} |\mathbf{E}^T(\mathbf{r}, \theta, \phi)|^2 r^2 \sin(\theta) d\theta d\phi$$

$$= (\mathbf{a}_{\alpha_T}^T)^H \left(\int_{-\pi}^{\pi} \int_0^{\pi} (\mathbf{F}^T(\theta, \phi))^H \mathbf{F}^T(\theta, \phi) \sin(\theta) d\theta d\phi\right) \mathbf{a}_{\alpha_T}^T = (\mathbf{b}_{\beta_T}^T)^H \mathbf{b}_{\beta_T}^T, \tag{4.38}$$

which through Eq. (4.35) can be further simplified to

$$P_{\alpha_T}^{T,\text{av}} = (\mathbf{b}_{\beta_T}^T)^H \mathbf{b}_{\beta_T}^T = (\mathbf{a}_{\alpha_T}^T)^H (\mathbf{S}_{\beta_T \alpha_T}^T)^H \mathbf{S}_{\beta_T \alpha_T}^T \mathbf{a}_{\alpha_T}^T \leq (\mathbf{a}_{\alpha_T}^T)^H (\mathbf{I} - (\mathbf{S}_{\alpha_T \alpha_T}^T)^H \mathbf{S}_{\alpha_T \alpha_T}^T) \mathbf{a}_{\alpha_T}^T \tag{4.39}$$

and where the equality applies to a lossless antenna.

Notice that a similar analysis can be conducted for a receiving MPA and its total received power. By virtue of the reciprocity theorem, one may obtain similar findings using such an analysis.

4.2.5   Canonical Minimum Scattering Antennas

When dealing with radiation parameters, one should guard against confusing the radiation pattern quantity $\mathbf{F}^A(\theta, \phi)$, as defined in Section 4.2.3, with the isolated radiation pattern ($\mathbf{F}^{A,\text{iso}}(\theta, \phi)$) or the open-circuit radiation pattern ($\mathbf{F}^{A,oc}(\theta, \phi)$) associated with a particular port of an MPA.

In the case of an MEA, the first one, $\mathbf{F}^{A,\text{iso}}(\theta, \phi)$, is that computed with no other antennas around the driven element, whereas the second one, $\mathbf{F}^{A,oc}(\theta, \phi)$, is computed by loading the surrounding antennas with an open-circuit termination. For example, when an open-circuited antenna is placed near an antenna connected to a generator, the electromagnetic boundary conditions are changed, leading to a change in the radiation behavior. However, because the coupled antenna is open-circuited, the impact on the pattern of the driven element is often relatively minor for most practical configurations and $\mathbf{F}^{A,\text{iso}}(\theta, \phi) \approx \mathbf{F}^{A,oc}(\theta, \phi)$. $\mathbf{F}^{A,\text{iso}}(\theta, \phi)$ and $\mathbf{F}^{A,oc}(\theta, \phi)$ are, however, different quantities, and although they are commonly assumed to be equal, they are, in fact, only equal under certain conditions. These conditions were investigated in Ref. [58], where the concept of canonical minimum scattering antennas (CMSA) [58] was introduced.

To understand the CMSA concept, we first need to introduce the definition of minimum scattering antennas (MSA) [58], which refers to a class of idealized single-port antennas (SPAs), or MPAs, the radiation and scattering properties of which are rigorously expressed solely in terms of their radiation patterns. By definition, an MSA satisfies $\mathbf{S}^A_{\alpha\alpha} = \mathbf{0}$; thus, MSAs sometimes incorporate external circuitry to make each accessible port reflectionless and decoupled from any of the other accessible ports. MSAs do not scatter; in other words, they become electromagnetically invisible when their ports are terminated in a particular set of reactive loads. If these reactive loads are pure open-circuits, the antennas are termed CMSA [58].

Therefore, in the case of an MEA, the open-circuit pattern of a particular radiating element will be the same as that of the isolated element, $\mathbf{F}^{A,\text{iso}}(\theta, \phi) \approx \mathbf{F}^{A,oc}(\theta, \phi)$, if and only if the MEA satisfies the CMSA concept. Except for certain types of antennas such as monopoles or dipoles, which reasonably approach the CMSA concept, it is not necessary true for all antenna types. Despite this, in Ref. [62], it was shown that antennas such as microstrip patches also approach the CMSA concept well.

4.2.6 Active Gain

As shown in Ref. [15], the active gain is defined as the gain associated with a particular port of the MPA when all the other ports are terminated with the characteristic reference impedance, Z^0. In an MEA, for example, as the surrounding antennas influence the pattern of the driven element, the gain in terms of directivity changes, and accordingly, we call it the active gain. Additionally, the energy radiated from an antenna may be directly absorbed by another closely spaced antenna; thus, the active gain is reduced. With this definition, the active gain is a property of each individual antenna within an MEA and depends on the MEA topology and the termination of the antennas. Notice that it is not a function of the incident field. The active gain associated with the ith port of a particular MPA can be defined in terms of the radiated electric field and the total radiated power, $P^{A,\text{av}}_{\alpha_i}$, using Ref. [10], as follows:

$$G^A_i = \frac{4\pi r^2 \max\{|\mathbf{W}^A_i(\mathbf{r}, \theta, \phi)|\}}{|\mathbf{a}^A_{\alpha_i}|^2} = \frac{4\pi r^2 \left(1 - (\mathbf{S}^A_{\alpha\alpha_{\cdot i}})^H \mathbf{S}^A_{\alpha\alpha_{\cdot i}}\right) \max\{|\mathbf{W}^A_i(\mathbf{r}, \theta, \phi)|\}}{P^{A,\text{av}}_{\alpha_i}}, \quad (4.40)$$

where $\mathbf{S}^A_{\alpha\alpha^*_i}$ is the ith column of $\mathbf{S}^A_{\alpha\alpha}$. Notice that from the previous definition, $\max\{\mathbf{F}^A_i(\theta,\phi)\} = \sqrt{\frac{G^A_i}{4\pi}}$. Notice that in the absence of mutual coupling effects ($\mathbf{S}^A_{\alpha\alpha^*_i} = \mathbf{0}$), the active gain reduces to

$$G^A_i = \frac{4\pi r^2 \max\{|\mathbf{W}^A_i(\mathbf{r},\theta,\phi)|\}}{P^{A,\text{av}}_{\alpha_i}} = \frac{4\pi r^2 \max\{|\mathbf{E}^A_i(\mathbf{r},\theta,\phi)|\}}{\eta_0 P^{A,\text{av}}_{\alpha_i}}. \qquad (4.41)$$

The upper-index A can either refer to the transmit (T) or receive (R) antennas.

Finally, we can define the power gain pattern associated with the ith port of an MPA, $G^A_i(\theta,\phi) \in \mathcal{C}^{1\times 1}$, as

$$G^A_i(\theta,\phi) = \frac{4\pi r^2 |\mathbf{W}^A_i(\mathbf{r},\theta,\phi)|}{|\mathbf{a}^A_{\alpha_i}|^2} = 4\pi|\mathbf{F}^A_i(\theta,\phi)|^2 = 4\pi\left(|\mathbf{F}^A_{i_\theta}(\theta,\phi)|^2 + |\mathbf{F}^A_{i_\phi}(\theta,\phi)|^2\right). (4.42)$$

Sometimes, it is interesting to define the power gain pattern for each of the polarizations. Keeping in mind that $\mathbf{F}^{A,0}_i(\theta,\phi) = \sqrt{\frac{4\pi}{G_i}}\mathbf{F}^A_i(\theta,\phi)$, we can define $\mathbf{G}^A_i(\theta,\phi) \in \mathcal{C}^{2\times 1}$ as

$$\mathbf{G}^A_i(\theta,\phi) = \sqrt{G^A_i}\mathbf{F}^{A,0}_i(\theta,\phi) = \sqrt{4\pi}\mathbf{F}^A_i(\theta,\phi)$$
$$= \sqrt{4\pi}\mathbf{F}^A_{i_\theta}(\theta,\phi)\widehat{\theta} + \sqrt{4\pi}\mathbf{F}^A_{i_\phi}(\theta,\phi)\widehat{\phi} = \mathbf{G}^A_{i_\theta}(\theta,\phi)\widehat{\theta} + \mathbf{G}^A_{i_\phi}(\theta,\phi)\widehat{\phi}, \qquad (4.43)$$

where it should be noticed that the θ component of the power gain patter is given by $\mathbf{G}^A_{i_\theta}(\theta,\phi) = \sqrt{4\pi}\mathbf{F}^A_{i_\theta}(\theta,\phi)$ and the ϕ component is given by $\mathbf{G}^A_{i_\phi}(\theta,\phi) = \sqrt{4\pi}\mathbf{F}^A_{i_\phi}(\theta,\phi)$.

4.3   COMPONENTS OF A MIMO SYSTEM NETWORK MODEL

As discussed in Section 4.1, to investigate the performance of a complete MIMO communication system, a network theory framework needs to be used. In this section, we describe each one of the components of the RF transmission chain by means of a scattering matrix representation. The basic components of the RF transmission chain of a MIMO communication system are the signal source (generator), the transmit MPA, the propagation channel, the receive MPA, and the signal drain (loads), as shown in Fig. 4.2. Afterward, in Section 4.4, these components are easily interconnected. Assume a transmit MPA with M transmit ports and a receive MPA with N receiving ports. We refer to such a system as an $M \times N$ MIMO system.

4.3.1   Signal Source (Generator)

The signal source is at the beginning of the transmission chain and determines the transmit power distribution among the transmit ports. The signal source has M output ports, according to the number of transmit ports of the transmit MPA. The internal impedances of the signal source

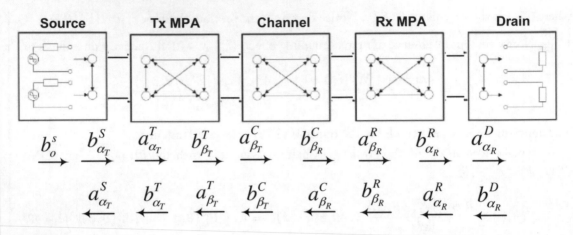

FIGURE 4.2: System model of the complete RF transmission chain including the source, drain, transmit and receive MPAs, and propagation channel. Block names and port references are included.

are characterized by the reflection matrix $\mathbf{\Gamma}^S_{\alpha_T} \in \mathcal{C}^{M \times M}$ (or simply $\mathbf{\Gamma}^S$), where the upper-index S denotes source. The outward propagating wave vector is

$$\mathbf{b}^S_{\alpha_T} = \mathbf{b}^S_0 + \mathbf{\Gamma}^S_{\alpha_T} \mathbf{a}^S_{\alpha_T} \tag{4.44}$$

where $\mathbf{b}^S_0 \in \mathcal{C}^{M \times 1}$ is the wave vector fed into the network, which determines the power distribution among the source ports. $\mathbf{a}^S_{\alpha_T}$ is the wave vector reflected by the transmit MPA.

4.3.2   Signal Drain (Loads)

The signal drain is at the end of the transmission chain and is described by the reflection matrix $\mathbf{\Gamma}^D_{\alpha_R} \in \mathcal{C}^{N \times N}$ (or simply $\mathbf{\Gamma}^D$), where the upper-index D denotes drain. It describes the reflections from the signal drain in case the receive ports are not perfectly matched to the loads. These reflections may arise as a result of mutual coupling effects among the ports of an MPAs. For example, in an MEA, the coupled currents among the antennas may change their input impedance, and mismatch may occur when the antennas are connected to the loads.

4.3.3   Transmit MPA

The transmit MPA is described by the scattering matrix, $\mathbf{S}^T$. The full scattering model associated with the transmit MPA is given by

$$\begin{pmatrix} \mathbf{b}^T_{\alpha_T} \\ \mathbf{b}^T_{\beta_T} \end{pmatrix} = \begin{pmatrix} \mathbf{S}^T_{\alpha_T \alpha_T} & \mathbf{S}^T_{\alpha_T \beta_T} \\ \mathbf{S}^T_{\beta_T \alpha_T} & \mathbf{S}^T_{\beta_T \beta_T} \end{pmatrix} \begin{pmatrix} \mathbf{a}^T_{\alpha_T} \\ \mathbf{a}^T_{\beta_T} \end{pmatrix}. \tag{4.45}$$

Notice that the upper-index T refers to the transmit MPA. The MPA model was described in Section 4.2.1. The power delivered to the input ports of the MPA can be expressed as

$$P_{\alpha_T}^S = (\mathbf{b}_{\alpha_T}^S)^H \mathbf{b}_{\alpha_T}^S - (\mathbf{a}_{\alpha_T}^S)^H \mathbf{a}_{\alpha_T}^S \tag{4.46}$$

or equivalently as

$$P_{\alpha_T}^T = (\mathbf{a}_{\alpha_T}^T)^H \mathbf{a}_{\alpha_T}^T - (\mathbf{b}_{\alpha_T}^T)^H \mathbf{b}_{\alpha_T}^T. \tag{4.47}$$

Unless specified, we will use $P_{\alpha_T}^S$ or simply P^S to designate the transmitted power.

4.3.4 Receive MPA

Similarly, the receive MPA is described using the scattering matrix, $\mathbf{S}^R$. The full scattering model associated with the transmit MPA is given by

$$\begin{pmatrix} \mathbf{b}_{\beta_R}^R \\ \mathbf{b}_{\alpha_R}^R \end{pmatrix} = \begin{pmatrix} \mathbf{S}_{\beta_R\beta_R}^R & \mathbf{S}_{\beta_R\alpha_R}^R \\ \mathbf{S}_{\alpha_R\beta_R}^R & \mathbf{S}_{\alpha_R\alpha_R}^R \end{pmatrix} \begin{pmatrix} \mathbf{a}_{\beta_R}^R \\ \mathbf{a}_{\alpha_R}^R \end{pmatrix}. \tag{4.48}$$

Notice that the upper-index R refers to the receive MPA. The MPA model is the same as that described in Section 4.2.1. However, in this case, the α and β ports have been permuted. The power delivered to the output ports of the MPA can be expressed as

$$P_{\alpha_R}^R = (\mathbf{b}_{\alpha_R}^R)^H \mathbf{b}_{\alpha_R}^R - (\mathbf{a}_{\alpha_R}^R)^H \mathbf{a}_{\alpha_R}^R \tag{4.49}$$

or equivalently as

$$P_{\alpha_R}^D = (\mathbf{a}_{\alpha_R}^D)^H \mathbf{a}_{\alpha_R}^D - (\mathbf{b}_{\alpha_R}^D)^H \mathbf{b}_{\alpha_R}^D. \tag{4.50}$$

Unless specified, we will use $P_{\alpha_R}^D$, or simply P^D to designate the received power.

4.3.5 Propagation Channel

The correct modeling of the channel is important to appropriately evaluate the impact of multi-antenna systems from a system-level perspective. A good channel model must have a clear physical meaning, and it must provide information on the characteristics of the propagation environment only, thus excluding the characteristics of the transmit and receive MPAs. The mathematical modeling of the channel must lead to its simple inclusion in the whole transmission chain of a MIMO communication system. In this way, the performance of different MPA geometries can be assessed for a given channel. To do this, one must model the propagation channel by a scattering matrix that takes into account how the transmitted spherical modes are modified by the propagation environment before reaching the receiving MPA. Also, it should express the incident field in terms of the spherical wave functions referred to a common coordinate system origin.

Following the aforementioned requirements, the propagation channel can be modeled by the scattering matrix $\mathbf{S}^C$. In this case, the full scattering model is given by

$$\begin{pmatrix} \mathbf{b}^C_{\beta_T} \\ \mathbf{b}^C_{\beta_R} \end{pmatrix} = \begin{pmatrix} \mathbf{S}^C_{\beta_T\beta_T} & \mathbf{S}^C_{\beta_T\beta_R} \\ \mathbf{S}^C_{\beta_R\beta_T} & \mathbf{S}^C_{\beta_R\beta_R} \end{pmatrix} \begin{pmatrix} \mathbf{a}^C_{\beta_T} \\ \mathbf{a}^C_{\beta_R} \end{pmatrix}, \tag{4.51}$$

where the upper-index C denotes channel. Assume for now that $\mathbf{S}^C$ expresses the relationship between the transmitted and received spherical modes. Notice that the coupling between the ports of the MPAs has already been included in the scattering matrices of the MPAs. The subblock matrices of $\mathbf{S}^C$ represent the following concepts: $\mathbf{S}^C_{\beta_T\beta_T} \in \mathcal{C}^{\infty\times\infty}$ is associated with the matching of the radiated spherical modes into the channel, $\mathbf{S}^C_{\beta_R\beta_T} \in \mathcal{C}^{\infty\times\infty}$ with the propagation of the radiated spherical modes from the transmitter to the receiver, $\mathbf{S}^C_{\beta_T\beta_R} \in \mathcal{C}^{\infty\times\infty}$ with the propagation of the backscattered spherical modes from the receiver to the transmitter, and $\mathbf{S}^C_{\beta_R\beta_R} \in \mathcal{C}^{\infty\times\infty}$ with the matching of the backscattered spherical modes into the channel.

In general, it can be assumed that there is no reflection from the far field into the channel. For example, for the transmitter side, nothing of the once-radiated transmit power is received by the transmit antennas. The same idea can be assumed for the receiver side, and thus, the submatrices $\mathbf{S}^C_{\beta_T\beta_T}$ and $\mathbf{S}^C_{\beta_R\beta_R}$ equal the zero matrix [15]. We refer to these types of channels as *matched channels*. On the other hand, the back transmission of signals through the propagation channel is subject to channel attenuation. Thus, the power reradiated by the receiver and received by the transmitter is twice as strongly attenuated as the signals at the receiver. Therefore, it is also justifiable to neglect the back transmission and set $\mathbf{S}^C_{\beta_T\beta_R} = \mathbf{0}$. We refer to these types of channels as *unilateral channels*.

4.4   ASSEMBLED MIMO SYSTEM NETWORK MODEL

4.4.1   Connection of Scattering Matrices

Before proceeding with the analysis of the MIMO system network model, it is necessary to review how two arbitrary scattering matrices can be interconnected. In particular, it is of interest to formulate the resultant scattering parameter matrix of the two concatenated networks. Without loss of generality, consider the cascade interconnection of two networks, as shown in Fig. 4.3, where a network of $M \times N$ ports is connected to a network of $N \times O$ ports. Obviously, the number of output ports of the first block must be equal to the number of input ports of the second block. The two original networks are described by the scattering matrices $\mathbf{S}^A \in \mathcal{C}^{M\times N}$ and $\mathbf{S}^B \in \mathcal{C}^{N\times O}$, given by

$$\mathbf{S}^A = \begin{pmatrix} \mathbf{S}^A_{11} & \mathbf{S}^A_{12} \\ \mathbf{S}^A_{21} & \mathbf{S}^A_{22} \end{pmatrix}, \mathbf{S}^B = \begin{pmatrix} \mathbf{S}^B_{11} & \mathbf{S}^B_{12} \\ \mathbf{S}^B_{21} & \mathbf{S}^B_{22} \end{pmatrix}, \tag{4.52}$$

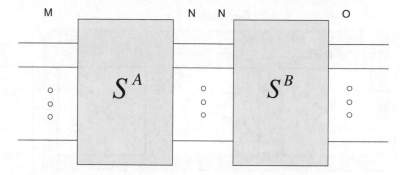

FIGURE 4.3: Block diagram of a network of $M \times N$ ports interconnected to a network of $N \times O$ ports.

where $\mathbf{S}_{11}^{A} \in \mathcal{C}^{M \times M}$, $\mathbf{S}_{12}^{A} \in \mathcal{C}^{M \times N}$, $\mathbf{S}_{21}^{A} \in \mathcal{C}^{N \times M}$, and $\mathbf{S}_{22}^{A} \in \mathcal{C}^{N \times N}$, and $\mathbf{S}_{11}^{B} \in \mathcal{C}^{N \times N}$, $\mathbf{S}_{12}^{B} \in \mathcal{C}^{N \times O}$, $\mathbf{S}_{21}^{B} \in \mathcal{C}^{O \times N}$, and $\mathbf{S}_{22}^{B} \in \mathcal{C}^{O \times O}$. The resultant scattering matrix after interconnecting the two aforementioned networks, $\mathbf{S}^{AB} \in \mathcal{C}^{M \times O}$, is given by

$$\mathbf{S}^{AB} = \mathbf{S}^{A} \odot \mathbf{S}^{B} = \begin{pmatrix} \mathbf{S}_{11}^{AB} & \mathbf{S}_{12}^{AB} \\ \mathbf{S}_{21}^{AB} & \mathbf{S}_{22}^{AB} \end{pmatrix}, \tag{4.53}$$

where $\odot$ is an operator used to describe the interconnection of two scattering matrices and where the subblock scattering matrices of $\mathbf{S}^{AB}$ are described by

$$\mathbf{S}_{11}^{AB} = \mathbf{S}_{11}^{A} + \mathbf{S}_{12}^{A} \mathbf{S}_{11}^{B} (\mathbf{I} - \mathbf{S}_{22}^{A} \mathbf{S}_{11}^{B})^{-1} \mathbf{S}_{21}^{A} \tag{4.54}$$

$$\mathbf{S}_{12}^{AB} = \mathbf{S}_{12}^{A} (\mathbf{I} - \mathbf{S}_{11}^{B} \mathbf{S}_{22}^{A})^{-1} \mathbf{S}_{12}^{B} \tag{4.55}$$

$$\mathbf{S}_{21}^{AB} = \mathbf{S}_{21}^{B} (\mathbf{I} - \mathbf{S}_{22}^{A} \mathbf{S}_{11}^{B})^{-1} \mathbf{S}_{21}^{A} \tag{4.56}$$

$$\mathbf{S}_{22}^{AB} = \mathbf{S}_{22}^{B} + \mathbf{S}_{21}^{B} \mathbf{S}_{22}^{A} (\mathbf{I} - \mathbf{S}_{11}^{B} \mathbf{S}_{22}^{A})^{-1} \mathbf{S}_{12}^{B} \tag{4.57}$$

Notice that $\mathbf{S}_{11}^{AB} \in \mathcal{C}^{M \times M}$, $\mathbf{S}_{12}^{AB} \in \mathcal{C}^{M \times O}$, $\mathbf{S}_{21}^{AB} \in \mathcal{C}^{O \times M}$, and $\mathbf{S}_{22}^{AB} \in \mathcal{C}^{O \times O}$.

4.4.2   Channel Matrix and Extended Channel Matrix

In this section, the components described in Section 4.3 are interconnected together in two steps. A formal definition for the channel matrix of a MIMO communication system is also given. To do so, notice first that, in our network model, the outward propagating wave vectors of a component are the inward propagating wave vectors of the adjacent component. In the first step, the inner three components: the transmit MPA, the propagation channel, and the receive MPA, are merged.

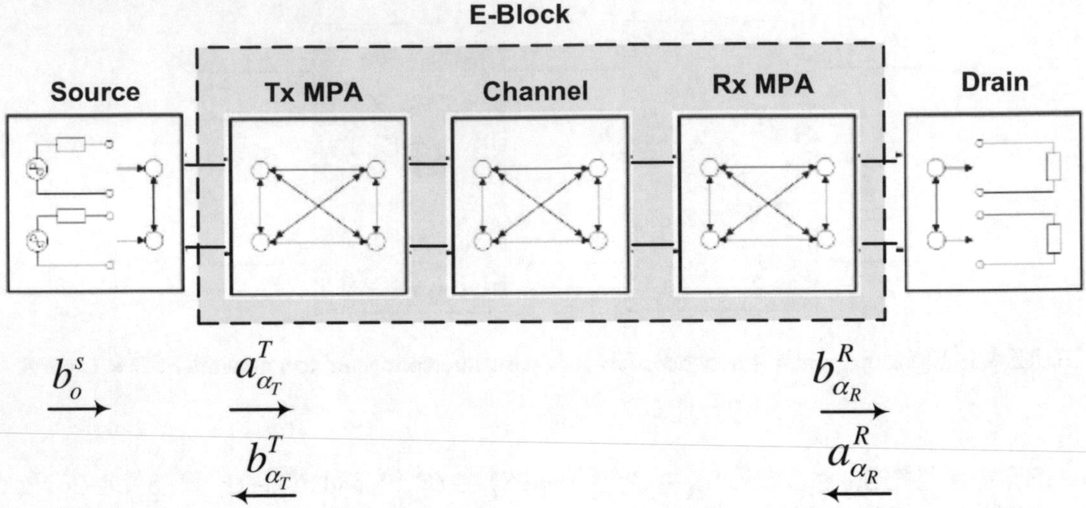

FIGURE 4.4: System model of the assembled RF transmission chain. Notice that the E block includes the transmit and receive MPAs and the propagation channel. This block itself constitutes an extended channel that includes the effects of the antennas.

As shown in Fig. 4.4, we refer to this merged block as the E block, described by a scattering matrix $\mathbf{S}^E$. The term $\mathbf{S}^E$ can be expanded into

$$\mathbf{S}^E = \begin{pmatrix} \mathbf{S}^E_{\alpha_T \alpha_T} & \mathbf{S}^E_{\alpha_T \alpha_R} \\ \mathbf{S}^E_{\alpha_R \alpha_T} & \mathbf{S}^E_{\alpha_R \alpha_R} \end{pmatrix}. \tag{4.58}$$

Notice that $\mathbf{S}^E$ is constructed as a result of performing the operation

$$\mathbf{S}^E = \mathbf{S}^T \odot \mathbf{S}^C \odot \mathbf{S}^R. \tag{4.59}$$

As shown in Ref. [15], the subblock matrices of $\mathbf{S}^E$ are given by

$$\mathbf{S}^E_{\alpha_T \alpha_T} = \mathbf{S}^T_{\alpha_T \alpha_T} + \mathbf{S}^T_{\alpha_T \beta_T} \mathbf{S}^C_{\beta_T \beta_R} \left(\mathbf{I} - \mathbf{S}^R_{\beta_R \beta_R} \mathbf{S}^C_{\beta_R \beta_T} \mathbf{S}^T_{\beta_T \beta_T} \mathbf{S}^C_{\beta_T \beta_R} \right)^{-1} \mathbf{S}^R_{\beta_R \beta_R} \mathbf{S}^C_{\beta_R \beta_T} \mathbf{S}^T_{\beta_T \alpha_T} \tag{4.60}$$

$$\mathbf{S}^E_{\alpha_T \alpha_R} = \mathbf{S}^T_{\alpha_T \beta_T} \mathbf{S}^C_{\beta_T \beta_R} \left(\mathbf{I} - \mathbf{S}^R_{\beta_R \beta_R} \mathbf{S}^C_{\beta_R \beta_T} \mathbf{S}^T_{\beta_T \beta_T} \mathbf{S}^C_{\beta_T \beta_R} \right)^{-1} \mathbf{S}^R_{\beta_R \alpha_R} \tag{4.61}$$

$$\mathbf{S}^E_{\alpha_R \alpha_T} = \mathbf{S}^R_{\alpha_R \beta_R} \left(\mathbf{I} - \mathbf{S}^C_{\beta_R \beta_T} \mathbf{S}^T_{\beta_T \beta_T} \mathbf{S}^C_{\beta_T \beta_R} \mathbf{S}^R_{\beta_R \beta_R} \right)^{-1} \mathbf{S}^C_{\beta_R \beta_T} \mathbf{S}^T_{\beta_T \alpha_T} \tag{4.62}$$

$$S^E_{\alpha_R \alpha_R} = S^R_{\alpha_R \alpha_R} + S^R_{\alpha_R \beta_R} \left(I - S^C_{\beta_R \beta_T} S^T_{\beta_T \beta_T} S^C_{\beta_T \beta_R} S^R_{\beta_R \beta_R} \right)^{-1} S^C_{\beta_R \beta_T} S^T_{\beta_T \beta_T} S^C_{\beta_T \beta_R} S^R_{\beta_R \alpha_R}. \quad (4.63)$$

Assuming a matched ($S^C_{\beta_T \beta_T} = 0$ and $S^C_{\beta_R \beta_R} = 0$) unilateral channel ($S^C_{\beta_T \beta_R} = 0$), S^E can be simplified and is given by

$$S^E = \begin{pmatrix} S^T_{\alpha_T \alpha_T} & 0 \\ S^R_{\alpha_R \beta_R} S^C_{\beta_R \beta_T} S^T_{\beta_T \alpha_T} & S^R_{\alpha_R \alpha_R} \end{pmatrix}. \quad (4.64)$$

Notice that the term $S^E_{\alpha_R \alpha_T} = S^R_{\alpha_R \beta_R} S^C_{\beta_R \beta_T} S^T_{\beta_T \alpha_T}$ describes the transmission of the signals from the input ports of the transmit MPA to the output ports of the receive MPA. On the other hand, the terms $S^E_{\alpha_T \alpha_T} = S^T_{\alpha_T \alpha_T}$ and $S^E_{\alpha_R \alpha_R} = S^R_{\alpha_R \alpha_R}$ contain the return loss and mutual coupling information among the ports of the transmit and receive MPAs, respectively.

It must be noted that the unilateral channel and the matched channel assumptions are not required because of limitations in the model, but rather because they are reasonably close to reality and because they simplify the model without compromising its accuracy in most situations. It is true that if these assumptions cannot be made, the subblock matrices $S^T_{\beta_T \beta_T}$ and $S^R_{\beta_R \beta_R}$ of the transmit and receive MPAs, respectively, must be known. These matrices relate the incident fields with the scattered fields of the MPA, and usually they are not known, simple to measure, or easy to relate to the MPA radiation patterns. Only for an MPA that was a good approximation of the CMSA concept could one easily compute the term $S^T_{\beta_T \beta_T}$ from the radiation patterns using $S^T_{\beta_T \beta_T} = I - S^T_{\beta_T \alpha_T} (S^T_{\beta_T \alpha_T})^H$ [58]. The same approach is valid for the computation of $S^R_{\beta_R \beta_R}$.

Coming back to our derivation, the (n, m)th entry of $S^E_{\alpha_R \alpha_T}$ can be computed as the ratio of the incoming wave at the mth transmit port to the outgoing wave at the nth receive port:

$$S^E_{\alpha_R \alpha_{T n,m}} = \sqrt{\frac{Z^0_{\alpha_{Tm}}}{Z^0_{\alpha_{Rn}}}} \frac{b_{\alpha_{Rn}}}{a_{\alpha_{Tm}}} \quad \text{when } a_{\alpha_{Tk}} = 0 \text{ for all } k \neq m, \quad (4.65)$$

where $Z^0_{\alpha_{Tm}}$ and $Z^0_{\alpha_{Rn}}$ are the characteristic reference impedances at the mth transmit and nth receive port, respectively. Notice that the channel matrix of a MIMO communication system, as used in previous sections, $H \in C^{N \times M}$, does not include the effects of the source and the drain and, by definition, is given by

$$H = S^E_{\alpha_R \alpha_T}. \quad (4.66)$$

Finally, notice that H is part of the scattering matrix S^E, as shown below:

$$S^E = \begin{pmatrix} S^T_{\alpha_T \alpha_T} & 0 \\ H & S^R_{\alpha_R \alpha_R} \end{pmatrix}. \quad (4.67)$$

The second step toward interconnecting the different components of the RF transmission chain of a MIMO communication system is to include the signal source and signal drain. Because the mutual coupling impacts the ports of an MPA by changing its input impedances, its effects strongly depend on the termination of the transmit and receive ports. In fact, there will be waves traveling between the MPA ports and the loads (source/drain) connected to them, which ultimately will affect the performance of the system. Therefore, it is important to include the termination effects of the signal source and drain in the calculations of the system capacity and other. To do this, let us define $\mathbf{E} \in \mathcal{C}^{N \times M}$ as the extended channel matrix of a MIMO communication system, which expresses the ratio of the voltages at the receive antennas $\mathbf{U}_{\alpha_R}^R$ to the voltages at the transmit antennas $\mathbf{U}_{\alpha_T}^T$, including the effects of source and drain terminations:

$$\mathbf{U}_{\alpha_R}^R = \mathbf{E}\mathbf{U}_{\alpha_T}^T. \tag{4.68}$$

Notice that the extended channel matrix $\mathbf{E}$ allows for more realistic capacity calculations than $\mathbf{H}$.

To compute $\mathbf{E}$ we use the flow diagram in Fig. 4.5. Notice that the wave vectors $\mathbf{a}_{\alpha_T}^T$, $\mathbf{a}_{\alpha_R}^R$, $\mathbf{b}_{\alpha_T}^T$, and $\mathbf{b}_{\alpha_R}^R$ represent the incident waves at the ports of the transmit MPA, the incident waves at the ports of the receive MPA, the reflected waves at the ports of the transmit MPA, and the reflected waves at the ports of the receive MPA, respectively. Using the Heavyside transformation, Eq. (4.68) leads to

$$Z_{\alpha_R}^0{}^{\frac{1}{2}} (\mathbf{a}_{\alpha_R}^R + \mathbf{b}_{\alpha_R}^R) = \mathbf{E} Z_{\alpha_T}^0{}^{\frac{1}{2}} (\mathbf{a}_{\alpha_T}^T + \mathbf{b}_{\alpha_T}^T). \tag{4.69}$$

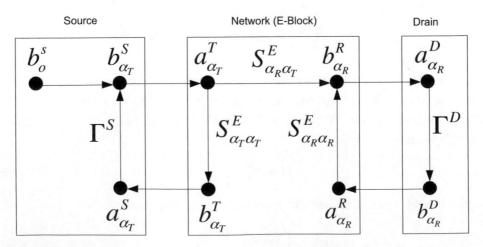

FIGURE 4.5: Signal flow graph for the assembled MIMO system network model.

Now, following the approach in Ref. [15], the aforementioned wave vectors can be expressed as a function of $\mathbf{b}_0^S$, as shown below:

$$\mathbf{a}_{\alpha_T}^T = \left(\mathbf{I} - \mathbf{\Gamma}_{\alpha_T}^S \mathbf{S}_{\alpha_T \alpha_T}^E\right)^{-1} \mathbf{b}_0^S \tag{4.70}$$

$$\mathbf{a}_{\alpha_R}^R = \left(\mathbf{I} - \mathbf{\Gamma}_{\alpha_R}^D \mathbf{S}_{\alpha_R \alpha_R}^E\right)^{-1} \mathbf{\Gamma}_{\alpha_R}^D \mathbf{S}_{\alpha_R \alpha_T}^E \left(\mathbf{I} - \mathbf{\Gamma}_{\alpha_T}^S \mathbf{S}_{\alpha_T \alpha_T}^E\right)^{-1} \mathbf{b}_0^S$$
$$= \mathbf{\Gamma}_{\alpha_R}^D \left(\mathbf{I} - \mathbf{S}_{\alpha_R \alpha_R}^E \mathbf{\Gamma}_{\alpha_R}^D\right)^{-1} \mathbf{S}_{\alpha_R \alpha_T}^E \left(\mathbf{I} - \mathbf{\Gamma}_{\alpha_T}^S \mathbf{S}_{\alpha_T \alpha_T}^E\right)^{-1} \mathbf{b}_0^S \tag{4.71}$$

$$\mathbf{b}_{\alpha_T}^T = \left(\mathbf{I} - \mathbf{S}_{\alpha_T \alpha_T}^E \mathbf{\Gamma}_{\alpha_T}^S\right)^{-1} \mathbf{S}_{\alpha_T \alpha_T}^E \mathbf{b}_0^S \tag{4.72}$$

$$\mathbf{b}_{\alpha_R}^R = \left(\mathbf{I} - \mathbf{S}_{\alpha_R \alpha_R}^E \mathbf{\Gamma}_{\alpha_R}^D\right)^{-1} \left(\mathbf{S}_{\alpha_R \alpha_T}^E + \mathbf{S}_{\alpha_R \alpha_T}^E \mathbf{\Gamma}_{\alpha_T}^S \left(\mathbf{I} - \mathbf{S}_{\alpha_T \alpha_T}^E \mathbf{\Gamma}_{\alpha_T}^S\right)^{-1} \mathbf{S}_{\alpha_T \alpha_T}^E\right) \mathbf{b}_0^S$$
$$= \left(\mathbf{I} - \mathbf{S}_{\alpha_R \alpha_R}^E \mathbf{\Gamma}_{\alpha_R}^D\right)^{-1} \mathbf{S}_{\alpha_R \alpha_T}^E \left(\mathbf{I} - \mathbf{\Gamma}_{\alpha_T}^S \mathbf{S}_{\alpha_T \alpha_T}^E\right)^{-1} \mathbf{b}_0^S, \tag{4.73}$$

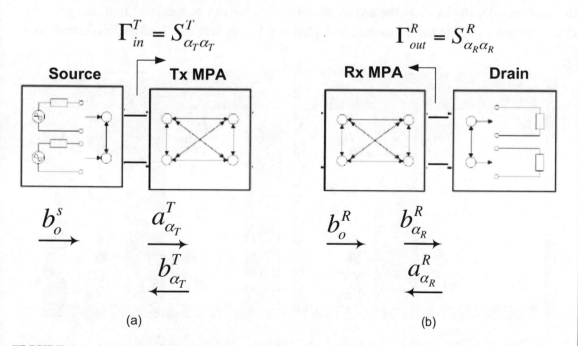

FIGURE 4.6: Separable MIMO system network model of the RF transmission chain: (a) transmitter and (b) receive side. If a unilateral channel can be assumed, the separation allows the analysis of the transmit and receive sides individually.

where we have used the fact that $(\mathbf{I} + \mathbf{AB})^{-1} = \mathbf{I} - \mathbf{A}(\mathbf{I} + \mathbf{BA})^{-1}\mathbf{B}$ and $(\mathbf{I} - \mathbf{AB})^{-1} = \mathbf{I} + \mathbf{A}(\mathbf{I} - \mathbf{BA})^{-1}\mathbf{B}$. We have also used the fact that $(\mathbf{I} - \mathbf{AB})^{-1}\mathbf{A} = \mathbf{A}(\mathbf{I} - \mathbf{BA})^{-1}$ if $\mathbf{B} \propto \mathbf{I}$ such as in the case of $\boldsymbol{\Gamma}^S_{\alpha_T}$ and $\boldsymbol{\Gamma}^D_{\alpha_R}$.

Finally, we can solve Eq. (4.69) for $\mathbf{E}$, knowing that the voltages $\mathbf{U}^R_{\alpha_R}$ can be expressed as a function of the voltages $\mathbf{U}^T_{\alpha_T}$ and using the above wave vector derivations. As a result, $\mathbf{E}$ is given by

$$
\begin{aligned}
\mathbf{E} &= \mathbf{Z}^{0\ \frac{1}{2}}_{\alpha_R} \left(\mathbf{I} + \boldsymbol{\Gamma}^D_{\alpha_R}\right) \left(\mathbf{I} - \mathbf{S}^E_{\alpha_R \alpha_R} \boldsymbol{\Gamma}^D_{\alpha_R}\right)^{-1} \mathbf{S}^E_{\alpha_R \alpha_T} \left(\mathbf{I} + \mathbf{S}^E_{\alpha_T \alpha_T}\right)^{-1} \mathbf{Z}^{0\ -\frac{1}{2}}_{\alpha_T} \\
&= \mathbf{Z}^{0\ \frac{1}{2}}_{\alpha_R} \left(\mathbf{I} + \boldsymbol{\Gamma}^D_{\alpha_R}\right) \left(\mathbf{I} - \mathbf{S}^E_{\alpha_R \alpha_R} \boldsymbol{\Gamma}^D_{\alpha_R}\right)^{-1} \mathbf{H} \left(\mathbf{I} + \mathbf{S}^E_{\alpha_T \alpha_T}\right)^{-1} \mathbf{Z}^{0\ -\frac{1}{2}}_{\alpha_T}.
\end{aligned}
\tag{4.74}
$$

4.4.3 Separation of the Transmitter and Receiver Side

Sometimes, when assessing the performance of an MPA within a MIMO communication system, one does not have information about all the components of the RF transmission chain. In fact, one does not need to have all this information to evaluate certain performance measures. In such a case, it would be desirable to assess the performance of the transmit and receive MPAs individually. Assuming a unilateral channel, it is possible to separate the network analysis of the transmitter and the receiver side. In that case, the system model in Fig. 4.4 can be separated into that given by Fig. 4.6a and 4.6b, for the transmitter and receive side, respectively. Similarly, the signal flow

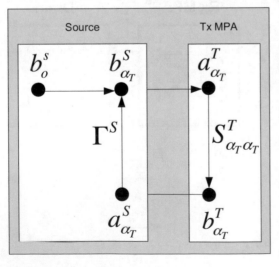
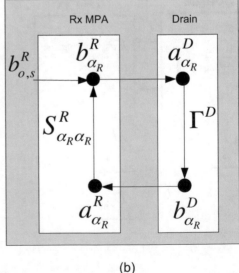

(a)   (b)

FIGURE 4.7: Signal flow graph for the separable MIMO system network model: (a) transmitter and (b) receive side.

diagram given in Fig. 4.5 can be separated into that given in Fig. 4.7a and 4.7b for the transmitter and receive side, respectively.

Using Eq. (4.44) and the fact that $\mathbf{a}_{\alpha_T}^S = \mathbf{b}_{\alpha_T}^T$, the voltages at the transmit MPA ports, $\mathbf{U}_{\alpha_T}^T$, are simply given by:

$$\mathbf{U}_{\alpha_T}^T = Z_{\alpha_T}^{0}{}^{\frac{1}{2}} \left(\mathbf{I} + \mathbf{S}_{\alpha_T \alpha_T}^E\right) \mathbf{a}_{\alpha_T}^T = Z_{\alpha_T}^{0}{}^{\frac{1}{2}} \left(\mathbf{I} + \mathbf{S}_{\alpha_T \alpha_T}^E\right) \left(\mathbf{I} - \mathbf{\Gamma}_{\alpha_T}^S \mathbf{S}_{\alpha_T \alpha_T}^E\right)^{-1} \mathbf{b}_0^S. \qquad (4.75)$$

Similarly, for the receiver we can express the outward propagating wave vector $\mathbf{b}_{\alpha_R}^R$ as

$$\mathbf{b}_{\alpha_R}^R = \mathbf{b}_0^R + \mathbf{S}_{\alpha_R \alpha_R}^R \mathbf{a}_{\alpha_R}^R, \qquad (4.76)$$

where $\mathbf{b}_0^R \in \mathcal{C}^{N \times 1}$ is the wave vector injected into the receive network blocks, generated by the impinging electromagnetic waves into the receive MPA. $\mathbf{b}_0^R$ then determines the power distribution among the receive ports, and it relates to $\mathbf{b}_0^S$ through the following expression:

$$\mathbf{b}_0^R = \mathbf{S}_{\alpha_R \alpha_T}^E \mathbf{a}_{\alpha_T}^T = \mathbf{S}_{\alpha_R \alpha_T}^E \left(\mathbf{I} - \mathbf{\Gamma}_{\alpha_T}^S \mathbf{S}_{\alpha_T \alpha_T}^E\right)^{-1} \mathbf{b}_0^S. \qquad (4.77)$$

Using the separate model, we can write the wave vectors $\mathbf{a}_{\alpha_R}^R$ and $\mathbf{b}_{\alpha_R}^R$ as a function of $\mathbf{b}_0^R$, as follows:

$$\mathbf{a}_{\alpha_R}^R = \mathbf{\Gamma}_{\alpha_R}^D \left(\mathbf{I} - \mathbf{S}_{\alpha_R \alpha_R}^E \mathbf{\Gamma}_{\alpha_R}^D\right)^{-1} \mathbf{b}_0^R \qquad (4.78)$$

$$\mathbf{b}_{\alpha_R}^R = \left(\mathbf{I} - \mathbf{S}_{\alpha_R \alpha_R}^E \mathbf{\Gamma}_{\alpha_R}^D\right)^{-1} \mathbf{b}_0^R. \qquad (4.79)$$

Finally, the voltages at the receive MPA ports, $\mathbf{U}_{\alpha_R}^R$, are given by

$$\mathbf{U}_{\alpha_R}^R = Z_{\alpha_R}^{0}{}^{\frac{1}{2}} \left(\mathbf{I} + \mathbf{\Gamma}_{\alpha_R}^D\right) \left(\mathbf{I} - \mathbf{S}_{\alpha_R \alpha_R}^E \mathbf{\Gamma}_{\alpha_R}^D\right)^{-1} \mathbf{b}_0^R. \qquad (4.80)$$

4.5 PHYSICAL REPRESENTATION OF THE PROPAGATION CHANNEL

This section deals with the physical representation of the propagation channel, which is described through the subblock matrix:

$$\mathbf{S}_{\alpha_R \alpha_T}^E = \mathbf{S}_{\alpha_R \beta_R}^R \mathbf{S}_{\beta_R \beta_T}^C \mathbf{S}_{\beta_T \alpha_T}^T. \qquad (4.81)$$

Except for the preliminary works in Refs. [63, 64], there is still very little in the literature of channel models that takes into account how the radiated spherical modes, from the transmit MPA, are modified by the propagation environment before reaching the receive MPA. Instead, a frequently used physical representation of $\mathbf{S}_{\alpha_R \alpha_T}^E$ is done through path models [2, 14, 15, 30]. Despite less generality than those based on the spherical modes, path models are better known at this moment. Therefore, we use them in this section to give a physical insight into the propagation channel.

Using a path model, each MPC p is described by a polarimetric transfer matrix $\mathbf{\Gamma}_{nm}^p \in \mathcal{C}^{2 \times 2}$, containing its polarization, gain, and phase information (among the mth transmit and nth receive antenna). $\mathbf{\Gamma}_{nm}^p$ is given by

$$\mathbf{\Gamma}_{nm}^p = \begin{pmatrix} \Gamma_{\theta\theta nm}^p & \Gamma_{\theta,\phi nm}^p \\ \Gamma_{\phi\theta nm}^p & \Gamma_{\phi\phi nm}^p \end{pmatrix}, \tag{4.82}$$

where $\Gamma_{\theta,\phi nm}^p$, for example, represents the percentage of the radiated field in the θ polarization that through propagation couples into the ϕ polarization. The entries of $\mathbf{\Gamma}_{nm}^p$ also take into account the free-space propagation term $\left(\frac{1}{r^2}\right)$ for each one of the polarizations and other losses that may occur because of a phenomenon, such as reflection, diffraction, or scattering, with objects present in the channel.

Let us recall the definition of the term $\mathbf{S}_{\alpha_R \alpha_{Tn,m}}^E$, given by

$$\mathbf{S}_{\alpha_R \alpha_{Tn,m}}^E = \sqrt{\frac{Z_{\alpha_{Tm}}^0}{Z_{\alpha_{Rn}}^0}} \frac{\mathbf{b}_{\alpha_{Rn}}}{\mathbf{a}_{\alpha_{Tm}}} \text{ when } \mathbf{a}_{\alpha_{Tk}} = 0 \text{ for all } k \neq m, \tag{4.83}$$

where $Z_{\alpha_{Tm}}^0$ and $Z_{\alpha_{Rn}}^0$ are the characteristic reference impedances at the mth transmit and nth receive α ports, respectively. Let us assume now that the mth transmit port is connected to a matched generator and the nth receive port to a matched load, whereas all the other transmit and receive ports are loaded with the characteristic reference impedance of the system, Z^0. In that case, using the Heavyside transformation, the term $\mathbf{S}_{\alpha_R \alpha_T}^E$ can be expressed as the ratio of the voltages at the mth transmit and nth receive antenna [15], as follows:

$$\mathbf{S}_{\alpha_R \alpha_{Tn,m}}^E = \sqrt{\frac{Z_{\alpha_{Tm}}^0}{Z_{\alpha_{Rn}}^0}} \frac{\mathbf{U}_{\alpha_{Rn}}^R}{\mathbf{U}_{\alpha_{Tm}}^T} \text{ when } \mathbf{a}_{\alpha_{Tk}} = 0, \ \mathbf{a}_{\alpha_{Ri}} = 0 \text{ for all } k \neq m \text{ and for all } i. \tag{4.84}$$

Then, using the Friis transmission formula [15, 65], one derives for the scattering parameter between transmit antenna m and receive antenna n, $\mathbf{S}_{\alpha_R \alpha_{Tn,m}}^E$, that

$$\mathbf{S}_{\alpha_R \alpha_{Tn,m}}^E = \lambda \sqrt{\frac{Z_{\alpha_{Tm}}^0}{Z_{\alpha_{Rn}}^0}} \sqrt{\frac{\Re(Z_{\alpha_{Rn}}^R)}{\Re(Z_{\alpha_{Tm}}^T)}} \sum_{p=1}^{P} (\mathbf{F}_n^R((\theta,\phi)_n^p))^\dagger \mathbf{\Gamma}_{nm}^p \mathbf{F}_m^T((\theta,\phi)_m^p), \tag{4.85}$$

where $Z_{\alpha_{Tm}}^T$ and $Z_{\alpha_{Rn}}^R$ are the antenna impedances of the transmit and receive ports, respectively. These impedances are modified as a result of the existing mutual coupling between ports. $(\theta,\phi)_m^p$ and $(\theta,\phi)_n^p$ are the directions of the multipath components associated with the mth transmit and

nth receive ports, respectively, and P is the number of relevant paths. Notice that $\mathbf{F}_m^T(\theta,\phi)$ and $\mathbf{F}_n^R(\theta,\phi)$ are the normalized far-field radiation patterns, as defined in Section 4.2.3, associated with the mth transmit and nth receiving port, respectively. Finally, notice that the upper-indexes T and R refers to the transmit and receive MPAs, respectively. Assuming that the antenna impedances and characteristic reference impedances are the same for all the ports, the complete subblock matrix $\mathbf{S}_{\alpha_R\alpha_T}^E$ can easily be expressed as

$$\mathbf{S}_{\alpha_R\alpha_T}^E = \lambda \sum_{p=1}^{P} (\mathbf{F}^{R,p}((\theta,\phi)))^{\dagger} \boldsymbol{\Gamma}_{nm}^p \mathbf{F}^{T,p}((\theta,\phi)), \tag{4.86}$$

where $\mathbf{F}^{A,p}((\theta,\phi)) = (\mathbf{F}_1^A((\theta,\phi)_1^p),\cdots,\mathbf{F}_n^A((\theta,\phi)_n^p))$ and where the upper-index A is used T here to refer to either T or R.

4.5.1 Separation of the Transmitter and Receiver Channel Characteristics

In the same way in which it is interesting to be able to conduct an analysis of the components at the transmitter separately from those at the receiver, or vice versa, as shown in Section 4.4.3, similar interests arise for the analysis of the propagation channel.

Starting from the statistical characteristics of the polarimetric channel matrix $\boldsymbol{\Gamma}_{nm}^p$ and the departing and arriving angles, it is always possible to define an outgoing power spectrum at the transmitter, $\boldsymbol{\Psi}^T(\theta,\phi) \in \mathcal{C}^{2\times 1}$, and an incoming power spectrum at the receiver, $\boldsymbol{\Psi}^R(\theta,\phi) \in \mathcal{C}^{2\times 1}$, representing the statistical distribution of incident and departing waves, respectively, in an environment. As shown in Fig. 4.8, from a system perspective, $\boldsymbol{\Psi}^T(\theta,\phi)$ represents the power spectrum over the angular intervals subtended by the scattering clusters of the channel being illuminated by the transmit array and that impact at the receiver, whereas $\boldsymbol{\Psi}^R(\theta,\phi)$ represents the power spectrum associated with the scattering intervals as observed from the receive array. Notice that if the random fading processes at the receiver are uncorrelated to those at the transmitter, which is a situation that occurs in most of the rich scattered NLOS propagation scenarios, then the statistics of the outgoing and incoming power spectrum are independent. Let us refer by $\boldsymbol{\Psi}^A(\theta,\phi)$ to any of the aforementioned power spectrum distributions, that is, $\boldsymbol{\Psi}^A(\theta,\phi)$ can be either $\boldsymbol{\Psi}^T(\theta,\phi)$ or $\boldsymbol{\Psi}^R(\theta,\phi)$. Then, $\boldsymbol{\Psi}^A(\theta,\phi)$ can be expanded into

$$\boldsymbol{\Psi}^A(\theta,\phi) = \boldsymbol{\Psi}_\theta^A(\theta,\phi)\widehat{\theta} + \boldsymbol{\Psi}_\phi^A(\theta,\phi)\widehat{\phi} = \begin{pmatrix} \boldsymbol{\Psi}_\theta^A(\theta,\phi) \\ \boldsymbol{\Psi}_\phi^A(\theta,\phi) \end{pmatrix}, \tag{4.87}$$

where $\boldsymbol{\Psi}_\theta^A(\theta,\phi) = \frac{\text{XPR}}{1+\text{XPR}}\boldsymbol{\Psi}_\theta^{A'}(\theta,\phi)$ is the elevation power spectrum, $\boldsymbol{\Psi}_\phi^A(\theta,\phi) = \frac{1}{1+\text{XPR}}\boldsymbol{\Psi}_\phi^{A'}(\theta,\phi)$ is the azimuth power spectrum, and XPR is the cross-polarization ratio, that is, the ratio of the

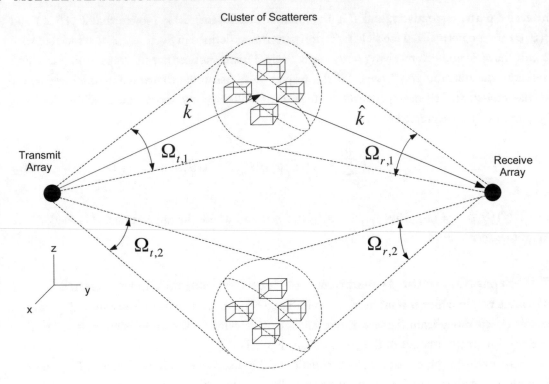

FIGURE 4.8: Graphical representation of the transmit and receive power spectra.

power in the θ polarization to the power in the ϕ polarization [6]. Measurements have shown that is it possible to decompose the azimuth and elevation power spectrum as follows:

$$\Psi^{A'}_{\theta}(\theta, \phi) = \Psi^{A'}_{\theta}(\theta)\Psi^{A'}_{\theta}(\phi) \tag{4.88}$$

$$\Psi^{A'}_{\phi}(\theta, \phi) = \Psi^{A'}_{\phi}(\theta)\Psi^{A'}_{\phi}(\phi), \tag{4.89}$$

where the power azimuth spectra, $\Psi^{A'}_{\theta}(\phi)$ and $\Psi^{A'}_{\phi}(\phi)$ are best modeled by a Laplacian function, for both polarizations. For the power elevation profiles, $\Psi^{A'}_{\theta}(\theta)$ and $\Psi^{A'}_{\phi}(\theta)$, a Gaussian function is normally assumed [6] for both polarizations. The azimuth and elevation power spectra must be normalized such that [6]:

$$\int_{-\pi}^{\pi}\int_{0}^{\pi}\Psi^{A'}_{\theta}(\theta, \phi)\sin(\theta)\mathrm{d}\theta\mathrm{d}\phi = 1 \tag{4.90}$$

$$\int_{-\pi}^{\pi}\int_{0}^{\pi}\Psi^{A'}_{\phi}(\theta, \phi)\sin(\theta)\mathrm{d}\theta\mathrm{d}\phi = 1. \tag{4.91}$$

Notice that in the case of an ideally scattered NLOS propagation scenario $\Psi^{A'}_{\theta}(\phi) = \Psi^{A'}_{\phi}(\phi) = \Psi^{A'}_{\theta}(\theta) = \Psi^{A'}_{\phi}(\theta) = \frac{1}{2\pi}$, and therefore, $\Psi^{A'}_{\theta}(\theta, \phi) = \frac{1}{4\pi}$ and $\Psi^{A'}_{\phi}(\theta, \phi) = \frac{1}{4\pi}$. Notice that after normalization of $\Psi^{A'}_{\theta}(\theta, \phi)$ and $\Psi^{A'}_{\phi}(\theta, \phi)$, these quantities no longer represent the real departing and arriving power spectra. They are normalized quantities so that it is possible to compare different spectrum distributions. Therefore, when using a separate modeling of the channel at the transmitter and at the receiver, with normalized power spectra, it is important to normalize the resultant channel matrix according to Section 1.5.5—that is, one must use the normalized version of the channel matrix $\mathbf{H}_0$ for the capacity and error probability calculations. This is because the capacity depends on receive SNR, and the channel matrices must be properly normalized for correct interpretation of results. The separate modeling of the transmitter and the receiver, if supported by the channel, is normally used in conjunction with the Kronecker channel model [18] described in Section 1.6.

4.6   MUTUAL COUPLING IN MPAs

The effect of mutual coupling between the ports of an MPA and its overall impact on the performance of a MIMO communication system has been a subject of significant study over the years. In general, mutual interaction between radiating structures is responsible for many, often undesirable effects. It has been reported in the literature [2] that mutual coupling may sometimes be beneficial. Whether mutual coupling is good or bad from a system-level perspective is a difficult question because the answer depends on other parameters such as the source and drain terminations, specific scattering characteristics propagation channels, and so forth. Mutual coupling may arise, for example, among closely located radiating structures, through surface waves in a dielectric, or through reflections on near field scatters. A graphical representation of the causes of mutual coupling in the particular case of a microstrip based MEA is given in Fig. 4.9.

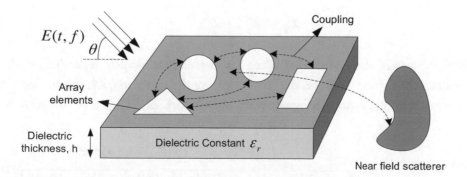

FIGURE 4.9: Mutual coupling in the particular case of a microstrip-based MEA.

To better understand the physical implications of mutual coupling, we here describe some of the mechanisms that are undergone by mutual coupling. These are indicated schematically on Fig. 4.10, through a three-element MEA. We can distinguish four main mechanisms:

- The first mechanism symbolizes the wanted radiation from the excited element when alone.
- Because of the interaction between the elements, there will be some scattering (second mechanism), either induced directly or indirectly through other elements. The scattered field may or may not have the same far-field pattern as the first mechanism but will in any case distort the primary field, in any or all of amplitude, phase, and polarization.
- The third mechanism represents the power coupled into the loads of adjacent antennas and is usually referred to as the coupling loss. This is also related to the input impedance of an MEA.
- The fourth mechanism represents the power coupled back into the source, leading to a mismatch.

Notice that mutual coupling is a property of the MPAs themselves and does not depend on the characteristics of the propagation channel. Its effects, however, strongly depend on the terminations associated with the antennas.

Based on the aforementioned mutual coupling mechanisms, we now conduct a more fundamental investigation of the problem. In Ref. [58], it was shown that the mutual impedance, $\mathbf{Z}_{ij}^{A}$, between any two generic antennas terminated on its characteristic reference impedance Z^0, may be expressed as a sum of two terms: a 0th-order term, $\mathbf{Z}_{ij}^{A,0}$, dependent exclusively on the radiation patterns ($\mathbf{S}_{\beta\alpha}^{A}$) of the radiating structures conforming the MPA, and a second term, $\mathbf{Z}_{ij}^{\widetilde{A}}$, involving the scattering properties, $\mathbf{S}_{\beta\beta}^{A}$, as well as the radiation patterns of the MPA. That is,

$$\mathbf{Z}_{ij}^{A} = \mathbf{Z}_{ij}^{A,0} + \mathbf{Z}_{ij}^{\widetilde{A}}. \qquad (4.92)$$

In Ref. [58], it is also shown that when the antennas are largely separated or when the MPA approaches the CMSA concept, the term involving the scattering, $\mathbf{Z}_{ij}^{\widetilde{A}}$, can be neglected. In particular, the mutual impedance among two ports of a CMSA, $\mathbf{Z}_{ij}^{\mathrm{CMSA}}$, can be computed as

$$\mathbf{Z}_{ij}^{\mathrm{CMSA}} = \mathbf{Z}_{ij}^{\mathrm{CMSA},0} \propto \frac{\int_{-\pi}^{\pi} \int_{0}^{\pi} (\mathbf{F}_{i}^{A}(\theta,\phi))^{H} \mathbf{F}_{j}^{A}(\theta,\phi) \sin(\theta) \mathrm{d}\theta \mathrm{d}\phi}{\int_{-\pi}^{\pi} \int_{0}^{\pi} |\mathbf{F}_{i}^{A}(\theta,\phi)|^{2} \sin(\theta) \mathrm{d}\theta \mathrm{d}\phi \int_{-\pi}^{\pi} \int_{0}^{\pi} |\mathbf{F}_{j}^{A}(\theta,\phi)|^{2} \sin(\theta) \mathrm{d}\theta \mathrm{d}\phi}. \qquad (4.93)$$

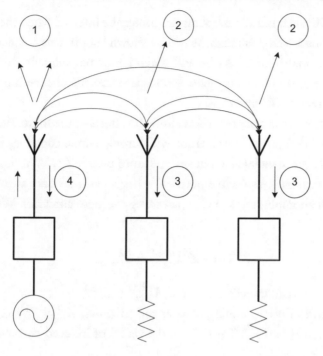

FIGURE 4.10: Schematic representation of interactions between antennas and the mechanisms invoked by mutual coupling.

In addition, because any lossless MPA satisfies the following condition $(\mathbf{S}_{\beta\alpha}^{A})^{H}\mathbf{S}_{\beta\alpha}^{A} = \mathbf{I} - (\mathbf{S}_{\alpha\alpha}^{A})^{H}\mathbf{S}_{\alpha\alpha}^{A}$ for lossless CMSAs, it is possible to rewrite Eq. (4.93) by means of the scattering parameter $\mathbf{S}_{\alpha\alpha}^{A}$ as follows:

$$\mathbf{Z}_{ij}^{\text{CMSA}} = \frac{(\mathbf{S}_{\alpha\alpha_i}^{A})^{H}\mathbf{S}_{\alpha\alpha_j}^{A}}{(1 - (\mathbf{S}_{\alpha\alpha_i}^{A})^{H}\mathbf{S}_{\alpha\alpha_i}^{A})(1 - (\mathbf{S}_{\alpha\alpha_j}^{A})^{H}\mathbf{S}_{\alpha\alpha_j}^{A})} \quad \text{for } i \neq j \qquad (4.94)$$

$$\mathbf{Z}_{ii}^{\text{CMSA}} \propto \frac{1}{1 - (\mathbf{S}_{\alpha\alpha_i}^{A})^{H}\mathbf{S}_{\alpha\alpha_i}^{A}} \quad \text{for } i = j, \qquad (4.95)$$

with $\mathbf{S}_{\alpha\alpha_i}^{A}$ being the ith column of $\mathbf{S}_{\alpha\alpha}^{A}$. For the remainder of this book, we will assume that the MPAs do a good job of approximating the CMSA concept. This is a convenient assumption because, in general, the term $\mathbf{S}_{\beta\beta}^{A}$ is complicated to measure. This has serious implications when investigating the impact of MPA on the performance of MIMO communication systems, because useful parameters such as the antenna correlation can easily be computed from the radiation pattern associated with the ports of the MPA and from the statistics of the channel.

Note also that although mutual coupling can usually be inferred when the matrix $\mathbf{S}_{\alpha\alpha}^A$ is not diagonal, its implications are far-reaching. As will be shown later in this chapter, mutual coupling impacts the correlation matrix of an MPA and may or may not degrade the performance of a MIMO system or its transmission power gain. Notice also that the upper-index A can either refer to the transmit (T) or receive (R) antennas.

Finally, we present an example that shows how the effects of the mutual coupling depend on the terminations at the MPA ports. In a situation with strong mutual coupling, $\mathbf{S}_{\alpha\alpha}^A$ is not diagonal and therefore neither is the associated antenna impedance matrix $\mathbf{Z}^A$. Without loss of generality, assume a receive MEA with n-accessible ports. We begin with a short analysis on the receive antenna voltages when an impinging signal is picked up on these antennas. We can establish the following relationship:

$$\mathbf{Z}^D \mathbf{i} = -\mathbf{Z}^R \mathbf{I}_{\alpha_R}^{R,\text{ind}} + \mathbf{U}_{\alpha_R}^{R,oc}, \tag{4.96}$$

where $\mathbf{Z}^R$ is the receive MEA impedance matrix, $\mathbf{U}_{\alpha_R}^{R,oc} \in \mathcal{C}^{n \times 1}$ is a vector describing the open-circuit voltage as a result of waves impinging over the antennas, $\mathbf{Z}^D = Z^D \mathbf{I} \in \mathcal{C}^{n \times n}$ is the load's impedance diagonal matrix, and $\mathbf{I}_{\alpha_R}^{R,\text{ind}} \in \mathcal{C}^{n \times 1}$ is the vector of induced currents on the antennas. Notice that $\mathbf{Z}^R$ can be expanded into

$$\mathbf{Z}^R = \begin{pmatrix} \mathbf{Z}_{11}^R & \cdots & \mathbf{Z}_{1n}^R \\ \vdots & \ddots & \vdots \\ \mathbf{Z}_{n1}^R & \cdots & \mathbf{Z}_{nn}^R \end{pmatrix}. \tag{4.97}$$

If we define the voltage at the load impedance as $\mathbf{U}_{\alpha_R}^R = Z^D \mathbf{i}$, then,

$$\mathbf{U}_{\alpha_R}^R = (\mathbf{C}^R)^{-1} \mathbf{U}_{\alpha_R}^{R,oc}, \tag{4.98}$$

where $\mathbf{U}_{\alpha_R}^R \in \mathcal{C}^{n \times 1}$, and $\mathbf{C}^R \in \mathcal{C}^{n \times n}$ is normally called the *coupling matrix*. $\mathbf{C}^R$ relates the open-circuit voltages with the voltages at the loads and is given by

$$\mathbf{C}^R = \begin{pmatrix} 1 + \dfrac{\mathbf{Z}_{11}^R}{Z^D} & \cdots & \dfrac{\mathbf{Z}_{1n}^R}{Z^D} \\ \vdots & \ddots & \vdots \\ \dfrac{\mathbf{Z}_{n1}^R}{Z^D} & \cdots & 1 + \dfrac{\mathbf{Z}_{nn}^R}{Z^D} \end{pmatrix}. \tag{4.99}$$

Observe the dependence of the coupling matrix $\mathbf{C}^R$ on the load termination at the ports Z^D. Notice that $\mathbf{C}^R$ will be diagonal if $\mathbf{S}_{\alpha_R \alpha_R}^R$ is also diagonal, and in the open-circuit case, $Z^D \to \infty$ and $\mathbf{U}_{\alpha_R}^R = \mathbf{U}_{\alpha_R}^{R,oc}$. A similar analysis can be performed for a transmit MEA.

Using Eq. (4.74), the aforementioned analysis can also be done in terms of scattering parameters. This analysis yields to the following relationship between the open-circuit voltage $\mathbf{U}_{\alpha_R}^{R,\text{oc}}$ and a generic voltage at the loads, $\mathbf{U}_{\alpha_R}^R$, depending on a particular termination:

$$\mathbf{U}_{\alpha_R}^R = \frac{1}{2}\left(\mathbf{I} + \mathbf{\Gamma}_{\alpha_R}^D\right)\left(\mathbf{I} - \mathbf{S}_{\alpha_R\alpha_R}^R \mathbf{\Gamma}_{\alpha_R}^D\right)^{-1}\left(\mathbf{I} - \mathbf{S}_{\alpha_R\alpha_R}^R\right)\mathbf{U}_{\alpha_R}^{R,\text{oc}}, \qquad (4.100)$$

where once again in the open-circuit case $Z^D \to \infty$, $\mathbf{\Gamma}_{\alpha_R}^D \to \mathbf{I}$, and $\mathbf{U}_{\alpha_R}^R = \mathbf{U}_{\alpha_R}^{R,\text{oc}}$.

4.7   PERFORMANCE MEASURES IN MIMO COMMUNICATION SYSTEMS

In this section, we describe the main performance measures that are used to assess the performance of MPAs within a MIMO communication system. System-level figures of merit are also presented.

4.7.1   Transmission Power Gain

Because most handheld devices are battery-driven, the efficiency of MPAs is important. For an MEA, for example, mutual coupling among closely spaced antennas not only influences the signal flow and the correlation properties, but it can also strongly reduce the efficiency of a MIMO communication system. We consider now the power transmission gain of the complete MIMO link, including the transmit and receive MPA, but independently of the excitation distribution $\mathbf{b}_0^S$.

To define the transmission power gain, we assume the extended channel defined in Section 4.4.2, whose channel matrix is given by $\mathbf{E}$. We define the transmission power gain as the ratio of the instantaneous real power delivered to the signal drain, $P^D = (\mathbf{b}_{\alpha_R}^R)^H\mathbf{b}_{\alpha_R}^R - (\mathbf{a}_{\alpha_R}^R)^H\mathbf{a}_{\alpha_R}^R$, to the instantaneous real power fed into the transmit antennas, $P^S = (\mathbf{a}_{\alpha_T}^T)^H\mathbf{a}_{\alpha_T}^T - (\mathbf{b}_{\alpha_T}^T)^H\mathbf{b}_{\alpha_T}^T$:

$$G_{\text{tp}} = \frac{P^D}{P^S} = \frac{(\mathbf{b}_{\alpha_R}^R)^H\mathbf{b}_{\alpha_R}^R - (\mathbf{a}_{\alpha_R}^R)^H\mathbf{a}_{\alpha_R}^R}{(\mathbf{a}_{\alpha_T}^T)^H\mathbf{a}_{\alpha_T}^T - (\mathbf{b}_{\alpha_T}^T)^H\mathbf{b}_{\alpha_T}^T}. \qquad (4.101)$$

The power gain allows us to draw conclusions about the efficiency of the whole transmission chain. Notice that the upper-index E refers to the extended channel matrix. Finally, we can also define the average transmit and receive power, $P^{S,\text{av}}$ and $P^{D,\text{av}}$, respectively, as

$$P^{S,\text{av}} = E\{P^S\} \qquad (4.102)$$

$$P^{D,\text{av}} = E\{P^D\}. \qquad (4.103)$$

By comparing the transmission power gain of MIMO systems with different transmit and receive MPAs within the same channel, one can draw conclusions on how specific MPAs perform on such systems. As a function of $\mathbf{b}_0^S$, the quantities P^D and P^S can be written as

$$P^S = (\mathbf{b}_0^S)^H (((\mathbf{I} - \mathbf{\Gamma}_{\alpha_T}^S \mathbf{S}_{\alpha_T \alpha_T}^E))^{-1})^H (\mathbf{I} - (\mathbf{S}_{\alpha_T \alpha_T}^E)^H \mathbf{S}_{\alpha_T \alpha_T}^E)(\mathbf{I} - \mathbf{\Gamma}_{\alpha_T}^S \mathbf{S}_{\alpha_T \alpha_T}^E)^{-1} \mathbf{b}_0^S \quad (4.104)$$

$$P^D = (\mathbf{b}_0^R)^H (((\mathbf{I} - \mathbf{S}_{\alpha_R \alpha_R}^E \mathbf{\Gamma}_{\alpha_R}^D))^{-1})^H (\mathbf{I} - (\mathbf{\Gamma}_{\alpha_R}^D)^H \mathbf{\Gamma}_{\alpha_R}^D)(\mathbf{I} - \mathbf{S}_{\alpha_R \alpha_R}^E \mathbf{\Gamma}_{\alpha_R}^D)^{-1} \mathbf{b}_0^R, \quad (4.105)$$

with $\mathbf{b}_0^R$ related to $\mathbf{b}_0^S$ in Eq. (4.77). Using Eqs. (4.104) and (4.105), one observes that the transmission power gain (G_{tp}) does not depend on the power distribution among the elements of the excitation vector $\mathbf{b}_0^S$; instead, it depends only on the radiation properties of the transmit and receive MPAs, the load and source terminations, and the scattering characteristics of the channel.

If needed, the transmission power gain G_{tp} can be separated into three components, as follows:

$$G_{\mathrm{tp}} = G_{\mathrm{tp}}^T G_{\mathrm{tp}}^C G_{\mathrm{tp}}^R, \quad (4.106)$$

where G_{tp}^T and G_{tp}^R represent the effective gain of the transmit and receive MPAs, respectively, on a system with an specific source and drain configuration. On the other hand, G_{tp}^C accounts for the channel propagation gains. These quantities are given by

$$G_{\mathrm{tp}}^T = \frac{(\mathbf{b}_{\beta_T}^T)^H \mathbf{b}_{\beta_T}^T - (\mathbf{a}_{\beta_T}^T)^H \mathbf{a}_{\beta_T}^T}{(\mathbf{a}_{\alpha_T}^T)^H \mathbf{a}_{\alpha_T}^T - (\mathbf{b}_{\alpha_T}^T)^H \mathbf{b}_{\alpha_T}^T}, \quad (4.107)$$

$$G_{\mathrm{tp}}^C = \frac{(\mathbf{a}_{\beta_R}^T)^H \mathbf{a}_{\beta_R}^R - (\mathbf{b}_{\beta_R}^R)^H \mathbf{b}_{\beta_R}^R}{(\mathbf{b}_{\beta_T}^T)^H \mathbf{b}_{\beta_T}^T - (\mathbf{a}_{\beta_T}^T)^H \mathbf{a}_{\beta_T}^T}, \quad (4.108)$$

$$G_{\mathrm{tp}}^R = \frac{(\mathbf{b}_{\alpha_R}^R)^H \mathbf{b}_{\alpha_R}^R - (\mathbf{a}_{\alpha_R}^R)^H \mathbf{a}_{\alpha_R}^R}{(\mathbf{a}_{\beta_R}^T)^H \mathbf{a}_{\beta_R}^R - (\mathbf{b}_{\beta_R}^R)^H \mathbf{b}_{\beta_R}^R}. \quad (4.109)$$

4.7.1.1 Transmit Matching Efficiency. In the optimal scenario in which one desires to maximize the power delivered from the source to the transmit MPA, one needs to satisfy the condition $\mathbf{\Gamma}_{\alpha_T}^S = (\mathbf{S}_{\alpha_T \alpha_T}^E)^H$ [66]. Thus, using Eq. (4.104), the maximum instantaneous transmitted power can be expressed as

$$P^{S,\mathrm{max}} = (\mathbf{b}_0^S)^H \left(\left(\mathbf{I} - (\mathbf{S}_{\alpha_T \alpha_T}^E)^H \mathbf{S}_{\alpha_T \alpha_T}^E \right)^{-1} \right)^H \mathbf{b}_0^S. \quad (4.110)$$

As a result, the power transmission gain is given by

$$G_{\text{tp}}^{S,\max} = \frac{(\mathbf{b}_{\alpha_R}^R)^H \mathbf{b}_{\alpha_R}^R - (\mathbf{a}_{\alpha_R}^R)^H \mathbf{a}_{\alpha_R}^R}{(\mathbf{b}_0^S)^H \left(\left(\mathbf{I} - (\mathbf{S}_{\alpha_T \alpha_T}^E)^H \mathbf{S}_{\alpha_T \alpha_T}^E \right)^{-1} \right)^H \mathbf{b}_0^S}. \tag{4.111}$$

On the other hand, in general, $\mathbf{\Gamma}_{\alpha_T}^S = \mathbf{0}$, and in the case that no maximum power transfer can be assured, which, for example, may happen if $\mathbf{S}_{\alpha_T \alpha_T}^E \neq \mathbf{0}$, then the transmission power gain will not be the maximum and is given by

$$G_{\text{tp}}^{S,0} = \frac{(\mathbf{b}_{\alpha_R}^R)^H \mathbf{b}_{\alpha_R}^R - (\mathbf{a}_{\alpha_R}^R)^H \mathbf{a}_{\alpha_R}^R}{(\mathbf{b}_0^S)^H (\mathbf{I} - (\mathbf{S}_{\alpha_T \alpha_T}^E)^H \mathbf{S}_{\alpha_T \alpha_T}^E) \mathbf{b}_0^S}, \tag{4.112}$$

where the upper-index 0 refers to the fact that we assumed $\mathbf{\Gamma}_{\alpha_T}^S = \mathbf{0}$. Notice that the radiated transmit power is limited by a factor that has to do with the coupling among antennas, $\mathbf{A} = (\mathbf{I} - (\mathbf{S}_{\alpha_T \alpha_T}^E)^H \mathbf{S}_{\alpha_T \alpha_T}^E)$. In this case, the radiated power is given by

$$P^{S,0} = (\mathbf{b}_0^S)^H \mathbf{A} \mathbf{b}_0^S. \tag{4.113}$$

Notice that for the case where $\mathbf{\Gamma}_{\alpha_T}^S = \mathbf{0}$, the radiated power will be maximum when the mutual coupling among the antennas is negligible, which is when $\mathbf{S}_{\alpha_T \alpha_T}^E = \mathbf{0}$ or when by means of specific antenna engineering techniques the MPAs can be decoupled and matched (as shown in Chapter 5). In that case, one obtains that the maximum radiated power is given by

$$P^{S,0,\max} = (\mathbf{b}_0^S)^H \mathbf{b}_0^S, \tag{4.114}$$

and the transmission power gain for the case of maximum radiated power and $\mathbf{\Gamma}_{\alpha_T}^S = \mathbf{0}$ is given by

$$G_{\text{tp}}^{S,0,\max} = \frac{(\mathbf{b}_{\alpha_R}^R)^H \mathbf{b}_{\alpha_R}^R - (\mathbf{a}_{\alpha_R}^R)^H \mathbf{a}_{\alpha_R}^R}{(\mathbf{b}_0^S)^H \mathbf{b}_0^S}. \tag{4.115}$$

We emphasize that in traditional analysis of MIMO systems, the total transmitted power is taken as $(\mathbf{b}_0^S)^H \mathbf{b}_0^S$, therefore ignoring the mutual coupling effects. Therefore, this illustrates one way in which mutual coupling impacts the system analysis. Using the above findings, and for the sake of completeness, the matching efficiency for the transmit MPA can be defined as

$$\eta^{T,0} = \frac{E\{P^{S,0}\}}{E\{P^{S,0,\max}\}}, \tag{4.116}$$

where, once again, the upper-index 0 refers to the fact that we assumed $\mathbf{\Gamma}_{\alpha_T}^S = \mathbf{0}$. Notice that with this definition of the efficiency, when the mutual coupling vanishes, $\eta^{T,0} \to 1$. We could also define the transmit efficiency independently of the source characteristic reference impedances, in which case, it is given by

$$\eta^T = \frac{E\{P^S\}}{E\{P^{S,\max}\}}.$$

(4.117)

4.7.1.2 Receive Matching Efficiency. Using the separable model for the receiver side, the power delivered to the load can be expressed as a function of $\mathbf{b}_0^R$, as shown in Eq. (4.105). The power delivered to the load is maximized when $\mathbf{\Gamma}_{\alpha_R}^D = (\mathbf{S}_{\alpha_R \alpha_R}^E)^H$, as shown in Ref. [14], and it can be written as

$$P^{D,\max} = (\mathbf{b}_0^R)^H \left(\left(\left(\mathbf{I} - \mathbf{S}_{\alpha_R \alpha_R}^E (\mathbf{S}_{\alpha_R \alpha_R}^E)^H\right)\right)^{-1}\right)^H \mathbf{b}_0^R.$$

(4.118)

Therefore, in this case, the power transmission gain is given by

$$G_{\mathrm{tp}}^{D,\max} = \frac{(\mathbf{b}_0^R)^H \left(\left(\left(\mathbf{I} - \mathbf{S}_{\alpha_R \alpha_R}^E (\mathbf{S}_{\alpha_R \alpha_R}^E)^H\right)\right)^{-1}\right)^H \mathbf{b}_0^R}{(\mathbf{a}_{\alpha_T}^T)^H \mathbf{a}_{\alpha_T}^T - (\mathbf{b}_{\alpha_T}^T)^H \mathbf{b}_{\alpha_T}^T}.$$

(4.119)

On the other hand, in general, $\mathbf{\Gamma}_{\alpha_R}^D = \mathbf{0}$, and in the case that no maximum power transfer can be assured, which, for example, may happen if $\mathbf{S}_{\alpha_R \alpha_R}^E \neq \mathbf{0}$, the power delivered to the loads is given by

$$P^{D,0} = (\mathbf{b}_0^R)^H \mathbf{b}_0^R,$$

(4.120)

where the upper-index 0 refers to the fact that we assumed $\mathbf{\Gamma}_{\alpha_R}^D = \mathbf{0}$. For the case of $\mathbf{\Gamma}_{\alpha_R}^D = \mathbf{0}$ and using specific antenna engineering techniques to decouple and match the ports of an MPA (as shown in Chapter 5), the maximum collected power is then given by

$$P^{D,0,\max} = (\mathbf{b}_0^R)^H \left(\left(\left(\mathbf{I} - \mathbf{S}_{\alpha_R \alpha_R}^E (\mathbf{S}_{\alpha_R \alpha_R}^E)^H\right)\right)^{-1}\right)^H \mathbf{b}_0^R.$$

(4.121)

Finally, at the receiver side, we can also define the receive matching efficiency η^R in the same manner as the relative collected power is defined in Ref. [14]. $\eta^{R,0}$ represents the ratio of the total received power by the MPA in the case that $\mathbf{\Gamma}_{\alpha_R}^D = \mathbf{0}$, to that of an optimally matched reference MPA in the same environment:

$$\eta^{R,0} = \frac{E\{P^{D,0}\}}{E\{P^{D,0,\max}\}}.$$

(4.122)

For a generic termination of the loads, the efficiency can be rewritten as follows:

$$\eta^R = \frac{E\{P^D\}}{E\{P^{D,\max}\}}.$$

(4.123)

4.7.2   Mean Effective Gain

Sometimes, one must compute the active gain associated with a particular port of an MPA by taking into account the statistics of a particular propagation scenario. To do this, we use the mean effective gain (MEG). The MEG is normally defined for a receive MPA as the ratio of the mean delivered power at a particular port of the receive MPA to the mean received power by a reference antenna, when the reference antenna is used in the same channel and with the same transmit MPA as the receive MPA under test. The MEG associated with the ith port of a receiving MPA is given by

$$G_i^{R,\text{MEG}} = \frac{E\{P_{\alpha_{R_i}}^D\}}{E\{P_{\text{ref}}^D\}} = \frac{P_{\alpha_{R_i}}^{D,\text{av}}}{P_{\text{ref}}^{D,\text{av}}}.$$

(4.124)

Notice that the MEG is a similar concept to the active gain. However, the MEG characteristics of a mobile MEA, for example, are determined by the mutual coupling relation between the antenna patterns and, in addition, by the statistical distribution of incident waves in an environment. By contrast, the distribution of incident waves is not taken into account when considering the active gain.

If an isotropic antennas is used as a reference antenna, the MEG associated with the ith port within an MPA can be calculated analytically as follows [6]:

$$G_i^{R,\text{MEG}} = \int_{-\pi}^{\pi} \int_0^{\pi} (\mathbf{G}_i^R(\theta,\phi))^\dagger \mathbf{\Psi}^R(\theta,\phi)\sin(\theta)\mathrm{d}\theta\mathrm{d}\phi,$$

(4.125)

where $\mathbf{\Psi}^R(\theta,\phi) \in \mathcal{C}^{2\times1}$ is the power spectrum describing the statistical distribution of the incoming waves in the azimuth and elevation profile, for both polarizations, as defined in Section 4.5.1. Eq. (4.125) can be also written expanding the integrands as follows:

$$G_i^{R,\text{MEG}} = \int_{-\pi}^{\pi} \int_0^{\pi} \left(\frac{\text{XPR}}{1+\text{XPR}}\mathbf{G}_{i,\theta}^R(\theta,\phi)\mathbf{\Psi}_\theta^{R'}(\theta,\phi)\right.$$

$$\left. + \frac{1}{1+\text{XPR}}\mathbf{G}_{i,\phi}^R(\theta,\phi)\mathbf{\Psi}_\phi^{R'}(\theta,\phi)\right)\sin(\theta)\mathrm{d}\theta\mathrm{d}\phi.$$

(4.126)

The MEG concept can be extended to assess complete MPAs. To do this, we define the mean effective array gain (MEAG) as the ratio of the mean received power by an MPA to the mean received power by a reference antenna within the same channel and using the same transmit MPA. The MEAG is given by

$$G^{R,\text{MEAG}} = \frac{E\{P_{\alpha_R}^D\}}{E\{P_{\text{ref}}^D\}} = \frac{E\{P^D\}}{P_{\text{ref}}^{D,\text{av}}} = \frac{P^{D,\text{av}}}{P_{\text{ref}}^{D,\text{av}}}.$$

(4.127)

Notice that both the MEG and MEAG depend on both the transmit and receive MPAs, the propagation channel, the load impedances (source/drain) that are connected to the MPAs, and the power distribution among the antennas $\mathbf{b}_0^S$.

Finally, notice that the MEG can also be defined for the ith port an MPA in transmission by using the following expression:

$$
G_i^{T,\text{MEG}} = \int_{-\pi}^{\pi} \int_0^{\pi} \left(\frac{\text{XPR}}{1+\text{XPR}} \mathbf{G}_{i,\theta}^T(\theta,\phi) \mathbf{\Psi}^{T'}_{\theta}(\theta,\phi) \right.
$$

$$
\left. + \frac{1}{1+\text{XPR}} \mathbf{G}_{i,\phi}^T(\theta,\phi) \mathbf{\Psi}^{T'}_{\phi}(\theta,\phi) \right) \sin(\theta)\mathrm{d}\theta\mathrm{d}\phi. \qquad (4.128)
$$

4.7.3 Channel Correlation

In this section, we define the main figures of merit associated with the correlation properties of a MIMO communication system. Let us begin by defining the channel correlation matrix $\mathcal{R}^H \in \mathcal{C}^{MN \times MN}$ given by

$$
\mathcal{R}^H = E\{\mathbf{vec}(\mathbf{H})\mathbf{vec}(\mathbf{H})^H\}, \qquad (4.129)
$$

where $\mathbf{vec}(\mathbf{H}) \in \mathcal{C}^{MN \times 1}$. We can also define the extended channel correlation matrix, $\mathcal{R}^E \in \mathcal{C}^{MN \times MN}$, as

$$
\mathcal{R}^E = E\{\mathbf{vec}(\mathbf{E})\mathbf{vec}(\mathbf{E})^H\}, \qquad (4.130)
$$

where $\mathbf{vec}(\mathbf{E}) \in \mathcal{C}^{MN \times 1}$. In both cases, these correlation matrices represent the correlation characteristics among different transmit and receive ports by taking into account the propagation of the signals over the channel and the radiation characteristics of the transmit and receive MPAs. In the latter case, $\mathcal{R}^E$ also takes into account the termination characteristics of the source and drain. The importance of these figures of merit is that they give us information of how correlation will influence the capacity of a MIMO communication system. These quantities can also be defined for the normalized versions of the channel matrix ($\mathbf{H}_0$) and extended channel matrix ($\mathbf{E}_0$), as follows:

$$
\mathcal{R}^{H_0} = E\{\mathbf{vec}(\mathbf{H}_0)\mathbf{vec}(\mathbf{H}_0)^H\} \qquad (4.131)
$$

$$
\mathcal{R}^{E_0} = E\{\mathbf{vec}(\mathbf{E}_0)\mathbf{vec}(\mathbf{E}_0)^H\}. \qquad (4.132)
$$

Notice that the number of correlation coefficients between all elements $\mathbf{H}_{ij}$ (or $\mathbf{E}_{ij}$) in $\mathbf{H}$ (or $\mathbf{E}$) is $N^2 M^2$. Thus, it is difficult to assess the correlation properties and to show a direct relationship between the capacity distribution and the correlation properties. In Ref. [67], a measure to describe the correlation among all elements $\mathbf{H}_{ij}$ (or $\mathbf{E}_{ij}$) is defined, and it is shown that the ergodic capacity of a MIMO system without CSI increases with decreasing correlation.

A simple way to assess whether the correlation of the system is high or low is to consider the transmit and receive correlation properties separately. Associated with the channel matrix $\mathbf{H}$, we can define the transmit antenna correlation matrix, $\mathbf{R}_H^T$, and the receive antenna correlation matrix, $\mathbf{R}_H^R$, as shown previously in Section 1.6:

$$\mathbf{R}_H^R = \frac{E\{\mathbf{H}(\mathbf{H})^H\}}{\sqrt{E\{||\mathbf{H}||_F^2\}}} \tag{4.133}$$

$$\mathbf{R}_H^T = \frac{E\{(\mathbf{H})^H\mathbf{H}\}}{\sqrt{E\{||\mathbf{H}||_F^2\}}}. \tag{4.134}$$

Notice that $\mathbf{R}_{H_{ij}}^T$ is the correlation value between the ith and jth ports of the transmit MPA, whereas and $\mathbf{R}_{H_{ij}}^R$ is the correlation value between the ith and jth ports of the receive MPA, without including the effects of source and drain termination. On the other hand, associated with the extended channel matrix $\mathbf{E}$, we can define then the extended transmit antenna correlation matrix, $\mathbf{R}_E^T$, and the extended receive antenna correlation matrix, $\mathbf{R}_E^R$, as follows:

$$\mathbf{R}_E^R = \frac{E\{\mathbf{E}(\mathbf{E})^H\}}{\sqrt{E\{||\mathbf{E}||_F^2\}}} \tag{4.135}$$

$$\mathbf{R}_H^T = \frac{E\{(\mathbf{E})^H\mathbf{E}\}}{\sqrt{E\{||\mathbf{E}||_F^2\}}}. \tag{4.136}$$

Notice that $\mathbf{R}_{E_{ij}}^T$ is the correlation value between the ith and jth ports of the transmit MPA, whereas $\mathbf{R}_{E_{ij}}^R$ is the correlation value between the ith and jth ports of the receive MPA, including the effects of source and drain termination. Notice that the entries of $\mathbf{H}$ (or $\mathbf{H}_0$) and $\mathbf{E}$ (or $\mathbf{E}_0$) have been assumed to have zero mean.

Notice that although this separation is always mathematically possible, in general, it may not be always possible to construct the resultant channel correlation matrix from the transmit and receive antenna correlation matrices, such as in the case that the channel does not supports this separability (because the channel statistics at the transmitter are dependent on those at the receiver, and vice versa). In most of the rich scattered NLOS propagation scenarios, however, the random fading processes at the receiver are uncorrelated to those at the transmitter. In that case, the separability of the transmitter and the receiver side of the channel is possible and, for example, a Kronecker channel model [18] can be used. As a result, it is much easier to evaluate how correlation may impact on the performance of a MIMO communication system. Let us assume then a Kronecker channel model for now to provide an intuitive physical insight on the correlation concept. Then, a good approximation for $\mathcal{R}^H$ is given by

$$\mathcal{R}^H = \mathbf{R}_H^T \otimes \mathbf{R}_H^R, \tag{4.137}$$

and the channel matrix $\mathbf{H}$ can be decomposed into

$$\mathbf{H} = (\mathbf{R}_H^R)^{\frac{1}{2}} \mathbf{W} ((\mathbf{R}_H^T)^{\frac{1}{2}})^\dagger, \tag{4.138}$$

where $\mathbf{W} \in \mathcal{C}^{N \times M}$ is a complex random matrix with all elements being iid Gaussian random variables with distribution $N(0,1)$. Notice that $\mathbf{R}_H^T = (\mathbf{R}_H^T)^{\frac{1}{2}}((\mathbf{R}_H^T)^{\frac{1}{2}})^H$ and $\mathbf{R}_H^R = (\mathbf{R}_H^R)^{\frac{1}{2}}((\mathbf{R}_H^R)^{\frac{1}{2}})^H$. The normalized version of the previous expressions can be found in Sections 1.6 and 3.2.4.

Similarly, if a Kronecker channel model can be assumed for the extended channel matrix case, a good approximation for $\mathcal{R}^E$ is given by

$$\mathcal{R}^E = \mathbf{R}_E^T \otimes \mathbf{R}_E^R, \tag{4.139}$$

and the extended channel matrix $\mathbf{E}$ can be decomposed into

$$\mathbf{E} = (\mathbf{R}_E^R)^{\frac{1}{2}} \mathbf{W} ((\mathbf{R}_E^T)^{\frac{1}{2}})^\dagger, \tag{4.140}$$

where $\mathbf{W} \in \mathcal{C}^{N \times M}$ is a complex random matrix with all elements being iid Gaussian random variables with distribution $N(0,1)$. Notice that $\mathbf{R}_E^T = (\mathbf{R}_E^T)^{\frac{1}{2}}((\mathbf{R}_E^T)^{\frac{1}{2}})^H$ and $\mathbf{R}_E^R = (\mathbf{R}_E^R)^{\frac{1}{2}}((\mathbf{R}_E^R)^{\frac{1}{2}})^H$.

4.7.4   Antenna Correlation

To express the antenna correlation matrices in terms of antenna parameters and the statistics of the propagation channel, we use the separated model for the transmitter and receiver described in Section 4.4.3. We start by analyzing the antenna correlation at the receiver. Later, we compute the transmit antenna correlation through the reciprocity theorem by solving the problem as a receiving structure. Let us start by defining the covariance of the receive signals, $\mathbf{X}_H^R$, as

$$\mathbf{X}_H^R = E\{\mathbf{b}_0^R (\mathbf{b}_0^R)^H\}. \tag{4.141}$$

For the transmit MPA acting as a receiving structure, we can define the outward propagating wave vector $\mathbf{b}_{\alpha_T}^T$ as

$$\mathbf{b}_{\alpha_T}^T = \mathbf{b}_0^T + \mathbf{S}_{\alpha_T \alpha_T}^T \mathbf{a}_{\alpha_T}^T, \tag{4.142}$$

where $\mathbf{b}_0^T \in \mathcal{C}^{N \times 1}$ is the wave vector injected into the network blocks, generated by the impinging electromagnetic waves into the transmit MPA, when acting as a receiving structure. $\mathbf{b}_0^T$ determines then the power distribution among the input ports of the transmit MPA. Then, we define the covariance of the transmitted signals, $\mathbf{X}_H^T$, as

$$\mathbf{X}_H^T = E\{\mathbf{b}_0^T(\mathbf{b}_0^T)^H\},$$

(4.143)

noticing that these matrices express the correlation of the impinging voltage waves on the transmit and receive antennas, respectively. Similarly, for the extended channel, we can define the covariance of the receive signals, $\mathbf{X}_E^R$, as

$$\mathbf{X}_E^R = E\{\mathbf{U}_{\alpha_R}^R(\mathbf{U}_{\alpha_R}^R)^H\}.$$

(4.144)

For the transmit MPA as a receiving structure, the voltages at the input ports of the transmit MPA, $\mathbf{U}_{\alpha_T}^T$, are given by

$$\mathbf{U}_{\alpha_T}^T = Z_{\alpha_T}^{0\,\frac{1}{2}}\left(\mathbf{I}+\boldsymbol{\Gamma}_{\alpha_T}^S\right)\left(\mathbf{I}-\mathbf{S}_{\alpha_T\alpha_T}^E\boldsymbol{\Gamma}_{\alpha_T}^S\right)^{-1}\mathbf{b}_0^T.$$

(4.145)

Then, we can also define the covariance of the transmitted signals, $\mathbf{X}_E^T$, as

$$\mathbf{X}_E^T = E\{\mathbf{U}_{\alpha_T}^T(\mathbf{U}_{\alpha_T}^T)^H\},$$

(4.146)

where one can easily observe that these last covariance terms depend on the termination values of the loads and sources.

4.7.4.1 Receive Antenna Correlation.

Using the aforementioned covariance terms, the (i,j)th entry of $\mathbf{R}_H^R$ can be computed as

$$\mathbf{R}_{H_{ij}}^R = \frac{\mathbf{X}_{H_{ij}}^R}{\sqrt{\mathbf{X}_{H_{ii}}^R\mathbf{X}_{H_{jj}}^R}}.$$

(4.147)

Let us express the covariance term $\mathbf{X}_{H_{ij}}^R$ in terms of antenna and channel parameters. To do this, recall that for an antenna in reception one has $\mathbf{b}_\alpha^R = \mathbf{S}_{\alpha\beta}^R\mathbf{a}_\beta^R$. Then,

$$\mathbf{X}_{H_{ij}}^R = E\left\{\mathbf{b}_{0_i}^R(\mathbf{b}_{0_j}^R)^*\right\} = E\{\mathbf{S}_{\alpha\beta_i}^R\mathbf{a}_\beta^R(\mathbf{S}_{\alpha\beta_j}^R)^*(\mathbf{a}_\beta^R)^*\}.$$

(4.148)

Using the fact that the receive antenna is reciprocal, we obtain

$$\mathbf{X}_{H_{ij}}^R = E\{(\mathbf{S}_{\beta\alpha_i}^R)^T\mathbf{a}_\beta^R(\mathbf{S}_{\beta\alpha_j}^R)^H(\mathbf{a}_\beta^R)^*\} = E\left\{\left(\sum_{k=1}^{\infty}\mathbf{S}_{\beta\alpha_{ki}}^R\mathbf{a}_{\beta_k}^R\right)\left(\sum_{k=1}^{\infty}(\mathbf{S}_{\beta\alpha_{kj}}^R)^*(\mathbf{a}_{\beta_k}^R)^*\right)\right\}.$$

(4.149)

Because the spherical modes are orthogonal among each other, the statistics of different arriving spherical modes can be assumed to be independent and uncorrelated, that is, $E\{\mathbf{a}_{\beta_k}^R(\mathbf{a}_{\beta_j}^R)^*\} = 0$ if $k \neq j$, where we have assumed that the modes have zero-mean statistic. In that case,

$$\mathbf{X}_{H_{ij}}^R = \sum_{k=1}^{\infty} \mathbf{S}_{\beta\alpha_{ki}}^R (\mathbf{S}_{\beta\alpha_{kj}}^R)^* E\{|\mathbf{a}_{\beta_k}^R|^2\}. \qquad (4.150)$$

Because $c_k = E\{|\mathbf{a}_{\beta_k}^R|^2\}$ is a real positive term, we can write $c_k = c_k^{\frac{1}{2}} c_k^{\frac{1}{2}}$. If we define the quantities $\mathbf{S'}_{\beta\alpha_{ki}}^R = \mathbf{S}_{\beta\alpha_{ki}}^R c_k^{\frac{1}{2}}$ and $\mathbf{S'}_{\beta\alpha_{kj}}^R = \mathbf{S}_{\beta\alpha_{kj}}^R c_k^{\frac{1}{2}}$, then, we obtain

$$\mathbf{X}_{H_{ij}}^R = \sum_{k=1}^{\infty} \mathbf{S'}_{\beta\alpha_{ki}}^R (\mathbf{S'}_{\beta\alpha_{kj}}^R)^* = \int_{-\pi}^{\pi}\int_0^{\pi} (\mathbf{F'}_i^R(\theta,\phi))^T (\mathbf{F'}_i^R(\theta,\phi))^* \sin(\theta)\mathrm{d}\theta\mathrm{d}\phi$$

$$= \int_{-\pi}^{\pi}\int_0^{\pi} \left(\mathbf{F'}_{i_\theta}^R(\theta,\phi)(\mathbf{F'}_{j_\theta}^R(\theta,\phi))^* + \mathbf{F'}_{i_\phi}^R(\theta,\phi)(\mathbf{F'}_{j_\phi}^R(\theta,\phi))^* \sin(\theta) \right) \mathrm{d}\theta\mathrm{d}\phi. \qquad (4.151)$$

Now, using the fact that we can write $\mathbf{F'}_{i_\theta}^R(\theta,\phi) = \mathbf{F}_{i_\theta}^R(\theta,\phi)\Xi_\theta^R(\theta,\phi)$ and $\mathbf{F'}_{j_\phi}^R(\theta,\phi) = \mathbf{F}_{j_\phi}^R(\theta,\phi)$ $\Xi_\phi^R(\theta,\phi)$, where $\Xi_\theta^R(\theta,\phi)(\Xi_\theta^R(\theta,\phi))^* = \Psi_\theta^R(\theta,\phi)$ and $\Xi_\phi^R(\theta,\phi)(\Xi_\phi^R(\theta,\phi))^* = \Psi_\phi^R(\theta,\phi)$, we can write

$$\mathbf{X}_{H_{ij}}^R = \int_{-\pi}^{\pi}\int_0^{\pi} \left(\mathbf{F'}_{i_\theta}^R(\theta,\phi)(\mathbf{F'}_{j_\theta}^R(\theta,\phi))^* + \mathbf{F'}_{i_\phi}^R(\theta,\phi)(\mathbf{F'}_{j_\phi}^R(\theta,\phi))^* \sin(\theta) \right) \mathrm{d}\theta\mathrm{d}\phi$$

$$= \int_{-\pi}^{\pi}\int_0^{\pi} \left(\mathbf{F}_{i_\theta}^R(\theta,\phi)(\mathbf{F}_{j_\theta}^R(\theta,\phi))^* \Psi_\theta^R(\theta,\phi) + \mathbf{F}_{i_\phi}^R(\theta,\phi)(\mathbf{F}_{j_\phi}^R(\theta,\phi))^* \Psi_\phi^R(\theta,\phi) \right) \sin(\theta)\theta\mathrm{d}\phi$$

$$= \int_{-\pi}^{\pi}\int_0^{\pi} \left(\frac{\mathrm{XPR}}{1+\mathrm{XPR}} \mathbf{F}_{i_\theta}^R(\theta,\phi)(\mathbf{F}_{j_\theta}^R(\theta,\phi))^* \Psi_\theta'^R(\theta,\phi) \right.$$

$$\left. + \frac{1}{1+\mathrm{XPR}} \mathbf{F}_{i_\phi}^R(\theta,\phi)(\mathbf{F}_{j_\phi}^R(\theta,\phi))^* \Psi_\phi'^R(\theta,\phi) \right) \sin(\theta)\theta\mathrm{d}\phi. \qquad (4.152)$$

It follows from this that the receive antenna correlation depends both on the mutual coupling characteristics of the receive MPA (observed through the coupled radiation patterns) and on the statistical characteristics of the propagation channel [6, 66].

4.7.4.2 Transmit Antenna Correlation. Using the aforementioned covariance terms, the (i,j)th entry of $\mathbf{R}_H^T$ can be computed as

$$\mathbf{R}_{H_{ij}}^T = \frac{\mathbf{X}_{H_{ij}}^T}{\sqrt{\mathbf{X}_{H_{ii}}^T \mathbf{X}_{H_{jj}}^T}}. \qquad (4.153)$$

Let us express the covariance term $\mathbf{X}_{H_{ij}}^T$ in terms of antenna parameters and channel parameters. As shown previously, the excitation vector and the radiated waves are related through the following expression $\mathbf{b}_\beta^T = \mathbf{S}_{\beta\alpha}^T \mathbf{a}_\alpha^T$. To compute the covariance of $\mathbf{a}_\alpha^T$, we must take into account what portion of the radiated waves contribute to the received signal. To do this, let us assume that $\mathbf{c}_\beta^T$ represents the set of transmit spherical modes subtended by the scattering clusters of the channel being illuminated by the transmit array and that impact at the receiver. Then, we can write $\mathbf{c}_\beta^T = \mathbf{S}_{\beta\alpha}^T \mathbf{b}_0^T$, where $\mathbf{b}_0^T$ is the excitation vector that produces the radiation $\mathbf{c}_\beta^T$. Using the fact that the transmit antenna is reciprocal, by virtue of the reciprocity theorem, we can analyze this problem from a receiving perspective. That is, we can compute $\mathbf{X}_{H_{ij}}^T$ from the covariance term $E\{\mathbf{b}_{0_i}^T(\mathbf{b}_{0_j}^T)^*\}$, where $\mathbf{b}_0^T$ is now the vector of received signals that are produced by the radiation $\mathbf{c}_\beta^T$, as shown in $\mathbf{b}_0^T = \mathbf{S}_{\alpha\beta}^T \mathbf{c}_\beta^T$. From a transmission point of view, $\mathbf{b}_0^T$ represents the excitation vector that is necessary to produce the radiation $\mathbf{c}_\beta^T$. In that case,

$$\mathbf{X}_{H_{ij}}^T = E\{\mathbf{b}_{0_i}^T(\mathbf{b}_{0_j}^T)^*\} = E\{\mathbf{S}_{\alpha\beta_{i^*}}^T \mathbf{c}_\beta^T(\mathbf{S}_{\alpha\beta_{j^*}}^T)^*(\mathbf{c}_\beta^T)^*\}. \tag{4.154}$$

Following the same approach as in Section 4.7.4.1, we finally obtain

$$\mathbf{X}_{H_{ij}}^T = \int_{-\pi}^{\pi}\int_0^\pi \left(\mathbf{F}_{i_\theta}^T(\theta,\phi)(\mathbf{F}_{j_\theta}^T(\theta,\phi))^*\mathbf{\Psi}_\theta^T(\theta,\phi) + \mathbf{F}_{i_\phi}^T(\theta,\phi)(\mathbf{F}_{j_\phi}^T(\theta,\phi))^*\mathbf{\Psi}_\phi^T(\theta,\phi)\right)\sin(\theta)\theta d\phi$$

$$= \int_{-\pi}^{\pi}\int_0^\pi (\frac{\text{XPR}}{1+\text{XPR}}\mathbf{F}_{i_\theta}^T(\theta,\phi)(\mathbf{F}_{j_\theta}^T(\theta,\phi))^*\mathbf{\Psi}_\theta'^T(\theta,\phi)$$

$$+ \frac{1}{1+\text{XPR}}\mathbf{F}_{i_\phi}^T(\theta,\phi)(\mathbf{F}_{j_\phi}^T(\theta,\phi))^*\mathbf{\Psi}_\phi'^T(\theta,\phi))\sin(\theta)\theta d\phi. \tag{4.155}$$

It follows that the transmit antenna correlation also depends both on the mutual coupling characteristics of the transmit MPA (observed through the coupled radiation patterns) and on the statistical characteristics of the propagation channel [6, 66].

4.7.4.3 Extended Transmit and Receive Antenna Correlation.

Regarding the extended channel model, the (i,j)th entry of the extended transmit antenna correlation matrix $\mathbf{R}_E^T$ can be computed as

$$\mathbf{R}_{E_{ij}}^T = \frac{\mathbf{X}_{E_{ij}}^T}{\sqrt{\mathbf{X}_{E_{ii}}^T \mathbf{X}_{E_{jj}}^T}}, \tag{4.156}$$

whereas the receive antenna correlation matrix $\mathbf{R}_E^R$ can be calculated using

$$\mathbf{R}_{E_{ij}}^R = \frac{\mathbf{X}_{E_{ij}}^R}{\sqrt{\mathbf{X}_{E_{ii}}^R \mathbf{X}_{E_{jj}}^R}}. \tag{4.157}$$

We proceed now to derive a general expression for the covariance term $\mathbf{X}_E^R$, given a generic load termination scenario in which $\mathbf{\Gamma}_{\alpha_R}^D \neq \mathbf{0}$. Consider again that

$$\mathbf{U}_{\alpha_R}^R = Z_{\alpha_R}^0{}^{\frac{1}{2}} \left(\mathbf{I} + \mathbf{\Gamma}_{\alpha_R}^D\right) \left(\mathbf{I} - \mathbf{S}_{\alpha_R \alpha_R}^E \mathbf{\Gamma}_{\alpha_R}^D\right)^{-1} \mathbf{b}_0^R. \qquad (4.158)$$

Let us define $\mathbf{Q} = Z_{\alpha_R}^0{}^{\frac{1}{2}} \left(\mathbf{I} + \mathbf{\Gamma}_{\alpha_R}^D\right) \left(\mathbf{I} - \mathbf{S}_{\alpha_R \alpha_R}^E \mathbf{\Gamma}_{\alpha_R}^D\right)^{-1}$, then using Eq. (4.144), $\mathbf{X}_E^R$, is given by

$$\mathbf{X}_E^R = \mathbf{Q} \mathbf{R}_H^R (\mathbf{Q})^H. \qquad (4.159)$$

We now compute the extended receive antenna covariance matrix $\mathbf{X}_E^R$ for two specific terminations of the loads: open-circuit termination and characteristic reference impedance terminations. In the case of an open circuit, $\mathbf{\Gamma}_{\alpha_R}^D = \mathbf{I}$, and $\mathbf{X}_E^R$ is given by

$$\mathbf{X}_E^R = 4 \left(\mathbf{I} - \mathbf{S}_{\alpha_R \alpha_R}^E\right)^{-1} \mathbf{X}_H^R \left(\mathbf{I} - (\mathbf{S}_{\alpha_R \alpha_R}^E)^H\right)^{-1}. \qquad (4.160)$$

In the case of characteristic reference impedance terminations, $\mathbf{\Gamma}_{\alpha_R}^D = \mathbf{0}$, and $\mathbf{X}_E^R$ is given by

$$\mathbf{X}_E^R = \mathbf{X}_H^R. \qquad (4.161)$$

A similar analysis can easily be derived for the extended transmit antenna correlation. We first derive a generic expression for $\mathbf{X}_E^T$ for a general source termination scenario in which $\mathbf{\Gamma}_{\alpha_T}^S \neq \mathbf{0}$. We begin considering once again the transmitter as a receiving structure and the fact that

$$\mathbf{U}_{\alpha_T}^T = Z_{\alpha_T}^0{}^{\frac{1}{2}} \left(\mathbf{I} + \mathbf{\Gamma}_{\alpha_T}^S\right) \left(\mathbf{I} - \mathbf{S}_{\alpha_T \alpha_T}^E \mathbf{\Gamma}_{\alpha_T}^S\right)^{-1} \mathbf{b}_0^T. \qquad (4.162)$$

Let us define $\mathbf{P} = Z_{\alpha_T}^0{}^{\frac{1}{2}} \left(\mathbf{I} + \mathbf{\Gamma}_{\alpha_T}^S\right) \left(\mathbf{I} - \mathbf{S}_{\alpha_T \alpha_T}^E \mathbf{\Gamma}_{\alpha_T}^S\right)^{-1}$, then using Eq. (4.146), $\mathbf{X}_E^T$ is given by

$$\mathbf{X}_E^T = \mathbf{P} \mathbf{X}_H^T (\mathbf{P})^H. \qquad (4.163)$$

Finally, also notice that in the case that the source internal impedance is that of the characteristic reference impedance, $\mathbf{\Gamma}_{\alpha_T}^S = \mathbf{0}$, then, the extended transmit antenna correlation matrix is given by

$$\mathbf{X}_E^T = \mathbf{X}_H^T. \qquad (4.164)$$

4.7.4.4 Antenna Correlation in Ideally Scattered Channels. In the special case that the channel is rich in scattering, the multipath arrival sector extends over the full angular range of the propagation environment, such that the power azimuth and elevation spectrum are uniform and a simplification over $\mathbf{X}_H^A$ can be introduced. The upper-index A can either refer to the transmit (t) or receive (R)

antennas. Let us assume then that $\Psi^{A'}_\theta(\theta,\phi) = \frac{1}{4\pi}$, $\Psi^{A'}_\phi(\theta,\phi) = \frac{1}{4\pi}$, and XPR $= 1$. In that case, X^A_H in Eqs. (4.155) and (4.152) can be rewritten as

$$X^A_H = \frac{1}{8\pi} \int_{-\pi}^{\pi} \int_0^\pi (\mathbf{F}^A(\theta,\phi))^\dagger (\mathbf{F}^A(\theta,\phi))^* \sin(\theta) d\theta d\phi, \qquad (4.165)$$

making the antenna correlation matrices only a function of the normalized far-field radiation patterns. Notice that the term $\frac{1}{8\pi}$ results from the used normalization convention of the power spectrums given in Section 4.5.1. Using a different normalization would result in a different value. In any case, this term depends and is proportional to the amount of radiated or received power. Notice also that Eq. (4.165) represents in fact the inner product of two matrices, that is,

$$X^A_H = \frac{1}{8\pi} \langle \mathbf{F}^A(\theta,\phi), \mathbf{F}^A(\theta,\phi) \rangle. \qquad (4.166)$$

The objective until the end of this sections is to link the MPA scattering matrix representation, $\mathbf{S}^A$, with Eq. (4.165). The fact is that in an ideally rich scattered environment, the concept of radiation pattern becomes partially insubstantial. The unpredictable DoA of the incoming wave, in these environments, turns the antenna radiation pattern into a second-order parameter. We proceed now to show this intuitive idea from a mathematical perspective. Using Eqs. (4.23) and (4.165), we can easily obtain

$$X^A_H = \frac{1}{8\pi} (\mathbf{S}^A_{\beta\alpha})^T (\mathbf{S}^A_{\beta\alpha})^*. \qquad (4.167)$$

On the other hand, from the unitarity property of lossless MPAs, the following condition must hold:

$$(\mathbf{S}^A_{\beta\alpha})^T (\mathbf{S}^A_{\beta\alpha})^* = \mathbf{I} - \mathbf{S}^A_{\alpha\alpha}(\mathbf{S}^A_{\alpha\alpha})^H, \qquad (4.168)$$

which clearly shows that any modification in the antenna input parameters, that is, in the subblock scattering matrix $\mathbf{S}^A_{\alpha\alpha}$, using external network and others, implies also a modification in the radiation pattern given by the matrix $\mathbf{S}^A_{\beta\alpha}$. Combining Eq. (4.167) and (4.168), we finally observe that in an ideally scattered environment, the correlation can be computed from the scattering parameters using

$$X^A_H = \frac{1}{8\pi} \left(\mathbf{I} - \mathbf{S}^A_{\alpha\alpha}(\mathbf{S}^A_{\alpha\alpha})^H \right). \qquad (4.169)$$

That is, as shown also in Refs. [66, 68], under the propagation conditions outlined above, the covariance matrix can also be computed without resorting to integration of the radiation patterns but only on the scattering characteristics measured at the input ports of the MPA. This also means that in ideally scattered environments, the correlation between the antenna ports of an MPA depends only on the mutual coupling characteristics. Finally, notice that although indoor and urban propagation environments are far from being ideally scattered environments, they are still very rich in scattering, in general, and some of the conclusions here can be extrapolated to more realistic environments.

Let us give now another physical insight on the implications of correlation in a MIMO communication system. Let us assume an MPA structure acting as a transmitter. The radiated power by the antenna system is given by

$$P^{S,0} = (\mathbf{b}_0^S)^H \left(\mathbf{I} - (\mathbf{S}_{\alpha_T \alpha_T}^E)^H \mathbf{S}_{\alpha_T \alpha_T}^E \right) \mathbf{b}_0^S. \tag{4.170}$$

Therefore, to maximize the transmitted power, one must impose the following condition upon the multiantenna system:

$$(\mathbf{S}_{\alpha_T \alpha_T}^E)^H \mathbf{S}_{\alpha_T \alpha_T}^E = \mathbf{0}. \tag{4.171}$$

The unique solution for this problem is $\mathbf{S}_{\alpha_T \alpha_T}^E = \mathbf{0}$. This constitutes a sufficient condition that will ensure that this MPA system will not degrade the wireless channel conditions (in an ideally scattered environment). Notice from Eq. (4.169) that this unique solution is also the one that

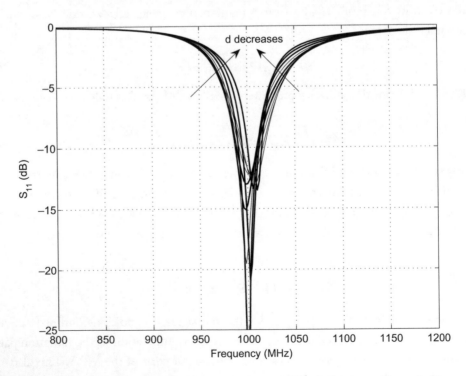

FIGURE 4.11: Representation of the scattering parameters $(\mathbf{S}_{\alpha \alpha_{ii}}^A)$ for two collinear half-wave dipoles separated by a distance, d, versus frequency.

provides minimum correlation. Therefore, antenna correlation will not only degrade the diversity order of the system but also the transmission power gain.

The conclusions derived from this analysis are that an n-port MPA, whose $\mathbf{S}^A$ parameter matrix is such that the correlation matrix $\mathbf{R}_{\mathbf{H}_0}^A$ (or $\mathbf{R}_{\mathbf{E}_0}^A$) is equal to the identity matrix, has its ports uncorrelated (decoupled, and matched) and will not degrade the performance of the MIMO system. In that case, the product $(\mathbf{S}_{\beta\alpha}^A)^H \mathbf{S}_{\beta\alpha}^A$ is also equal to the identity matrix, which means that the associated radiation patterns are orthogonal to each other. For the sake of completeness, notice the difference between a decoupled and a decorrelated MPA. In the first case, $\mathbf{S}_{ij}^A = 0$ for $i \neq j$, and in the second, $\mathbf{R}_{\mathbf{H}_{0_{ij}}}^A = 0$ (or $\mathbf{R}_{\mathbf{E}_{0_{ij}}}^A = 0$) for $i \neq j$.

4.7.4.5 Examples. To understand the different concepts and its relation with antenna parameters, let us consider a very simple case. Consider two collinear half-wave dipoles separated by a distance d, as shown in Fig. 5.1. With an electromagnetic analysis code, it is simple to obtain the scattering parameters of the two-port system formed by the two antennas as a function of the distance (d). Figs. 4.11 and 4.12 show the scattering parameters as a function of the frequency, at distances that

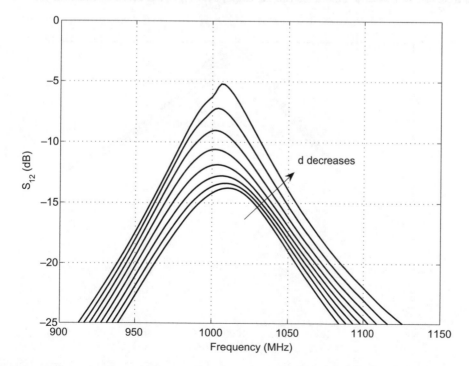

FIGURE 4.12: Representation of the scattering parameters ($\mathbf{S}_{\alpha\alpha_{ij}}^A$) for two collinear half-wave dipoles separated by a distance, d, versus frequency.

range from 0.15λ to 0.5λ in steps of 0.05λ. The wavelength (λ) is defined for the central frequency of 1000 MHz. Notice that the scattering parameters are defined for the antenna ports loaded with a reference impedance that in this case is 50 Ω. Also notice that in this antenna system, in addition to exhibit reciprocity (i.e., $\mathbf{S}^A_{\alpha\alpha_{12}} = \mathbf{S}^A_{\alpha\alpha_{21}}$), it also has symmetry, and thus $\mathbf{S}^A_{\alpha\alpha_{11}} = \mathbf{S}^A_{\alpha\alpha_{22}}$. The plots show that regardless of mutual effects, the antennas are matched around 1000 MHz and that mutual coupling increases as the antennas are closer. Mutual coupling measured as the magnitude of $\mathbf{S}^A_{\alpha\alpha_{21}}$ ranges from -14 dB when the antennas are spaced 0.5λ to approximately -5 dB when the dipoles are 0.15λ far apart. The scattering parameters of an MPA are usually computed and measured in the antenna design process. These parameters, although relevant and useful in the design process, provide little insight on the potential behavior of the MPA in a MIMO system. Other parameters that give us more information from a system-level perspective are, for example, the matching efficiency, and the correlation of the system. Using Eqs. (4.169) and (4.147)/(4.153), we can find the elements of the channel covariance/correlation matrix, assuming a perfectly scattered propagation channel (NLOS). On the other hand, in Eq. (4.170), it is shown that a very meaningful quantity to represent directly is $\mathbf{X}^A = (\mathbf{I} - (\mathbf{S}^A_{\alpha\alpha})^H \mathbf{S}^A_{\alpha\alpha})$ because it gives us the ratio of the power that is radiated to the power available to the antenna, that is, the matching efficiency. Notice that

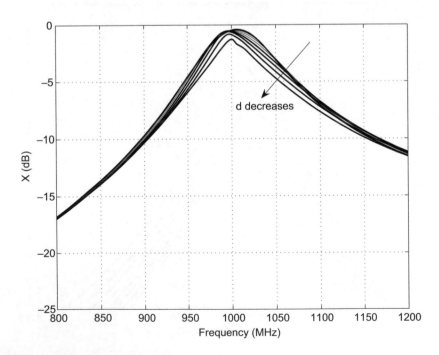

FIGURE 4.13: Representation of the $\mathbf{X}^A_{ii}$ for two collinear half-wave dipoles separated by a distance, d, versus frequency.

we have defined $\mathbf{X}^A$ such that it is equal to the covariance term $\mathbf{X}_H^A$ without the factor $\frac{1}{8\pi}$. Fig. 4.13 shows the quantity $\mathbf{X}_{ii}^A$ and Fig. 4.14 the quantity $\mathbf{X}_{ij}^A$. From the representation of $\mathbf{X}_{ij}^A$, it is clear that in all cases, at the central frequency and for the case of a perfect scattered NLOS propagation scenario, the values are below -10 dB. On the other hand, the observation of the term $\mathbf{X}_{ii}^A$ shows a variation of the values, at the central frequency, ranging from -0.5 dB when the antennas are at its maximum distance to -1.5 dB when they are at the closest distance. This quantity provides a clear physical interpretation when Eqs. (4.170) and (4.116) are recalled. Let us also recall that the receive matching efficiency is given by

$$\eta^{T,0} = \frac{E\{(\mathbf{b}_0^S)^H\left(\mathbf{I} - (\mathbf{S}_{\alpha\alpha}^A)^H\mathbf{S}_{\alpha\alpha}^A\right)\mathbf{b}_0^S\}}{E\{(\mathbf{b}_0^S)^H\mathbf{b}_0^S\}}, \qquad (4.172)$$

where $\mathbf{b}_0^S$ is the excitation vector. Therefore, what Fig. 4.13 shows is the fact that as mutual coupling between antennas increases, a larger fraction of the power radiated by one antenna is actually coupled to the second antenna and is not delivered into the radioelectric channel. In our example, this accounts for about 1.5 dB in power losses when the antennas are at their closest distance. Notice

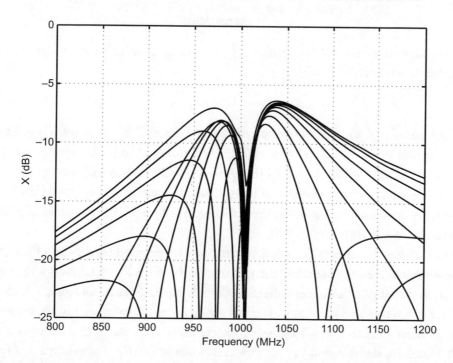

FIGURE 4.14: Representation of the $\mathbf{X}_{ij}^A$ for two collinear half-wave dipoles separated by a distance, d, versus frequency.

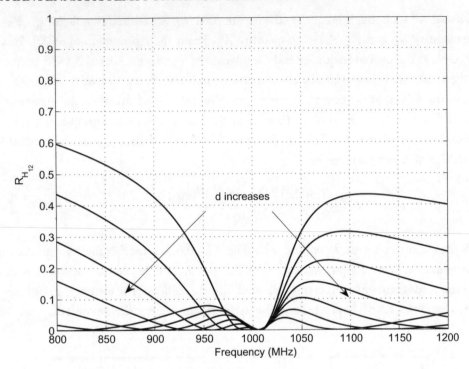

FIGURE 4.15: Representation of the correlation $\mathbf{R}_{H_{ij}}^A$ for two collinear half-wave dipoles separated by a distance, d, versus frequency.

that this means that if 1 W is effectively dissipated in port 1, only 0.7 W are effectively radiated and 0.3 W are dissipated in the 50-Ω load of port 2. Finally, from the correlation perspective, notice that the values of the term $\mathbf{X}_{ij}^A$ are low, thus the cross-correlation values will also be small, and as a consequence, we can expect the existence of independent subchannels when this MPA is used in a highly scattering environment. However, there will be actually a 1.5 dB reduction of SNR ratio compared with the case in which $\mathbf{X}_{ii}^A = 0$ dB.

We proceed now to compute the actual antenna correlation matrix as defined in Eq. (4.147)/(4.153). The entries of the envelope correlation matrix are shown in Fig. 4.15. Notice that we have plotted the cross-correlation values normalized to the self-correlation. In this case, it is clear that assuming a rich scattering environment with uniformly distributed power spectrums, the antennas exhibit a correlation close to 0 in a bandwidth that tends to decrease as the antennas are close together. In Fig. 4.16, the values of the envelop cross-correlation ($\mathbf{R}_{H_{ij}}^A$) are shown as a function of the antenna separation, and compared with the analytic expression (see Eq. (5.6)) developed in Section 5.2.1, which corresponds to the ideal case of point sources and no mutual coupling between

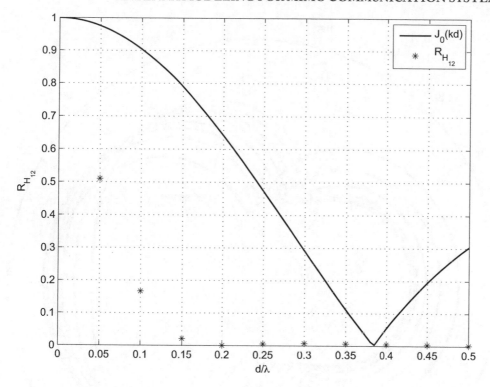

FIGURE 4.16: Representation of the correlation $\mathbf{R}^{A}_{H_{ij}}$ for two collinear half-wave dipoles separated by a distance, d, versus antenna separation.

them. Interestingly, the correlation in the practical implementation of the MPA (thus, including mutual coupling effects) is lower than in the case that mutual coupling is neglected. Sometimes, in the literature, mutual coupling between antennas is identified as a beneficial effect because it reduces the antenna correlation. Nevertheless, it must be recalled that as discussed previously mutual coupling is also responsible for the losses in radiated power, which has a negative impact in the performance of the system. In any case, it is clear that mutual coupling between ports of the MPA cannot be neglected for an accurate assessment of the performance of the MIMO system.

Finally, to understand the role of mutual coupling in antenna decorrelation, it is interesting to look at the antenna patterns. In Fig. 4.17, the resulting patterns at the central frequency in the H plane of the antenna system (plane xy), assuming that the dipoles are aligned in the z direction according to Fig. 5.1, are shown. Notice that because of the symmetry of the antenna system, a rotation of $180°$ results when the other port is excited. Because of antenna coupling, the patterns are not omnidirectional, but are heavily distorted. It is clear that a set of disjoint patterns results.

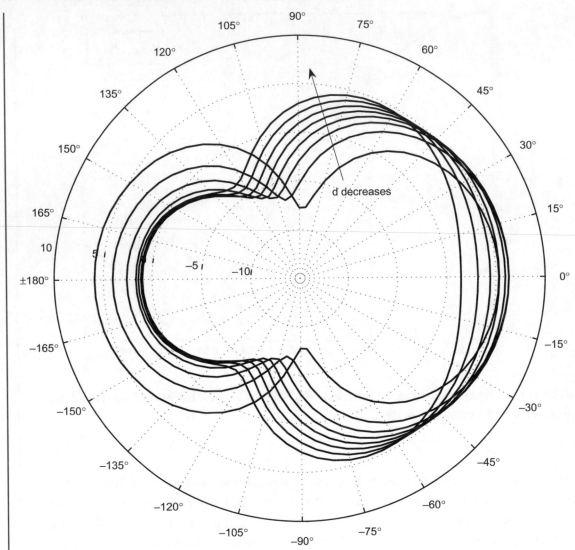

FIGURE 4.17: Representation of the radiation patterns of two collinear half-wave dipoles separated by a distance, *d*.

That is, when one port is fed, radiation tends to happen in one half-space, and when the other port is fed, radiation concentrates in the opposite half-space. This helps to keep the correlation low even when the antennas are very close. However, in general, because channel correlation can also be computed using expression Eq. (4.165), the result is that because of mutual coupling, the radiation patterns lose their orthogonality.

4.7.5   System Capacity

The capacity of a MIMO communication system, C, is a measure of the maximum transmission rate for a reliable communication on a certain channel. It is "reliable" in the sense that it is possible to transmit at such a speed with an arbitrarily small error probability. The concept of capacity was extensively defined in Chapter 1. However, we recall now how to relate the concept of antenna correlation and capacity. In the particular case where the transmitter does not know the channel $\mathbf{E}_0$, the transmitted power is equally distributed among the ports of the transmit MPA, that is, $\mathbf{R}_s = \frac{P^{S,av}}{M}\mathbf{I}$, and the capacity expression is given by

$$C = \log_2 \det\left(\mathbf{I} + \frac{\rho_0}{M}\mathbf{E}_0\mathbf{E}_0{}^H\right),   \text{(4.173)}$$

where $\rho_0 = \frac{P^{D,av}}{N_0}$. Notice that the previous equation does not impose any constraints of linear combining of the MPA outputs—that is, it is a general expression of the capacity, which may or not be attained with current channel coding techniques. Notice also that ρ_0 is the SNR at the receiver.

To give a physical insight into the impact of correlation upon the capacity expression, we assume again a Kronecker channel model [18]. Using this model, the capacity expression for a MIMO communication system can be rewritten including the correlation effects at the MPAs, as follows [22]:

$$C = \log_2 \det\left(\mathbf{I} + \frac{\rho_0}{M}\mathbf{R}_{E_0}^R \mathbf{W} \mathbf{R}_{E_0}^T \mathbf{W}^H\right),   \text{(4.174)}$$

where $\mathbf{W} \in C^{N \times M}$ is a complex random matrix with all elements being iid Gaussian random variables with distribution $N(0,1)$. The ports of an MPA are said to be uncorrelated when $\mathbf{R}_E^T$ and $\mathbf{R}_E^R$ are equal to the identity matrix. In that case, the capacity of the MIMO communication system is not degraded because of the characteristics of the used MPAs. In other words, the MPAs become transparent to the system from a performance perspective. Notice, however, that Eq. (4.174) allows us to evaluate the performance of any MPA, from a transmission rate perspective, within a MIMO wireless communication system. Notice, too, that to conduct numerical simulations for capacity, one can, for example, use the Monte-Carlo methodology, which consists in generating a large number of channel realizations and computing the associated channel matrix for each of them. Finally, for example, the CDF can be plotted.

4.7.6   Diversity Order

The diversity gain (DG) is a figure of merit used to quantify how resistant our system is to multipath fading in terms of the receive signal strength. The concept of DG (or diversity order) was extensively defined in Chapter 1 and basically is defined as the slope of the error probability curve in terms of the received SNR in a log–log scale. For a system with M transmit and N receive antennas, the maximum diversity order is MN. The diversity order is maximum if the used STC is full rank and the ports of the transmit and receive MPAs are decorrelated; in general, it tends to decrease as the correlation increases.

As with capacity, it is important to relate the diversity order of a MIMO system to the correlation properties of the antennas. In this manner, we can predict and evaluate the performance of any MPA, from a diversity perspective, in a MIMO wireless communication system. Different space–time architectures have distinct expressions for the diversity order. Therefore, to illustrate, we give here the expression for a system using STCs (such as OSTBCs, QOSTBC, etc.). We assume again a Kronecker channel model [18] and an STC designed such that it can achieve full diversity in an ideal NLOS situation. Such a code is called a full-rank code. Using Ref. [45] and the notation in Chapter 3, the PEP, $P(\mathbf{C}^1 \rightarrow \mathbf{C}^2)$, of a MIMO communication system using STBCs can be written as

$$P(\mathbf{C}^1 \rightarrow \mathbf{C}^2) \leq \left(\frac{\rho_0}{4}\right)^{r(\mathbf{R}_{E_0}^T)r(\mathbf{R}_{E_0}^R)} \prod_{l=1}^{r(\mathbf{R}_{E_0}^R)-1} (\lambda_l(\mathbf{R}_{E_0}^R))^{-r(\mathbf{R}_{E_0}^T)}$$

$$\cdot \prod_{i=1}^{r(\mathbf{R}_{E_0}^T)-1} (\lambda_i((\mathbf{C}^1 - \mathbf{C}^2)^\dagger \mathbf{R}_{E_0}^T (\mathbf{C}^1 - \mathbf{C}^2)^*)^{-r(\mathbf{R}_{E_0}^R)}, \qquad (4.175)$$

where $r(\mathbf{A})$ denotes the rank of $\mathbf{A}$, and $\lambda_i(\mathbf{A})$ denotes the ith eigenvalue of the matrix $\mathbf{A}$. Notice that the above expression gives us an upper bound on the PEP for such a system using a particular MPA, such that the transmit and receive correlation matrices are given by $\mathbf{R}_{E_0}^T$ and $\mathbf{R}_{E_0}^R$, respectively. As shown also in Ref. [45], the diversity order of this system is $G_d = r(\mathbf{R}_{E_0}^T)r(\mathbf{R}_{E_0}^R)$.

To numerically estimate the diversity order, it is necessary to plot the BER versus SNR curve and compute the slope of that curve for the high SNR regime [26]. To conduct numerical simulations one can, for example, use the Monte-Carlo methodology, which consists in generating a large number of channel realizations and computing the number of errors generated by the system.

4.7.7 System Bandwidth

In the extant literature, there is no single definition of the bandwidth of a single antenna. This is because the term *bandwidth* simply represents a frequency range wherein the antenna operates within certain specifications. In wireless communications, it is commonly defined by the frequency range over which the return loss is less than a threshold level, e.g., -6 dB (or 75% efficiency). The -6-dB threshold is a rule of thumb for the design of mobile terminal antennas.

As with matching efficiency, the bandwidth (B) of MIMO communication system using MPAs differs from that of a communication system using SPAs, especially when the mutual coupling effect becomes significant. To provide a formal definition for it, let us first define $\mathbf{\Gamma}_{\text{in}}^T$ as the reflection coefficient looking into the input ports of the transmit RF chain, after the source, and $\mathbf{\Gamma}_{\text{out}}^R$ as the reflection coefficient looking into the output ports of the receive RF chain, before the loads. In Ref. [69], a definition for the bandwidth of MIMO communication system using MPAs is proposed. The system bandwidth is defined using the input coefficients $\mathbf{\Gamma}_{in_{ij}}^T$ and the output coefficients $\mathbf{\Gamma}_{out_{ij}}^R$, which for now are given by $\mathbf{\Gamma}_{\text{in}}^T = \mathbf{S}_{\alpha_T \alpha_T}^E$ and $\mathbf{\Gamma}_{\text{out}}^R = \mathbf{S}_{\alpha_R \alpha_R}^E$. The system bandwidth is then the frequency range in which both its self-reflections $|\mathbf{\Gamma}_{in_{ii}}^T|$ (and $|\mathbf{\Gamma}_{out_{ii}}^R|$) and mutual reflections $|\mathbf{\Gamma}_{in_{ij}}^T|$ (and $|\mathbf{\Gamma}_{out_{ij}}^R|$) with $i \neq j$ satisfy a specified maximum return loss threshold. That is,

$$B = 2\min\left(\min_{\forall i,j}(f_{ii}^U, f_{ij}^U) - f_c, f_c - \max_{\forall i,j}(f_{ii}^L, f_{ij}^L) \right), \qquad (4.176)$$

where f_{ij}^U and f_{ij}^L are the first frequency higher and lower, respectively, than the center frequency f_c at which $|\mathbf{\Gamma}_{in_{ij}}^T|$ (and $|\mathbf{\Gamma}_{out_{ij}}^R|$) coincides with the specified return loss threshold, given an initial level that is below that threshold at f_c. Where this condition is not met by all f_{ij}^Ls and f_{ij}^Us, except when $|\mathbf{\Gamma}_{in_{ij}}^T|$ (and $|\mathbf{\Gamma}_{out_{ij}}^R|$) is strictly less than the threshold, the bandwidth is then undefined. In other words, this implies that the bandwidth is specified by the dominant antenna reflection. The outermost min operator ensures that the bandwidth is symmetrical around the center frequency. This is a straightforward extension of the single antenna case, which is commonly defined by its return loss. We shall henceforth refer to the bandwidth as the efficiency bandwidth because it is the range of frequency in which the matching efficiency is acceptable. Finally, it is important to notice that although narrowband matching can be always be attained, in theory and practice, wideband matching has fundamental theoretical as well as practical limits. Therefore, it is crucial to study the wideband performance of such MPAs for successful implementations in future systems.

4.7.8 Effective Degrees of Freedom

In Ref. [70], the concept of effective degrees of freedom (EDOF) of the system is proposed as a metric used to measure the number of subchannels effectively active on a particular MIMO communication system. According to this proposal, the EDOF is defined as

$$EDOF = \frac{dC(x)}{d\log_2(x)}\bigg|_{x=\rho}, \qquad (4.177)$$

where EDOF should converge to $\min(M,N)$ in the case of uncorrelated channels, when the SNR increases. Notice that, the EDOF is also often defined as the number of significantly large eigenvalues of the channel correlation matrix, $\mathcal{R}^H$ (or $\mathcal{R}^E$), as a result of the SVD, defining it as a measure of the number of uncorrelated signals. Finally, other similar quantities to the EDOF are proposed in the literature, such as in Ref. [71].

4.8   REVIEW OF THE FUNDAMENTAL LIMITS OF MPAs

In a radio communication system, a particular MPA poses specific limitations on its potential maximum performance. We briefly review these fundamental limits. A general requirement in MIMO communication systems is to reduce the size of the MPA structures to accommodate them into smaller terminals. For simplicity in this discussion, let us consider a SISO communication system using one single antenna at the transmitter and at the receiver. The absolute capacity expression of such system, having an operational bandwidth B, is given by

$$C = B\log_2\left(1 + \rho_0\right) \qquad \text{bits/s (bps)}, \qquad (4.178)$$

where $\rho_0 = \frac{P^{D,\text{av}}}{N_0}$ is the average SNR at the receiver, N_0 is the noise power of the additive noise, and $P^{D,\text{av}}$ is the average received power. Notice that, as opposed to Eq. (1.2), this expression of capacity explicitly includes the bandwidth term. Both parameters, B and ρ_0, are heavily determined by the size of the antenna, and specifically by the size compared with the wavelength. There exists a fundamental limit that restricts the performance of antennas enclosed in a given volume. This limit is given in terms of a figure of merit called *quality factor* (Q), which is the ratio of time-average nonpropagating energy to the radiated power of an antenna [72] and is given by

$$Q = \frac{1}{ka} + \frac{1}{(ka)^3}, \qquad (4.179)$$

where k is the wave number at the operating wavelength and a is the radius of the smallest virtual sphere that encloses the antenna. In all the cases, reaching such an extreme quality factor requires that just a TE_{01} or TM_{01} spherical mode is excited (or a combined $\text{TE}_{01}/\text{TM}_{01}$ mode for circular polarization). If the antenna is modeled as a resonant RLC circuit (series or parallel), the quality factor can be related with the fractional bandwidth B through

$$Q \approx \frac{1}{B} = \frac{f_c}{f_{\max} - f_{\min}}. \qquad (4.180)$$

This approximation is quite good for $Q \gg 1$ but is rather inaccurate for $Q < 2$, although it useful to predict the potential broadband behavior of the antenna. When $Q \gg 1$, the antenna has a narrow bandwidth, and therefore, it has great frequency sensitivity.

It is clear that as the electrical size of the antenna is reduced, the Q of the antenna increases, and consequently, the bandwidth of the antenna shrinks as well. This bandwidth reduction has a direct impact on the maximum capacity that a SISO or MIMO communication system can achieve.

Practical antenna designs are far from approaching the fundamental limit. A realistic bound for the Q of small antennas has been proposed by Thiele et al. [73]. This prediction is based on the assumption of a sinusoidal current distribution along an electrically small antenna. There is approximately one order of magnitude between the fundamental limit for the Q of small antennas

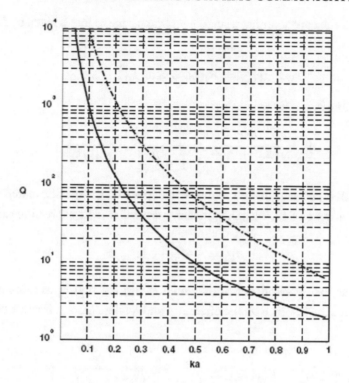

FIGURE 4.18: Fundamental limit (solid line) and realistic limit (dashed line) for the Q of a lossless antenna.

and the actual Q that is achieved in practical designs. This idea can be observed in Fig. 4.18, where the fundamental limit and the realistic limit for the Q of a lossless antenna are graphically compared.

In addition to the bandwidth limitations, there is also a reduction of ohmic efficiency. For an antenna in transmission, the ohmic efficiency is defined as

$$\eta_\Omega^T = \frac{E\{P^S\}}{E\{P^S + P_\Omega^T\}},\tag{4.181}$$

where P^S is the radiated power and P_Ω^T is the dissipated power due to the ohmic losses. Let us consider a conductor surface S with all of its dimensions smaller than λ, so $S \ll \lambda^2$. For simplicity, let us assume a uniform current density distribution in the conductor directed along the z axis. Given these assumptions, the total radiated power is approximately proportional to

$$P^{S,\mathrm{av}} \propto \frac{\pi \eta}{2} \frac{S^2}{\lambda^2} |J_0|^2.\tag{4.182}$$

On the other hand, if the conductor has a surface resistance due to losses given by R_s, the dissipated power is

$$P_\Omega^{T,av} = R_s|J_0|^2 S \propto S|J_0|^2, \tag{4.183}$$

and the antenna efficiency can be written as

$$\eta_\Omega^T = \frac{P^{S,av}}{P^{S,av} + P_\Omega^{T,av}} = \frac{1}{1 + \frac{P_\Omega^{T,av}}{P^{S,av}}} = \frac{1}{1 + C\frac{\lambda^2}{S}}, \tag{4.184}$$

where C is some positive constant that ultimately will depend on the specific material. It is clear that, as the surface of the antenna S shrinks, the efficiency is smaller. The expression can be rewritten as

$$\eta_\Omega^T = \frac{(ka)^2}{(ka)^2 + C'}. \tag{4.185}$$

It is also clear that as the electrical size of the antenna shrinks, that is, as a becomes smaller, the efficiency reduces. In Fig. 4.19, the ohmic efficiency and lossless Q for a family of electrically small

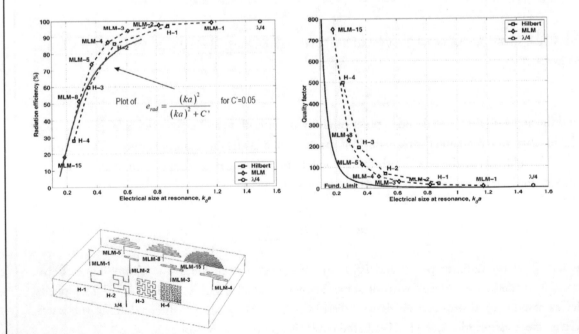

FIGURE 4.19: Radiation efficiency and lossless quality factor for a family of wire antennas. The approximate expression for the efficiency is in blue.

wire antennas is shown. The measured efficiency is compared with the approximate expression. The reduction of efficiency implies a reduction in SNR (ρ_0) with the corresponding implications in the capacity of the communication system.

In the case of MPA systems, orthogonality between radiation patterns requires that high-order spherical modes are radiated. Specifically, for an N-port MPA, at least, N spherical modes must be radiated to ensure orthogonality. For a compact MPA, satisfying $ka < 1$, where a is the radius of the virtual sphere that encloses the MPA, results in a supergain condition and very high Q [74]. For example, for $N = 5$ and $ka = 1$, the Q of a lossless MPA is of the order of 105.

Also, in the case of MPA, in addition to ohmic losses, the radiation efficiency is reduced because of mutual coupling between closely spaced elements. As MPA dimension is reduced, the coupling between elements increases, and in addition to a decrease of ohmic efficiency due to the MPA size reduction, a higher fraction of power is coupled into adjacent ports instead of being radiated.

Therefore, MPA miniaturization is practically constrained by a fundamental limit that sets a bound for the maximum attainable bandwidth for a given electrical dimension. For an MEA, for example, when element spacing is less than half a wavelength, pattern orthogonality is achieved at the expense of attaining a supergain condition. As elements are placed closer together, this supergain condition is increased. In practical applications, the supergain condition results in narrow bandwidth and high losses. Therefore, a careful assessment is required in reducing the size of MPA in MIMO systems to determine whether a real advantage results from increasing the number of antennas at the expense of placing them closer together.

·　·　·　·

CHAPTER 5

Design of MPAs for MIMO Communication Systems

5.1 INTRODUCTION

Antennas have traditionally been designed from a radioelectric perspective. However, with the introduction of MIMO communication systems, multiantenna systems now need to be designed so that they take into account both system-level considerations and the statistical characteristics of the propagation environment.

Ultimately, one would like to design the transmit and receive MPAs so that, for example, the capacity of the system is maximized or the channel correlation is minimized or maximum transmission power gain is achieved between the source and the drain. One might even wish these three qualities to be satisfied simultaneously. However, especially when the coupling between the ports of the MPAs become important, such as for closely spaced antennas in an MEA, the input impedances at the input ports of the MPAs will change and become unmatched to those at the source and drain. As a consequence, the transmission power gain may degrade as well as the transmit and receive antenna correlation (see Section 4.7.1). Ultimately, the capacity of the system or its diversity order will also degrade (see Section 4.7.5).

Using well-known microwave analysis techniques, in Ref. [14], it is shown that assuming an ideally rich scattering NLOS propagation channel, to satisfy any (or all) of the three requirements, one must satisfy the multiport-conjugate matching condition [56] at the interface between the source and the transmit MPA and at the interface between the receive MPA and the loads. That is,

$$\mathbf{\Gamma}^S = (\mathbf{S}_{\alpha_T \alpha_T}^T)^H \tag{5.1}$$

$$\mathbf{\Gamma}^D = (\mathbf{S}_{\alpha_R \alpha_R}^E)^H. \tag{5.2}$$

On the other hand, notice that, normally, $\mathbf{\Gamma}^S = \mathbf{0}$ and $\mathbf{\Gamma}^D = \mathbf{0}$ because the internal impedance of the generator and the terminal loads are normally equal to the characteristic reference impedances of the system. Therefore, the aforementioned conditions are normally only satisfied when the ports of the MPAs suffer only from negligible mutual coupling effects, in which case, $\mathbf{S}_{\alpha_T \alpha_T}^T = \mathbf{0}$

and $\mathbf{S}^E_{\alpha_R \alpha_R} = \mathbf{0}$. However, if the ports of the MPAs undergo a strong mutual coupling phenomenon, then $\mathbf{S}^T_{\alpha_T \alpha_T} \neq \mathbf{0}$ and $\mathbf{S}^E_{\alpha_R \alpha_R} \neq \mathbf{0}$, and the optimal conditions are not satisfied, that is, $\mathbf{\Gamma}^S \neq (\mathbf{S}^T_{\alpha_T \alpha_T})^H$ and $\mathbf{\Gamma}^D \neq (\mathbf{S}^E_{\alpha_R \alpha_R})^H$. Therefore, one needs some design mechanism to satisfy the above design criteria.

The approach followed in this book is to design the MPAs as if they were intended to be used in an ideally rich scattering NLOS propagation channel. As shown in Sections 4.7.1 and 4.7.4, in this case, the aforementioned optimization goals converge to the single requirement of minimizing the mutual coupling phenomena among the ports of the MPA while requiring matched ports with the source/loads. There exist two main families of techniques that can be used to optimize the design of an MPA for MIMO systems:

- antenna diversity
- decorrelating networks (DNs).

Both techniques are based on the reduction of mutual coupling phenomena. Afterward, if the multiantenna system is intended to be used in other types of channel, channel coding techniques can be used to tune the antenna to better fit the characteristics of the propagation channel, as shown in Sections 1.5.4 and 3.6.3. Notice that the antennas can also be tuned to fit a realistic propagation environment by means of using DNs, as explained later in this chapter. The drawbacks of this approach are the following: first, the design becomes a static solution—that is, a different network needs to be used for each propagation scenario; second, these networks become remarkably more complex to implement than if they had been designed for an ideal NLOS scenario. The advantage, however, is that this solution does not need to use specific channel coding techniques. Notice that whether channel coding techniques or DNs are used to adapt the design to a nonideal NLOS propagation channel, from a physical perspective, the goal is to decorrelate the ports of an MPA by means of creating directive patterns that adapt to the channel.

For the remainder of this book, unless otherwise specified, let us assume an ideally rich scattering NLOS propagation channel.

5.2   MUTUAL COUPLING REDUCTION USING ANTENNA DIVERSITY

Antenna diversity techniques are one of the fundamental design tools that engineers can use to reduce the mutual coupling associated with the ports of an MPA. These techniques are based on the fact that mutual coupling reduction can be accomplished by providing orthogonality among the radiation patterns of the MPA. This fact was already observed in Section 4.7.4.4 through the expression:

$$(\mathbf{S}_{\beta\alpha}^{A})^{T}(\mathbf{S}_{\beta\alpha}^{A})^{*} = \mathbf{I} - \mathbf{S}_{\alpha\alpha}^{A}(\mathbf{S}_{\alpha\alpha}^{A})^{H} = \langle \mathbf{F}^{A}(\theta,\phi), \mathbf{F}^{A}(\theta,\phi)\rangle. \tag{5.3}$$

Designing an MPA under the pattern orthogonality criteria means that the MPA must be designed such that

$$\langle \mathbf{F}^{A}(\theta,\phi), \mathbf{F}^{A}(\theta,\phi)\rangle = \mathbf{I}, \tag{5.4}$$

which implies that the equality

$$\mathbf{S}_{\alpha\alpha}^{A}(\mathbf{S}_{\alpha\alpha}^{A})^{H} = \mathbf{0} \tag{5.5}$$

must hold. Finally, the unique solution to the above condition is $\mathbf{S}_{\alpha\alpha}^{A} = \mathbf{0}$. When considering the complete MIMO communication system, this solution tells us that one must design the system such that the ports of the transmit and receive MPAs are decoupled and matched, that is, $\mathbf{S}_{\alpha_T\alpha_T}^{E} = \mathbf{0}$ and $\mathbf{S}_{\alpha_R\alpha_R}^{E} = \mathbf{0}$. This is a desired feature as long as $\mathbf{\Gamma}^{S} = \mathbf{0}$ and $\mathbf{\Gamma}^{D} = \mathbf{0}$, because then, the multiport conjugate match condition is satisfied.

Because in an ideally scattered environment, mutual coupling and antenna correlation are totally equivalent, satisfying pattern orthogonality will also produce decorrelated antenna ports. On the other hand, in a generic environment, the pattern orthogonality design criteria can also be used as long as one weighs the patterns of the antennas by the power spectrum of the incoming or departing waves, as shown in Eq. (4.152).

Now that the fundamentals of pattern orthogonality have been revised, we describe the most commonly used antenna diversity techniques. Antenna diversity techniques can be classified according to which physical mechanism is used to produce orthogonal radiation patterns among the ports of an MPA. From an electromagnetic point of view, orthogonal patterns can be produced using space [26], polarization [4], pattern diversity techniques [5, 6], or any combination of these. In particular, space diversity imposes orthogonality by using identical antennas spaced apart, polarization diversity imposes orthogonality by creating two orthogonal polarizations, and pattern diversity imposes orthogonality by producing spatially disjoint radiation patterns.

This classification is rather arbitrary, and for most cases, a combination of all these sources is present on antenna diversity systems. Using the abovementioned diversity techniques, multiple and uncorrelated versions of the transmitted signals may appear at the receiver. Afterward, to improve the performance of a particular MIMO communication system, those signals need to be combined.

5.2.1 Space Diversity

Using space diversity, equal antennas are spaced apart, so that the magnitude of the radiation pattern at each port is essentially the same, but the phase patterns relative to a common coordinate system are such that the resultant radiation patterns among different ports of the MEA become orthogonal [33]. As a result, the received signals are statistically uncorrelated. In general, a minimum antenna separation of 0.5λ is used in space diversity schemes.

Let us illustrate this concept with a simple graphical example. Imagine the situation given in Fig. 5.1, where two antennas are placed next to each other at a certain distance. For a plane wave arriving from direction **k**, the received signals at the two antennas will have approximately the same amplitude but different phase, because the wavefront is not parallel to the x axis and the antennas are not in the same exact position. Such phase difference, $\Delta\gamma$, is, in fact, the parameter that one needs to optimize such that the chances of having a fade in both antennas, in multipath channels, is minimized.

As shown in Section 2.7 for the simplified case of two isotropic antennas and considering only incident waves with DoA uniformly distributed in the xy plane (see Fig. 5.1) and neglecting any mutual effects between antennas, the correlation coefficient of the received signals is [33]

$$\mathbf{R}_{H_{ji}} = J_0(kd_{ij}), \tag{5.6}$$

where J_0 is the 0th-order Bessel function, d_{ij} is the distance between antennas, and k is the wave number. In Fig. 5.2, the correlation is plotted as a function of the distance. It is clear that for this

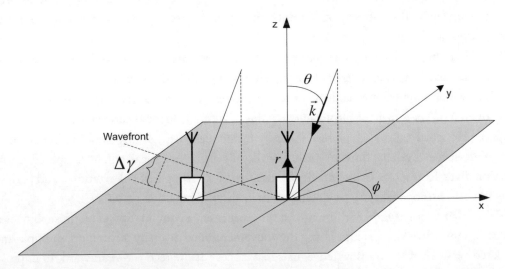

FIGURE 5.1: Graphical representation of the space diversity concept.

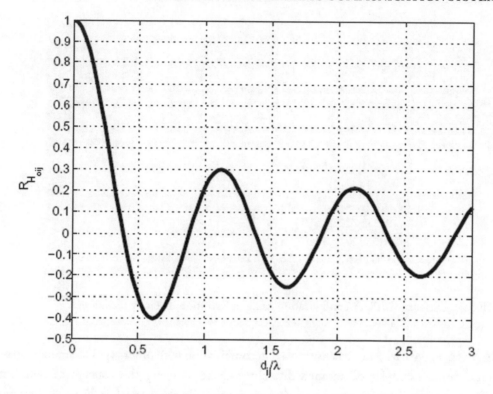

FIGURE 5.2: Jakes' model [33] for the correlation between two isotropic antennas in a uniformly distributed power azimuth spectrum.

simplified model, distances between antennas of the order of half a wavelength (0.5λ) are necessary to attain low correlation between antenna ports.

5.2.2 Polarization Diversity

Polarization diversity is another well-known antenna diversity technique. Some references on polarization diversity are [4, 12, 33]. In the case of polarization diversity, the integrand in Eq. (4.165) vanishes because the pattern scalar product is zero for each direction of space due to the vectorial orthogonal characteristic of the field radiated by the antennas, as a result of using two orthogonal polarizations.

For example, one possible scenario of polarization diversity would be that in which for two collocated antennas, their associated radiation pattern is $\mathbf{F}_1^A(\theta,\phi) = \mathbf{F}_{1_\theta}^A(\theta,\phi)\widehat{\theta}$ and $\mathbf{F}_2^A(\theta,\phi) = \mathbf{F}_{2_\phi}^A(\theta,\phi)\widehat{\phi}$. A practical implementation of this example would be using a linear dipole having the radiation pattern $\mathbf{F}_1^A(\theta,\phi)$ and a loop antenna having $\mathbf{F}_2^A(\theta,\phi)$ as a radiation pattern.

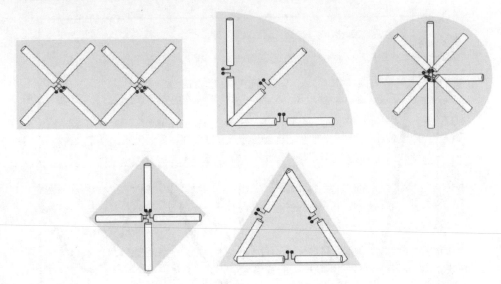

FIGURE 5.3: Different MPOA configuration using polarization diversity techniques.

In practice, polarization diversity is used in combination with other types of antenna diversity techniques. Some examples of antenna diversity techniques using the concept of polarization diversity (in combination with space diversity) are graphically represented in Fig. 5.3. As one can observe, not all of these configurations completely satisfy orthogonality on the radiation patterns. However, they are of interest in real applications because they become very compact designs and can still provide significant DGs.

5.2.3 Pattern Diversity

Pattern diversity imposes orthogonality by producing spatially (angularly) disjoint radiation patterns, as shown in Fig. 5.4. This is done by shaping the radiation patterns associated with different ports, such that the integrand in Eq. (4.165) tends to vanish. To generate pattern diversity, one can use MMAs [5, 6] or external beam-forming networks, such as a Butler matrix [13] (see Fig. 5.4).

5.3   MUTUAL COUPLING REDUCTION USING DNs

If the mutual coupling cannot be reduced by means of antenna diversity techniques—for example, because the antennas are very close together or if by design conditions $\Gamma^S \neq \mathbf{0}$ and $\Gamma^D \neq \mathbf{0}$ are needed—it is then necessary to include some additional components on the RF chain of a MIMO communication system. These additional components are know as DNs and have the capability to decorrelate the ports MPAs (assuming an ideally rich scattered NLOS environment)

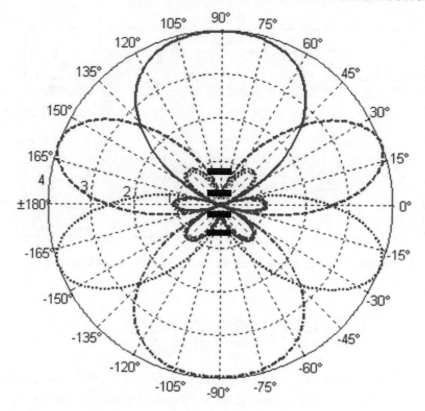

FIGURE 5.4: Possible radiation patterns of an MMA employing pattern diversity techniques. Notice that each beam is spatially disjoint from the others.

by compensating for the effects of mutual coupling through linear combinations of the received signals and by using the structural symmetries in the MPAs. Fig. 5.5 shows the components of the RF chain of a complete MIMO system, including the transmit and receive DNs.

These DNs can be described as well using a scattering matrix representation. Using the aforementioned formulation, the extended scattering model of the transmit DN, $\mathbf{S}^{MT}$, is given by

$$\begin{pmatrix} \mathbf{b}_{\alpha_1}^{MT} \\ \mathbf{b}_{\alpha_2}^{MT} \end{pmatrix} = \begin{pmatrix} \mathbf{S}_{11}^{MT} & \mathbf{S}_{12}^{MT} \\ \mathbf{S}_{21}^{MT} & \mathbf{S}_{22}^{MT} \end{pmatrix} \begin{pmatrix} \mathbf{a}_{\alpha_1}^{MT} \\ \mathbf{a}_{\alpha_2}^{MT} \end{pmatrix}, \qquad (5.7)$$

where we use the upper-index MT to refer to the transmit DN block within the RF chain. Similarly, the extended scattering model of the receive DN, $\mathbf{S}^{MR}$, is given by:

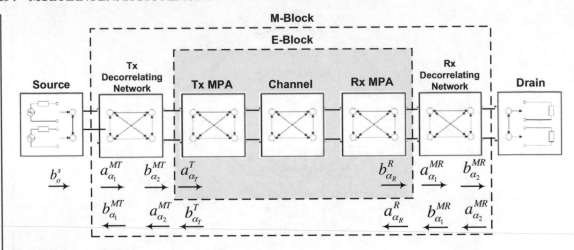

FIGURE 5.5: System model of the complete RF transmission chain, including the DNs.

$$\begin{pmatrix} \mathbf{b}_{\alpha_1}^{MR} \\ \mathbf{b}_{\alpha_2}^{MR} \end{pmatrix} = \begin{pmatrix} \mathbf{S}_{11}^{MR} & \mathbf{S}_{12}^{MR} \\ \mathbf{S}_{21}^{MR} & \mathbf{S}_{22}^{MR} \end{pmatrix} \begin{pmatrix} \mathbf{a}_{\alpha_1}^{MR} \\ \mathbf{a}_{\alpha_2}^{MR} \end{pmatrix}, \tag{5.8}$$

where we use the upper-index MR to refer to the receive DN block within the RF chain. In the next sections, we extend the system analysis given in Chapter 4 to include the effects of transmit and receive DNs.

5.4 ASSEMBLED MIMO SYSTEM NETWORK MODEL INCLUDING DNs

5.4.1 Superextended Channel Matrix

In this section, we derive an expression for the superextended channel matrix, $\mathbf{M}$, which includes the input and output DNs and all the effects included in the extended channel matrix $\mathbf{E}$. For any value of $\mathbf{\Gamma}^S$ and $\mathbf{\Gamma}^D$, the equivalent scattering matrix of the superextended channel, including the transmit and receive antennas, and the transmit and receive DNs, denoted by $\mathbf{S}^M$, is constructed as a result of performing the following operation:

$$\mathbf{S}^M = \mathbf{S}^T \odot \mathbf{S}^{MT} \odot \mathbf{S}^C \odot \mathbf{S}^{MR} \odot \mathbf{S}^R. \tag{5.9}$$

Therefore, assuming once again a unilateral and matched channel, $\mathbf{S}^M$ is given by

$$\mathbf{S}^M = \begin{pmatrix} \mathbf{S}_{11}^M & \mathbf{0} \\ \mathbf{S}_{21}^M & \mathbf{S}_{22}^M \end{pmatrix}, \tag{5.10}$$

where the subblock matrices of $\mathbf{S^M}$ are

$$\mathbf{S}_{11}^M = \mathbf{S}_{11}^{MT} + \mathbf{S}_{12}^{MT}\left(\mathbf{I} - \mathbf{S}_{\alpha_T\alpha_T}^T\mathbf{S}_{22}^{MT}\right)^{-1}\mathbf{S}_{\alpha_T\alpha_T}^T\mathbf{S}_{21}^{MT} \tag{5.11}$$

$$\mathbf{S}_{21}^M = \mathbf{S}_{21}^{MR}\left(\mathbf{I} - \mathbf{S}_{\alpha_R\alpha_R}^E\mathbf{S}_{11}^{MR}\right)^{-1}\mathbf{S}_{\alpha_R\alpha_T}^E\left(\mathbf{I} - \mathbf{S}_{22}^{MT}\mathbf{S}_{\alpha_T\alpha_T}^E\right)^{-1}\mathbf{S}_{21}^{MT} \tag{5.12}$$

$$\mathbf{S}_{22}^M = \mathbf{S}_{22}^{MR} + \mathbf{S}_{21}^{MR}\left(\mathbf{I} - \mathbf{S}_{\alpha_R\alpha_R}^E\mathbf{S}_{11}^{MR}\right)^{-1}\mathbf{S}_{\alpha_R\alpha_R}^E\mathbf{S}_{12}^{MR}. \tag{5.13}$$

Notice that the expression in Eq. (5.10) is a generalization of that in Eq. (4.67), which includes the effects of the transmit and receive DNs in the system. For this superextended channel model, including the source and drain terminations as well as the transmit and receive DNs, we can now derive the superextended channel matrix $\mathbf{M}$, which expresses the ratio of the voltages at the output ports of the receive DN ($\mathbf{U}_{\alpha_2}^{MR}$) to the voltages at the input ports of the transmit DN ($\mathbf{U}_{\alpha_1}^{MT}$), that is,

$$\mathbf{U}_{\alpha_2}^{MR} = \mathbf{M}\mathbf{U}_{\alpha_1}^{MT}. \tag{5.14}$$

Using the Heavyside transformation over the previous expression, we obtain

$$Z_{\alpha_2}^{0,MR\frac{1}{2}}\left(\mathbf{a}_{\alpha_2}^{MR} + \mathbf{b}_{\alpha_2}^{MR}\right) = \mathbf{M}Z_{\alpha_1}^{0,MT\frac{1}{2}}\left(\mathbf{a}_{\alpha_1}^{MT} + \mathbf{b}_{\alpha_1}^{MT}\right). \tag{5.15}$$

Now, following the approach in Ref. [15], the aforementioned wave vectors can be expressed as a function of $\mathbf{b}_0^S$, as shown below:

$$\mathbf{a}_{\alpha_1}^{MT} = \left(\mathbf{I} - \mathbf{\Gamma}_{\alpha_T}^S\mathbf{S}_{11}^M\right)^{-1}\mathbf{b}_0^S \tag{5.16}$$

$$\begin{aligned}\mathbf{a}_{\alpha_2}^{MR} &= \left(\mathbf{I} - \mathbf{\Gamma}_{\alpha_R}^D\mathbf{S}_{22}^M\right)^{-1}\mathbf{\Gamma}_{\alpha_R}^D\mathbf{S}_{21}^M\left(\mathbf{I} - \mathbf{\Gamma}_{\alpha_T}^S\mathbf{S}_{11}^M\right)^{-1}\mathbf{b}_0^S \\ &= \mathbf{\Gamma}_{\alpha_R}^D\left(\mathbf{I} - \mathbf{S}_{22}^M\mathbf{\Gamma}_{\alpha_R}^D\right)^{-1}\mathbf{S}_{21}^M\left(\mathbf{I} - \mathbf{\Gamma}_{\alpha_T}^S\mathbf{S}_{11}^M\right)^{-1}\mathbf{b}_0^S \end{aligned} \tag{5.17}$$

$$\mathbf{b}_{\alpha_1}^{MT} = \left(\mathbf{I} - \mathbf{S}_{11}^M\mathbf{\Gamma}_{\alpha_T}^S\right)^{-1}\mathbf{S}_{11}^M\mathbf{b}_0^S \tag{5.18}$$

$$\begin{aligned}\mathbf{b}_{\alpha_2}^{MR} &= \left(\mathbf{I} - \mathbf{S}_{22}^M\mathbf{\Gamma}_{\alpha_R}^D\right)^{-1}\left(\mathbf{S}_{21}^M + \mathbf{S}_{21}^M\mathbf{\Gamma}_{\alpha_T}^S\left(\mathbf{I} - \mathbf{S}_{11}^M\mathbf{\Gamma}_{\alpha_T}^S\right)^{-1}\mathbf{S}_{11}^M\right)\mathbf{b}_0^S \\ &= \left(\mathbf{I} - \mathbf{S}_{22}^M\mathbf{\Gamma}_{\alpha_R}^D\right)^{-1}\mathbf{S}_{21}^M\left(\mathbf{I} - \mathbf{\Gamma}_{\alpha_T}^S\mathbf{S}_{11}^M\right)^{-1}\mathbf{b}_0^S. \end{aligned} \tag{5.19}$$

Finally, $\mathbf{M}$ can be expressed as

$$
\begin{aligned}
\mathbf{M} &= Z_{\alpha_2}^{0,MR\frac{1}{2}} \left(\mathbf{I} + \boldsymbol{\Gamma}_{\alpha_R}^{D}\right) \left(\mathbf{I} - \mathbf{S}_{22}^{M}\boldsymbol{\Gamma}_{\alpha_R}^{D}\right)^{-1} \\
&\quad \mathbf{S}_{21}^{MR}\left(\mathbf{I} - \mathbf{S}_{\alpha_R\alpha_R}^{E}\mathbf{S}_{11}^{MR}\right)^{-1} \mathbf{S}_{\alpha_R\alpha_T}^{E} \left(\mathbf{I} - \mathbf{S}_{22}^{MT}\mathbf{S}_{\alpha_T\alpha_T}^{E}\right)^{-1} \mathbf{S}_{21}^{MT}\left(\mathbf{I} + \mathbf{S}_{11}^{M}\right)^{-1} Z_{\alpha_1}^{0,MT-\frac{1}{2}} \\
&= Z_{\alpha_2}^{0,MR\frac{1}{2}} \left(\mathbf{I} + \boldsymbol{\Gamma}_{\alpha_R}^{D}\right) \left(\mathbf{I} - \mathbf{S}_{22}^{M}\boldsymbol{\Gamma}_{\alpha_R}^{D}\right)^{-1} \\
&\quad \mathbf{S}_{21}^{MR}\left(\mathbf{I} - \mathbf{S}_{\alpha_R\alpha_R}^{E}\mathbf{S}_{11}^{MR}\right)^{-1} \mathbf{H} \left(\mathbf{I} - \mathbf{S}_{22}^{MT}\mathbf{S}_{\alpha_T\alpha_T}^{E}\right)^{-1} \mathbf{S}_{21}^{MT}\left(\mathbf{I} + \mathbf{S}_{11}^{M}\right)^{-1} Z_{\alpha_1}^{0,MT-\frac{1}{2}}. \quad (5.20)
\end{aligned}
$$

5.4.2 Separation of the Transmitter and Receiver Side Including DNs

By assuming an unilateral superextended channel, and for the sake of simplicity, it is possible to separate the network analysis of the transmitter and the receiver side, even when including the DNs. In that case, the system model in Fig. 4.4 can be separated into that given by Fig. 5.6a and 5.6b, for the transmitter and receiver sides, respectively. Similarly, the signal flow diagram given in Fig. 4.5 can be separated into that given in Fig. 5.7a and 5.7b for the transmitter and receiver sides, respectively. The input and output reflection coefficients for the input and output DN, as shown in Fig. 5.6, can be expressed as follows:

$$
\boldsymbol{\Gamma}_{\text{in}}^{T} = \mathbf{S}_{11}^{M} = \mathbf{S}_{11}^{MT} + \mathbf{S}_{12}^{MT}\left(\mathbf{I} - \mathbf{S}_{\alpha_T\alpha_T}^{T}\mathbf{S}_{22}^{MT}\right)^{-1} \mathbf{S}_{\alpha_T\alpha_T}^{T}\mathbf{S}_{21}^{MT} \quad (5.21)
$$

$$
\boldsymbol{\Gamma}_{\text{out}}^{T} = \mathbf{S}_{22}^{MT} + \mathbf{S}_{21}^{MT}\left(\mathbf{I} - \boldsymbol{\Gamma}^{S}\mathbf{S}_{11}^{MT}\right)^{-1} \boldsymbol{\Gamma}^{S}\mathbf{S}_{12}^{MT} \quad (5.22)
$$

$$
\boldsymbol{\Gamma}_{\text{in}}^{R} = \mathbf{S}_{11}^{MR} + \mathbf{S}_{12}^{MR}\left(\mathbf{I} - \boldsymbol{\Gamma}^{D}\mathbf{S}_{22}^{MR}\right)^{-1} \boldsymbol{\Gamma}^{D}\mathbf{S}_{21}^{MR} \quad (5.23)
$$

$$
\boldsymbol{\Gamma}_{\text{out}}^{R} = \mathbf{S}_{22}^{M} = \mathbf{S}_{22}^{MR} + \mathbf{S}_{21}^{MR}\left(\mathbf{I} - \mathbf{S}_{\alpha_R\alpha_R}^{E}\mathbf{S}_{11}^{MR}\right)^{-1} \mathbf{S}_{\alpha_R\alpha_R}^{E}\mathbf{S}_{12}^{MR}. \quad (5.24)
$$

For the receiver side, the outward propagating wave vector can be defined as

$$
\mathbf{b}_{\alpha_1}^{MR} = \mathbf{b}_0^{R} + \mathbf{S}_{11}^{M}\mathbf{a}_{\alpha_1}^{MR}, \quad (5.25)
$$

where $\mathbf{b}_0^{R} \in \mathcal{C}^{N\times 1}$ is the wave vector fed into the network, which determines the power distribution among the receiving antenna ports and relates to $\mathbf{b}_0^{S}$ through the following expression:

$$
\mathbf{b}_0^{R} = \mathbf{S}_{21}^{M}\mathbf{a}_{\alpha_1}^{MT} = \mathbf{S}_{21}^{M}(\mathbf{I} - \boldsymbol{\Gamma}_{\alpha_T}^{S}\mathbf{S}_{11}^{M})^{-1}\mathbf{b}_0^{S}. \quad (5.26)
$$

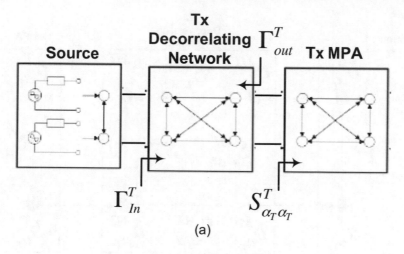

(a)

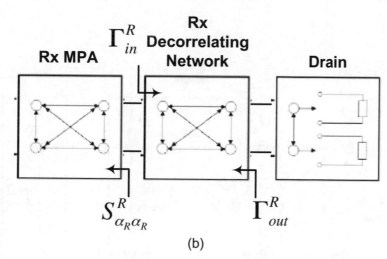

(b)

FIGURE 5.6: Separable system model of the complete RF transmission chain, including the DNs.

Using the separate model, we can write the wave vectors $\mathbf{a}_2^{MR}$ and $\mathbf{b}_2^{MR}$ as a function of $\mathbf{b}_0^R$, as follows:

$$
\begin{aligned}
\mathbf{a}_{\alpha_2}^{MR} &= \left(\mathbf{I} - \mathbf{\Gamma}_{\alpha_R}^{D}\mathbf{S}_{22}^{MR}\right)^{-1}\mathbf{\Gamma}_{\alpha_R}^{D}\mathbf{S}_{21}^{MR}\left(\mathbf{I} - \mathbf{S}_{\alpha_R\alpha_R}^{E}\mathbf{S}_{11}^{MR}\right)^{-1}\mathbf{b}_0^R \\
&= \mathbf{\Gamma}_{\alpha_R}^{D}\left(\mathbf{I} - \mathbf{S}_{22}^{MR}\mathbf{\Gamma}_{\alpha_R}^{D}\right)^{-1}\mathbf{S}_{21}^{MR}\left(\mathbf{I} - \mathbf{S}_{\alpha_R\alpha_R}^{E}\mathbf{S}_{11}^{MR}\right)^{-1}\mathbf{b}_0^R
\end{aligned}
\tag{5.27}
$$

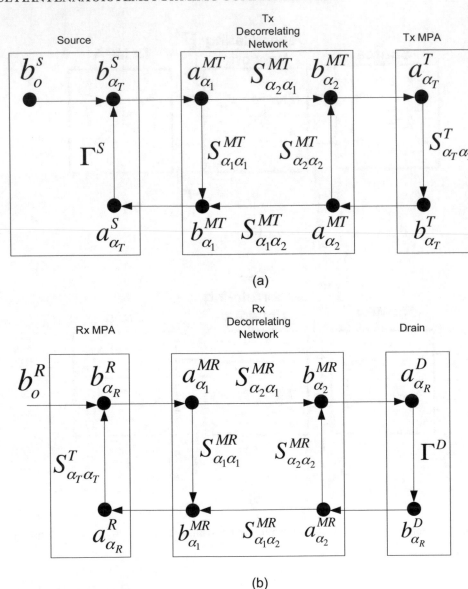

FIGURE 5.7: Signal flow graph for the separable MIMO system network model, including the DNs.

$$\mathbf{b}_{\alpha_2}^{MR} = \left(\mathbf{I} - \mathbf{S}_{22}^{MR}\mathbf{\Gamma}_{\alpha_R}^{D}\right)^{-1}\left(\mathbf{S}_{21}^{MR} + \mathbf{S}_{21}^{MR}\mathbf{S}_{\alpha_R\alpha_R}^{E}\left(\mathbf{I} - \mathbf{S}_{11}^{MR}\mathbf{S}_{\alpha_R\alpha_R}^{E}\right)^{-1}\mathbf{S}_{11}^{MR}\right)\mathbf{b}_0^{R}$$

$$= \left(\mathbf{I} - \mathbf{S}_{22}^{MR}\mathbf{\Gamma}_{\alpha_R}^{D}\right)^{-1}\mathbf{S}_{21}^{MR}\left(\mathbf{I} - \mathbf{S}_{\alpha_R\alpha_R}^{E}\mathbf{S}_{11}^{MR}\right)^{-1}\mathbf{b}_0^{R}. \tag{5.28}$$

For the superextended channel model, the voltages at the output ports of the receive DN ($\mathbf{U}_{\alpha_2}^{MR}$) are given by

$$\mathbf{U}_{\alpha_2}^{MR} = Z_{\alpha_2}^{0,MR\frac{1}{2}} \left(\mathbf{I} + \mathbf{\Gamma}_{\alpha_R}^{D}\right) \left(\mathbf{I} - \mathbf{S}_{22}^{MR}\mathbf{\Gamma}_{\alpha_R}^{D}\right)^{-1} \mathbf{S}_{21}^{MR} \left(\mathbf{I} - \mathbf{S}_{\alpha_R\alpha_R}^{E}\mathbf{S}_{11}^{MR}\right)^{-1} \mathbf{b}_0^{R}. \qquad (5.29)$$

Similarly, for the transmitter side, we defined previously that the outward propagating wave vector was

$$\mathbf{b}_{\alpha_T}^{S} = \mathbf{b}_0^{S} + \mathbf{\Gamma}_{\alpha_T}^{S}\mathbf{a}_{\alpha_T}^{S}, \qquad (5.30)$$

where $\mathbf{b}_0^{S} \in \mathcal{C}^{M\times1}$ is the wave vector fed into the system, which determines the power distribution among the source ports. $\mathbf{a}_{\alpha_T}^{S}$ is now the wave vector reflected by the input ports of the transmit DN, that is, $\mathbf{a}_{\alpha_T}^{S} = \mathbf{b}_{\alpha_{MT}}^{MT}$. For the superextended channel model, the voltages at the input ports of the transmit DNs ($\mathbf{U}_{\alpha_1}^{MT}$) are given by

$$\mathbf{U}_{\alpha_1}^{MT} = Z_{\alpha_1}^{0,MT\frac{1}{2}} \left(\mathbf{I} + \mathbf{\Gamma}_{\text{in}}^{T}\right) \mathbf{a}_{\alpha_1}^{MT} = Z_{\alpha_1}^{0,MT\frac{1}{2}} \left(\mathbf{I} + \mathbf{\Gamma}_{\text{in}}^{T}\right) \left(\mathbf{I} - \mathbf{\Gamma}_{\alpha_T}^{S}\mathbf{\Gamma}_{\text{in}}^{T}\right)^{-1} \mathbf{b}_0^{S}. \qquad (5.31)$$

5.5   DESIGN CRITERIA OF DNs

In this section, we present the design criteria of lossless DNs (also valid for reciprocal lossless DNs) to maximize the capacity and transmission power gain of a MIMO system and minimize the antenna correlation at the transmitter and receiver.

5.5.1   Introduction to Matching Conditions

In the following, we introduce three different matching conditions, which are widely used in the literature. One may choose one matching condition or another depending on the design degrees of freedom and the characteristics of the source and drain. For simplicity, $\mathbf{S}^{MA}$ refers to either the transmit ($\mathbf{S}^{MT}$) or receive ($\mathbf{S}^{MR}$) DNs.

5.5.1.1 Characteristic Impedance Match. Characteristic impedance match happens when the ports are terminated with the characteristic reference impedances, at either the source or at the loads, $\mathbf{\Gamma}^{S} = \mathbf{0}$ and $\mathbf{\Gamma}^{D} = \mathbf{0}$. Therefore, there are no DNs. This can be modeled either by removal of the DN or by setting $\mathbf{S}_{11}^{MA} = \mathbf{S}_{22}^{MA} = \mathbf{0}$ and $\mathbf{S}_{21}^{MA} = \mathbf{S}_{12}^{MA} = \mathbf{I}$. In this case, the degree of mismatch depends on the difference between the MPA port impedances and the characteristic reference impedance. Thus, $\mathbf{M} = \mathbf{E}$.

5.5.1.2 Self-Impedance Match. The self-impedance matching can be defined in two ways. In Ref. [66], it is defined as $\mathbf{S}_{11}^{MA} = diag(\mathbf{S}_{\alpha\alpha}^{A})^{H}$, where $diag(\cdot)$ retains only the diagonal elements of the matrix operand. On the other hand, in Ref. [69], a more common definition of the self-impedance match is used, given by

$$\mathbf{S}_{11}^{MA} = diag(\mathbf{Z}_{\alpha\alpha}^{A})^{H}. \tag{5.32}$$

The self-impedance match has the same basic property of being an individual-port match. That is, for an isolated antenna, the self-impedance match is also known as the complex conjugate match. It facilitates maximum power transfer to the load when there is no mutual coupling among the ports, such as, for example, when the elements of an MEA are infinitely far apart. At finite antenna separations, however, the goodness of the match depends on the behavior of the mutual impedance, which is not taken into account. Therefore, it is not a very good method when the mutual coupling is strongly present among the ports of the MPA. Individual-port match is well-studied, and many different implementations are possible for a given matching impedance value at the center frequency. One can, for example, use the transmission-line open-circuited stub configuration [56].

5.5.1.3 Multiport Conjugate Match. The multiport conjugate match, or optimal Hermitian match, takes into account the mutual coupling among the ports of the MPA. It allows the interconnections between all ports on the two sides of the network. The multiport conjugate match requires one side of the DN to be conjugate-matched to the antennas and the other side to the load or source impedance. That is, for the case where $\mathbf{\Gamma}^{S} = \mathbf{0}$ and $\mathbf{\Gamma}^{D} = \mathbf{0}$, the multiport conjugate match is given by

$$\mathbf{S}_{22}^{MT} = (\mathbf{S}_{\alpha_T\alpha_T}^{T})^{H} \tag{5.33}$$

$$\mathbf{S}_{11}^{MR} = (\mathbf{S}_{\alpha_R\alpha_R}^{E})^{H} \tag{5.34}$$

$$\mathbf{S}_{11}^{MT} = \mathbf{0} \tag{5.35}$$

$$\mathbf{S}_{22}^{MR} = \mathbf{0}. \tag{5.36}$$

5.5.2 Optimal Matching Conditions

In this section, we present the optimal design criteria when no other elements than those described above are considered. If receiving amplifiers or other components are considered in the MIMO system, as described in Ref. [75], then the design criteria varies slightly. Using [14, 66], it is easy to observe that the maximum transmission power gain, the maximum capacity, and the minimum antenna correlation, interestingly happen simultaneously when it is satisfied that

$$\Gamma_{in}^T = (\Gamma^S)^H \tag{5.37}$$

$$\Gamma_{out}^T = (\mathbf{S}_{\alpha_T \alpha_T}^T)^H \tag{5.38}$$

$$\Gamma_{in}^R = (\mathbf{S}_{\alpha_R \alpha_R}^E)^H \tag{5.39}$$

$$\Gamma_{out}^R = (\Gamma^D)^H. \tag{5.40}$$

Notice that the bilateral multiport conjugate matching condition holds if one uses lossless DNs, that is,

$$\Gamma_{in}^T = (\Gamma^S)^H \Leftrightarrow \Gamma_{out}^T = (\mathbf{S}_{\alpha_T \alpha_T}^T)^H \tag{5.41}$$

$$\Gamma_{out}^R = (\Gamma^D)^H \Leftrightarrow \Gamma_{in}^R = (\mathbf{S}_{\alpha_R \alpha_R}^E)^H. \tag{5.42}$$

Because normally we assume $\Gamma^S = \mathbf{0}$ and $\Gamma^D = \mathbf{0}$, some of the expressions given in Section 5.4.1 can be simplified to

$$\Gamma_{in}^T = \mathbf{S}_{11}^{MT} + \mathbf{S}_{12}^{MT} \left(\mathbf{I} - \mathbf{S}_{\alpha_T \alpha_T}^T \mathbf{S}_{22}^{MT}\right)^{-1} \mathbf{S}_{\alpha_T \alpha_T}^T \mathbf{S}_{21}^{MT} \tag{5.43}$$

$$\Gamma_{out}^T = \mathbf{S}_{22}^{MT} \tag{5.44}$$

$$\Gamma_{in}^R = \mathbf{S}_{11}^{MR} \tag{5.45}$$

$$\Gamma_{out}^R = \mathbf{S}_{22}^{MR} + \mathbf{S}_{21}^{MR} \left(\mathbf{I} - \mathbf{S}_{\alpha_R \alpha_R}^E \mathbf{S}_{11}^{MR}\right)^{-1} \mathbf{S}_{\alpha_R \alpha_R}^E \mathbf{S}_{12}^{MR}. \tag{5.46}$$

In this case, as will be shown later, to maximize the capacity and the transmission power gain of the system and minimize antenna correlation, it is enough to satisfy the following two conditions:

$$\Gamma_{out}^T = \mathbf{S}_{22}^{MT} = (\mathbf{S}_{\alpha_T \alpha_T}^T)^H \tag{5.47}$$

$$\Gamma_{in}^R = \mathbf{S}_{11}^{MR} = (\mathbf{S}_{\alpha_R \alpha_R}^E)^H. \tag{5.48}$$

Notice that, in this case, it is also satisfied that $\Gamma_{in}^T = \mathbf{0}$ and $\Gamma_{out}^R = \mathbf{0}$, and this happens to be independently of the chosen value for $\mathbf{S}_{11}^{MT}$ and $\mathbf{S}_{22}^{MR}$, if the DNs are lossless. This can be proved as follows. For the receiver side, if one must have $\mathbf{S}_{11}^{MR} = (\mathbf{S}_{\alpha_R\alpha_R}^E)^H$, then it must also hold $\mathbf{S}_{\alpha_R\alpha_R}^E = (\mathbf{S}_{11}^{MR})^H$. Then

$$
\begin{aligned}
\mathbf{\Gamma}_{\text{out}}^R &= \mathbf{S}_{22}^{MR} + \mathbf{S}_{21}^{MR} \left(\mathbf{I} - \mathbf{S}_{\alpha_R\alpha_R}^E \mathbf{S}_{11}^{MR} \right)^{-1} \mathbf{S}_{\alpha_R\alpha_R}^E \mathbf{S}_{12}^{MR} \\
&= \mathbf{S}_{22}^{MR} + \mathbf{S}_{21}^{MR} \left(\mathbf{I} - (\mathbf{S}_{11}^{MR})^H \mathbf{S}_{11}^{MR} \right)^{-1} (\mathbf{S}_{11}^{MR})^H \mathbf{S}_{12}^{MR} \\
&= \mathbf{S}_{22}^{MR} + \mathbf{S}_{21}^{MR} \left((\mathbf{S}_{21}^{MR})^H \mathbf{S}_{21}^{MR} \right)^{-1} (\mathbf{S}_{11}^{MR})^H \mathbf{S}_{12}^{MR} \\
&= \mathbf{S}_{22}^{MR} + \left((\mathbf{S}_{21}^{MR})^H \right)^{-1} (\mathbf{S}_{11}^{MR})^H \mathbf{S}_{12}^{MR} \\
&= \mathbf{S}_{22}^{MR} - \left((\mathbf{S}_{21}^{MR})^H \right)^{-1} (\mathbf{S}_{21}^{MR})^H \mathbf{S}_{22}^{MR} \\
&= \mathbf{S}_{22}^{MR} - \mathbf{S}_{22}^{MR} \\
&= \mathbf{0}.
\end{aligned}
\tag{5.49}
$$

For the transmitter side, if one must have $\mathbf{S}_{11}^{MT} = (\mathbf{S}_{\alpha_T\alpha_T}^E)^H$, then it must also hold $\mathbf{S}_{\alpha_T\alpha_T}^E = (\mathbf{S}_{22}^{MR})^H$. Then

$$
\begin{aligned}
\mathbf{\Gamma}_{\text{in}}^T &= \mathbf{S}_{11}^{MT} + \mathbf{S}_{12}^{MT} \left(\mathbf{I} - \mathbf{S}_{\alpha_T\alpha_T}^E \mathbf{S}_{22}^{MT} \right)^{-1} \mathbf{S}_{\alpha_T\alpha_T}^E \mathbf{S}_{21}^{MT} \\
&= \mathbf{S}_{11}^{MT} + \mathbf{S}_{12}^{MT} \left(\mathbf{I} - (\mathbf{S}_{22}^{MT})^H \mathbf{S}_{22}^{MT} \right)^{-1} (\mathbf{S}_{22}^{MT})^H \mathbf{S}_{21}^{MT} \\
&= \mathbf{S}_{11}^{MT} + \mathbf{S}_{12}^{MT} \left((\mathbf{S}_{12}^{MT})^H \mathbf{S}_{12}^{MT} \right)^{-1} (\mathbf{S}_{22}^{MT})^H \mathbf{S}_{21}^{MT} \\
&= \mathbf{S}_{11}^{MT} + \left((\mathbf{S}_{12}^{MT})^H \right)^{-1} (\mathbf{S}_{22}^{MT})^H \mathbf{S}_{21}^{MT} \\
&= \mathbf{S}_{11}^{MT} - \left((\mathbf{S}_{12}^{MT})^H \right)^{-1} (\mathbf{S}_{12}^{MT})^H \mathbf{S}_{11}^{MT} \\
&= \mathbf{S}_{11}^{MT} - \mathbf{S}_{11}^{MT} \\
&= \mathbf{0}.
\end{aligned}
\tag{5.50}
$$

Therefore, to achieve this optimal condition, one needs to make sure that multiport conjugate match is provided at the output ports of the transmit DN and at the input ports of the receive DN. Notice that the multiport conjugate matching explained in Section 5.5.1.3 is just a particular solution to this problem in which it is set $\mathbf{S}_{11}^{MT} = \mathbf{0}$ and $\mathbf{S}_{22}^{MR} = \mathbf{0}$. In the above derivations, we have used the fact that lossless DN networks must satisfy the unitarity condition given by

$$
(\mathbf{S}^{MA})^H \mathbf{S}^{MA} = \mathbf{I},
\tag{5.51}
$$

where the upper index MA refers to either the transmit (MT) or receive (MR) DN. See Section 5.7 for additional properties on losses DNs.

Because multiport conjugate matching is the optimal method to minimize antenna correlation, the DNs designed based upon this methodology are referred to in this book as DNs. Interestingly, the aforementioned matching condition is optimal independently of the characteristics of the channel, as long as the degrees of freedom offered by an arbitrary design of the $\mathbf{S}_{11}^{MT}$ and $\mathbf{S}_{22}^{MR}$ subblock matrices are appropriately taken into account and exploited.

5.5.3 Proof of Optimal Transmission Power Gain

In this section, we prove that the transmission power gain is maximized when $\mathbf{S}_{11}^{MR} = (\mathbf{S}_{\alpha_R \alpha_R}^E)^H = (\mathbf{S}_{\alpha_R \alpha_R}^R)^H$, in the case that $\mathbf{\Gamma}_{\alpha_R}^D = \mathbf{0}$ (following the approach in Ref. [14]). The proof focuses on the receiver side, although by virtue of the reciprocity theorem, it also applies for the transmit side, for which the transmission power gain is maximized when $\mathbf{S}_{22}^{MT} = (\mathbf{S}_{\alpha_T \alpha_T}^E)^H = (\mathbf{S}_{\alpha_T \alpha_T}^T)^H$, in the case that $\mathbf{\Gamma}_{\alpha_T}^S = \mathbf{0}$. If $\mathbf{\Gamma}_{\alpha_R}^D = \mathbf{0}$ can be assumed, we can consider the equivalent problem depicted in Fig. 5.8. Everything to the right of the reference plane associated with input ports of the receive DN has been replaced with a block with input reflection coefficient $\mathbf{S}_{11}^{MR}$ (because $\mathbf{\Gamma}_{\alpha_R}^D = \mathbf{0}$). We can easily write the following set of equations:

$$\mathbf{b}_{\beta_R}^R = \mathbf{S}_{\beta_R \alpha_R}^R \mathbf{a}_{\alpha_R}^R + \mathbf{S}_{\beta_R \beta_R}^R \mathbf{a}_{\beta_R}^R \tag{5.52}$$

$$\mathbf{b}_{\alpha_R}^R = \mathbf{S}_{\alpha_R \alpha_R}^R \mathbf{a}_{\alpha_R}^R + \mathbf{S}_{\alpha_R \beta_R}^R \mathbf{a}_{\beta_R}^R \tag{5.53}$$

$$\mathbf{a}_{\alpha_R}^R = \mathbf{S}_{11}^{MR} \mathbf{b}_{\alpha_R}^R \tag{5.54}$$

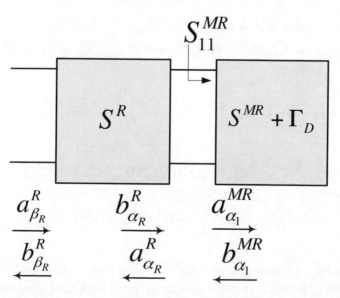

FIGURE 5.8: Network model for the equivalent receive impedance matching problem.

$$\mathbf{b}_0^R = \mathbf{S}_{\alpha_R \beta_R}^R \mathbf{a}_{\beta_R}^R. \tag{5.55}$$

After some algebraic manipulations of the above expressions, we can express the backscattered field $\mathbf{b}_{\beta_R}^R$ as follows:

$$\mathbf{b}_{\beta_R}^R = \left(\mathbf{S}_{\beta_R \alpha_R}^R \mathbf{S}_{11}^{MR} \left(\mathbf{I} - \mathbf{S}_{\alpha_R \alpha_R}^R \mathbf{S}_{11}^{MR} \right)^{-1} + \mathbf{S}_{\beta_R \beta_R}^R (\mathbf{S}_{\alpha_R \beta_R}^R)^{-1} \right) \mathbf{b}_0^R. \tag{5.56}$$

Because the receive antenna is assumed to be lossless, the power available to the load is $(\mathbf{a}_{\beta_R}^R)^H \mathbf{a}_{\beta_R}^R = ||\mathbf{a}_{\beta_R}^R||^2 = ||(\mathbf{S}_{\alpha_R \beta_R}^R)^{-1} \mathbf{b}_0^R||^2$. The load will collect all of this available power if we can choose $\mathbf{S}_{11}^{MR}$, such that the reflection $\mathbf{b}_{\beta_R}^R = \mathbf{0}$. In such a case, the antenna does not scatter. To solve this problem, we require an expression that relates the subblocks of the scattering matrix $\mathbf{S}^R$ of the lossless receive MPA. The relations are given by the SVD of the subblocks (see Section 5.7):

$$\mathbf{S}_{\alpha_R \alpha_R}^R = \mathbf{U}_{\alpha\alpha}^R (\mathbf{\Lambda}_{\alpha\alpha}^R)^{\frac{1}{2}} (\mathbf{V}_{\alpha\alpha}^R)^H \tag{5.57}$$

$$\mathbf{S}_{\alpha_R \beta_R}^R = -\mathbf{U}_{\alpha\alpha}^R (\mathbf{\Theta}^R)^H (\mathbf{I} - \mathbf{\Lambda}_{\alpha\alpha}^R)^{\frac{1}{2}} (\mathbf{V}_{\beta\beta}^R)^H \tag{5.58}$$

$$\mathbf{S}_{\beta_R \alpha_R}^R = \mathbf{U}_{\beta\beta}^R \mathbf{\Theta}^R (\mathbf{I} - \mathbf{\Lambda}_{\alpha\alpha}^R)^{\frac{1}{2}} (\mathbf{V}_{\alpha\alpha}^R)^H \tag{5.59}$$

$$\mathbf{S}_{\beta_R \beta_R}^R = \mathbf{U}_{\beta\beta}^R (\mathbf{\Lambda}_{\alpha\alpha}^R)^{\frac{1}{2}} (\mathbf{V}_{\beta\beta}^R)^H, \tag{5.60}$$

where $\mathbf{U}_{\alpha\alpha}^R$, $\mathbf{V}_{\alpha\alpha}^R$, $\mathbf{U}_{\beta\beta}^R$, and $\mathbf{V}_{\beta\beta}^R$ are unitary matrices and $\mathbf{\Theta}^R$ is a diagonal matrix with complex unit magnitude entries. Letting $\mathbf{S}_{11}^{MR} = (\mathbf{S}_{\alpha_R \alpha_R}^R)^H = (\mathbf{S}_{\alpha_R \alpha_R}^R)^H$, where $\mathbf{S}_{\alpha_R \alpha_R}^R = \mathbf{U}_{\alpha\alpha}^R (\mathbf{\Lambda}_{\alpha\alpha}^R)^{\frac{1}{2}} (\mathbf{V}_{\alpha\alpha}^R)^H$ (and therefore, $(\mathbf{S}_{\alpha_R \alpha_R}^R)^H = \mathbf{V}_{\alpha\alpha}^R (\mathbf{\Lambda}_{\alpha\alpha}^R)^{\frac{1}{2}} (\mathbf{U}_{\alpha\alpha}^R)^H$ and $(\mathbf{S}_{\alpha_R \alpha_R}^R (\mathbf{S}_{\alpha_R \alpha_R}^R)^H)^{-1} = \mathbf{U}_{\alpha\alpha}^R (\mathbf{I} - \mathbf{\Lambda}_{\alpha\alpha}^R)^{-1} (\mathbf{U}_{\alpha\alpha}^R)^H$), results in $\mathbf{b}_{\beta_R}^R$ given by

$$\mathbf{b}_{\beta_R}^R = \left(\mathbf{S}_{\beta_R \alpha_R}^R (\mathbf{S}_{\alpha_R \alpha_R}^R)^H \left(\mathbf{I} - \mathbf{S}_{\alpha_R \alpha_R}^R (\mathbf{S}_{\alpha_R \alpha_R}^R)^H \right)^{-1} + \mathbf{S}_{\beta_R \beta_R}^R (\mathbf{S}_{\alpha_R \beta_R}^R)^{-1} \right) \mathbf{b}_0^R, \tag{5.61}$$

and finally,

$$
\begin{aligned}
\mathbf{b}_{\beta_R}^R &= (\mathbf{U}_{\beta\beta}^R \mathbf{\Theta}^R (\mathbf{I} - \mathbf{\Lambda}_{\alpha\alpha}^R)^{\frac{1}{2}} (\mathbf{V}_{\alpha\alpha}^R)^H \mathbf{V}_{\alpha\alpha}^R (\mathbf{\Lambda}_{\alpha\alpha}^R)^{\frac{1}{2}} (\mathbf{U}_{\alpha\alpha}^R)^H \mathbf{U}_{\alpha\alpha}^R (\mathbf{\Lambda}_{\alpha\alpha}^R)^{-1} (\mathbf{U}_{\alpha\alpha}^R)^H \\
&\quad - \mathbf{U}_{\beta\beta}^R (\mathbf{\Lambda}_{\alpha\alpha}^R)^{\frac{1}{2}} \mathbf{\Theta}^R (\mathbf{I} - \mathbf{\Lambda}_{\alpha\alpha}^R)^{-\frac{1}{2}} (\mathbf{U}_{\alpha\alpha}^R)^H) \mathbf{b}_0^R \\
&= \left(\mathbf{U}_{\beta\beta}^R (\mathbf{\Lambda}_{\alpha\alpha}^R)^{\frac{1}{2}} \mathbf{\Theta}^R (\mathbf{I} - \mathbf{\Lambda}_{\alpha\alpha}^R)^{-\frac{1}{2}} (\mathbf{U}_{\alpha\alpha}^R)^H - \mathbf{U}_{\beta\beta}^R (\mathbf{\Lambda}_{\alpha\alpha}^R)^{\frac{1}{2}} \mathbf{\Theta}^R (\mathbf{I} - \mathbf{\Lambda}_{\alpha\alpha}^R)^{-\frac{1}{2}} (\mathbf{U}_{\alpha\alpha}^R)^H \right) \mathbf{b}_0^R = \mathbf{0}.
\end{aligned}
\tag{5.62}
$$

We have ensured that $\mathbf{b}_{\beta_R}^R = \mathbf{0}$ for any choice of $\mathbf{b}_0^R$. Therefore, the assignment $\mathbf{S}_{11}^{MR} = (\mathbf{S}_{\alpha_R \alpha_R}^E)^H = (\mathbf{S}_{\alpha_R \alpha_R}^R)^H$ ensures that all available power is dissipated in the load, thus maximizing the collected receive power and the transmission power gain.

5.5.4   Proof of Optimal Capacity

We prove now that the above condition $\mathbf{S}_{11}^{MR} = (\mathbf{S}_{\alpha_R \alpha_R}^{E})^H = (\mathbf{S}_{\alpha_R \alpha_R}^{R})^H$, in the case that $\mathbf{\Gamma}_{\alpha_R}^{D} = \mathbf{0}$, also maximizes the capacity of the MIMO system [14]. For simplicity, we assume the uninformed transmitter capacity expression given in Section 1.5.3. The proof focuses on the receiver side, although by virtue of the reciprocity theorem, it also applies for the transmit side. For the superextended channel model, the voltages at the output ports of the receive DN ($\mathbf{U}_{\alpha_2}^{MR}$), are now given by

$$\mathbf{U}_{\alpha_2}^{MR} = Z_{\alpha_2}^{0,MR\frac{1}{2}} \mathbf{S}_{21}^{MR} \left(\mathbf{I} - \mathbf{S}_{\alpha_R \alpha_R}^{E} \mathbf{S}_{11}^{MR} \right)^{-1} \mathbf{S}_{21}^{MT} \mathbf{a}_{\alpha_2}^{MT}. \tag{5.63}$$

Let us denote $\mathbf{H} = Z_{\alpha_2}^{0,MR\frac{1}{2}} \mathbf{S}_{21}^{MR} \left(\mathbf{I} - \mathbf{S}_{\alpha_R \alpha_R}^{E} \mathbf{S}_{11}^{MR} \right)^{-1} \mathbf{S}_{21}^{MT}$, $\mathbf{s} = \mathbf{a}_{\alpha_2}^{MT}$. If we include the effects of AWGN ($\mathbf{n}$), the received signal $\mathbf{r}$ can be expressed as

$$\mathbf{r} = \mathbf{Hs} + \mathbf{n}. \tag{5.64}$$

In this case, the mutual information expression is

$$I(\mathbf{r}; \mathbf{s}) = \det \left(\mathbf{I} + \frac{\mathbf{H} \mathbf{R}_s \mathbf{H}^H}{N_0} \right) = \det \left(\mathbf{I} + \frac{\mathbf{W}(\mathbf{S}_{11}^{MR}) \mathbf{M}}{N_0} \right), \tag{5.65}$$

where we have defined

$$\mathbf{W}(\mathbf{S}_{11}^{MR}) = \left((\mathbf{I} - \mathbf{S}_{\alpha_R \alpha_R}^{E} \mathbf{S}_{11}^{MR})^{-1} \right)^H \left(\mathbf{I} - (\mathbf{S}_{11}^{MR})^H \mathbf{S}_{11}^{MR} \right) (\mathbf{I} - \mathbf{S}_{\alpha_R \alpha_R}^{E} \mathbf{S}_{11}^{MR})^{-1} \tag{5.66}$$

and where the noise is AWGN, such that the entries of $\mathbf{n}$ are iid random variables with equal variance N_0, with $\mathbf{M} = \mathbf{S}_{21}^{MT} \mathbf{R}_s (\mathbf{S}_{21}^{MT})^H$. In general, $\mathbf{M}$ is a Hermitian-positive semidefinite matrix, so that we can use the EVD of $\mathbf{M}$ to write it in the form $\mathbf{M} = \mathbf{M}^{\frac{1}{2}} \left(\mathbf{M}^{\frac{1}{2}} \right)^H$. Thus, maximization of the mutual information for a fixed (but arbitrary) $\mathbf{R}_s$ requires maximization of

$$I(\mathbf{r}; \mathbf{s}) = \det \left(\mathbf{I} + \frac{\left(\mathbf{M}^{\frac{1}{2}} \right)^H \mathbf{W}(\mathbf{S}_{11}^{MR}) \mathbf{M}^{\frac{1}{2}}}{N_0} \right) \tag{5.67}$$

over all positive values of $\mathbf{S}_{11}^{MR}$ and $\mathbf{R}s$. This multivariate maximization is simplified by the fact that a simple conjugate match will always maximize Eq. (5.67) for fixed but arbitrary $\mathbf{R}_s$. To show this, we use the result from Section 5.5.3, for which the received power is maximized when $\mathbf{S}_{11}^{MR} = (\mathbf{S}_{\alpha_R \alpha_R}^{E})^H$ (see also Section 5.6.1.2), and therefore,

$$\mathbf{u}^H \mathbf{W}((\mathbf{S}_{\alpha_R \alpha_R}^{E})^H) \mathbf{u} \geq \mathbf{u}^H \mathbf{W}(\mathbf{S}_{11}^{MR}) \mathbf{u} \tag{5.68}$$

for all possible values of $\mathbf{S}_{11}^{MR}$ and $\mathbf{u}$. Letting $\mathbf{u} = \mathbf{M}^{\frac{1}{2}}\mathbf{w}$ and $\mathbf{W}'(\mathbf{S}) = \left(\mathbf{M}^{\frac{1}{2}}\right)^{H}\mathbf{W}(\mathbf{S})\mathbf{M}^{\frac{1}{2}}$, we obtain

$$\mathbf{w}^{H}\mathbf{W}'((\mathbf{S}_{\alpha_R \alpha_R}^{E})^{H})\mathbf{w} \geq \mathbf{w}^{H}\mathbf{W}(\mathbf{S}_{11}^{\prime MR})\mathbf{w}, \tag{5.69}$$

and therefore,

$$\mathbf{w}^{H}\left(\frac{\mathbf{W}'((\mathbf{S}_{\alpha_R \alpha_R}^{E})^{H})}{N_0} + \mathbf{I}\right)\mathbf{w} \geq \mathbf{w}^{H}\left(\frac{\mathbf{W}(\mathbf{S}_{11}^{\prime MR})}{N_0}\mathbf{I}\right)\mathbf{w}. \tag{5.70}$$

For two positive definite matrices, $\mathbf{A}$ and $\mathbf{B}$,

$$\mathbf{u}^{H}\mathbf{A}\mathbf{u} \geq \mathbf{u}^{H}\mathbf{B}\mathbf{u} \tag{5.71}$$

for all $\mathbf{u}$, if and only if the singular values of $\mathbf{B}(\mathbf{A})^{-1}$ are all less than or equal to 1 [17]. Hence, the previous relation implies that

$$\det(\mathbf{A}) \geq \det(\mathbf{B}), \tag{5.72}$$

leading to the conclusion that

$$\det\left(\frac{\mathbf{W}'((\mathbf{S}_{\alpha_R \alpha_R}^{E})^{H})}{N_0} + \mathbf{I}\right) \geq \det\left(\frac{\mathbf{W}(\mathbf{S}_{11}^{\prime MR})}{N_0}\mathbf{I}\right). \tag{5.73}$$

Therefore, for arbitrary $\mathbf{R}_s$, the condition $\mathbf{S}_{11}^{MR} = (\mathbf{S}_{\alpha_R \alpha_R}^{E})^{H}$ will maximize Eq. (5.67) and the mutual-information expression reduces to

$$I(\mathbf{r};\mathbf{s}) = \det\left(\mathbf{I} + Z^{0,MR}\frac{(\mathbf{I} - \mathbf{S}_{\alpha_R \alpha_R}^{R}(\mathbf{S}_{\alpha_R \alpha_R}^{R})^{H})^{-1}\mathbf{S}_{21}^{MT}\mathbf{R}_s(\mathbf{S}_{21}^{MT})^{H}}{N_0}\right). \tag{5.74}$$

Finding the value of $\mathbf{R}_s$ that maximizes this equation will, therefore, lead to an expression of the channel capacity.

5.6   PERFORMANCE MEASURES IN MIMO COMMUNICATION SYSTEMS INCLUDING DNs

In this section, we describe the main performance measures that are used to asses the performance of MPAs within a MIMO communication system based on the superextended channel model. System-level figures of merit are also presented.

5.6.1   Transmission Power Gain

Similarly as before, the transmission power gain of the superextended channel including DNs is the ratio of the real power delivered to the signal drain, $P^{D} = P_{\alpha_2}^{MR} = (\mathbf{b}_{\alpha_2}^{MR})^{H}\mathbf{b}_{\alpha_2}^{MR} - (\mathbf{a}_{\alpha_2}^{MR})^{H}\mathbf{a}_{\alpha_2}^{MR}$, to the real power fed into the transmit DN, $P^S = P_{\alpha_1}^{MT} = (\mathbf{a}_{\alpha_1}^{MT})^H \mathbf{a}_{\alpha_1}^{MT} - (\mathbf{b}_{\alpha_1}^{MT})^H \mathbf{b}_{\alpha_1}^{MT}$. It allows us to draw conclusions about the efficiency of the whole transmission chain, and it is given by

$$G_{tp} = \frac{P^D}{P^S} = \frac{(\mathbf{b}_{\alpha_2}^{MR})^H \mathbf{b}_{\alpha_2}^{MR} - (\mathbf{a}_{\alpha_2}^{MR})^H \mathbf{a}_{\alpha_2}^{MR}}{(\mathbf{a}_{\alpha_1}^{MT})^H \mathbf{a}_{\alpha_1}^{MT} - (\mathbf{b}_{\alpha_1}^{MT})^H \mathbf{b}_{\alpha_1}^{MT}}. \tag{5.75}$$

As a function of $\mathbf{b}_0^S$, P^D and P^S can now be written as

$$P^S = (\mathbf{b}_0^S)^H \left(\left(\mathbf{I} - \mathbf{\Gamma}_{\alpha_T}^S \mathbf{\Gamma}_{in}^T \right)^{-1} \right)^H (\mathbf{I} - (\mathbf{\Gamma}_{in}^T)^H \mathbf{\Gamma}_{in}^T)(\mathbf{I} - \mathbf{\Gamma}_{\alpha_T}^S \mathbf{\Gamma}_{in}^T)^{-1} \mathbf{b}_0^S \tag{5.76}$$

$$P^D = (\mathbf{b}_0^R)^H \left((\mathbf{I} - \mathbf{S}_{\alpha_R \alpha_R}^E \mathbf{S}_{11}^{MR})^{-1} \right)^H (\mathbf{S}_{21}^{MR})^H \left(\mathbf{I} - \mathbf{S}_{22}^{MR} \mathbf{\Gamma}_{\alpha_R}^D \right)^H$$
$$\cdot (\mathbf{I} - (\mathbf{\Gamma}_{\alpha_R}^D)^H \mathbf{\Gamma}_{\alpha_R}^D)(\mathbf{I} - \mathbf{S}_{22}^{MR} \mathbf{\Gamma}_{\alpha_R}^D) \mathbf{S}_{21}^{MR} (\mathbf{I} - \mathbf{S}_{\alpha_R \alpha_R}^E \mathbf{S}_{11}^{MR})^{-1} \mathbf{b}_0^R, \tag{5.77}$$

with $\mathbf{b}_0^R$ related to $\mathbf{b}_0^S$ in Eq. (5.26).

5.6.1.1 Transmit Matching Efficiency.
In the optimal scenario in which we set $\mathbf{\Gamma}_{\alpha_T}^S = (\mathbf{\Gamma}_{in}^T)^H$, the power delivered by the source is maximized, that is,

$$P^{S,\max} = (\mathbf{b}_0^S)^H \left(\left(\mathbf{I} - (\mathbf{\Gamma}_{in}^T)^H \mathbf{\Gamma}_{in}^T \right)^{-1} \right)^H \mathbf{b}_0^S. \tag{5.78}$$

On the other hand, in general, $\mathbf{\Gamma}_{\alpha_T}^S = \mathbf{0}$, and the radiated power is given by

$$P^{S,0} = (\mathbf{b}_0^S)^H \mathbf{A} \mathbf{b}_0^S, \tag{5.79}$$

where the upper-index 0 refers to the fact that we assumed $\mathbf{\Gamma}_{\alpha_T}^S = \mathbf{0}$. Notice that the radiated transmit power is limited by a factor that has to do with the coupling among the input ports of the transmit DN, $\mathbf{A} = (\mathbf{I} - (\mathbf{\Gamma}_{in}^T)^H \mathbf{\Gamma}_{in}^T)$. Now, besides the possible occurrence of $\mathbf{S}_{\alpha_T \alpha_T}^E \neq \mathbf{0}$, if the transmit DN is designed according to the optimal criteria with $\mathbf{S}_{22}^{MT} = (\mathbf{S}_{\alpha_T \alpha_T}^T)^H$ (and therefore, $\mathbf{\Gamma}_{in}^T = \mathbf{0}$), then the radiated power is maximized and given by

$$P^{S,0,\max} = (\mathbf{b}_0^S)^H \mathbf{b}_0^S. \tag{5.80}$$

The matching efficiency for the transmit multiantenna system associated with the superextended channel can be defined as for the extended channel, that is,

$$\eta^{T,0} = \frac{E\{P^{S,0}\}}{E\{P^{S,0,\max}\}}, \tag{5.81}$$

where once again the upper-index 0 refers to the fact that we assumed $\mathbf{\Gamma}_{\alpha_T}^S = \mathbf{0}$. Notice that with this definition of the efficiency, when the mutual coupling vanishes because of the use of the DN, that is, when $\mathbf{\Gamma}_{in}^{T} = \mathbf{0}$, then $\eta^{T,0} \to 1$. We could also define the transmit efficiency independently of the source characteristic reference impedances, here given by

$$\eta^{T} = \frac{E\{P^{S}\}}{E\{P^{S,\max}\}}. \tag{5.82}$$

5.6.1.2 Receive Matching Efficiency.

Using the separable model for the receiver side, the power delivered to the load can be expressed as a function of $\mathbf{b}_{0}^{R}$, as shown in Eq. (4.105). The power delivered to the load is maximized when $\mathbf{\Gamma}_{\alpha_{R}}^{D} = (\mathbf{\Gamma}_{out}^{R})^{H}$, as shown in Ref. [14]. On the other hand, in general, $\mathbf{\Gamma}_{\alpha_{R}}^{D} = \mathbf{0}$, and in the case that no maximum power transfer can be assured, which, for example, may happen if $\mathbf{\Gamma}_{out}^{R} \neq \mathbf{0}$ (i.e., $\mathbf{S}_{\alpha_{R}\alpha_{R}}^{E} \neq \mathbf{0}$), then the power delivered to the loads is given by

$$\begin{aligned}
P^{D,0} &= (\mathbf{b}_{0}^{R})^{H} \left((\mathbf{I} - \mathbf{S}_{\alpha_{R}\alpha_{R}}^{E} \mathbf{S}_{11}^{MR})^{-1} \right)^{H} (\mathbf{S}_{21}^{MR})^{H} \mathbf{S}_{21}^{MR} (\mathbf{I} - \mathbf{S}_{\alpha_{R}\alpha_{R}}^{E} \mathbf{S}_{11}^{MR})^{-1} \mathbf{b}_{0}^{R} \\
&= (\mathbf{b}_{0}^{R})^{H} \left((\mathbf{I} - \mathbf{S}_{\alpha_{R}\alpha_{R}}^{E} \mathbf{S}_{11}^{MR})^{-1} \right)^{H} \left(\mathbf{I} - (\mathbf{S}_{11}^{MR})^{H} \mathbf{S}_{11}^{MR} \right) (\mathbf{I} - \mathbf{S}_{\alpha_{R}\alpha_{R}}^{E} \mathbf{S}_{11}^{MR})^{-1} \mathbf{b}_{0}^{R}.
\end{aligned} \tag{5.83}$$

Now, besides the possibility that $\mathbf{S}_{\alpha_{R}\alpha_{R}}^{E} \neq \mathbf{0}$, if the receive DN is designed according to the optimal criteria with $\mathbf{S}_{11}^{MR} = (\mathbf{S}_{\alpha_{R}\alpha_{R}}^{E})^{H}$ (and therefore, $\mathbf{\Gamma}_{out}^{R} = \mathbf{0}$), then the collected power is maximized and given by

$$\begin{aligned}
P^{D,0,\max} &= (\mathbf{b}_{0}^{R})^{H} \left(\left(\mathbf{I} - \mathbf{S}_{\alpha_{R}\alpha_{R}}^{E} (\mathbf{S}_{\alpha_{R}\alpha_{R}}^{E})^{H} \right)^{-1} \right)^{H} \mathbf{b}_{0}^{R} \\
&= (\mathbf{b}_{0}^{R})^{H} \mathbf{W}((\mathbf{S}_{\alpha_{R}\alpha_{R}}^{E})^{H}) \mathbf{b}_{0}^{R}.
\end{aligned} \tag{5.84}$$

Finally, the matching efficiency for the receive multiantenna system, η^{R}, can be defined in the same manner as the relative collected power in Ref. [14], that is, as the ratio of the total received load power for the multiple antenna system in the case that $\mathbf{\Gamma}_{\alpha_{R}}^{D} = \mathbf{0}$ to that of an optimally matched reference single antenna in the same environment:

$$\eta^{R,0} = \frac{E\{P^{D,0}\}}{E\{P^{D,0,\max}\}}. \tag{5.85}$$

Similarly, as with the transmitter, if the receive DN is optimally designed, then $\eta^{R,0} \to 1$. For a generic termination of the loads, the efficiency can be rewritten as follows:

$$\eta^{R} = \frac{E\{P^{D}\}}{E\{P^{D,\max}\}}. \tag{5.86}$$

5.6.2   Superextended Channel Correlation

As in Section 4.7.3, we can define the superextended channel matrix correlation matrix $\mathcal{R}^M \in \mathcal{C}^{MN \times MN}$ as

$$\mathcal{R}^M = E\{\mathbf{vec}(\mathbf{M})\mathbf{vec}(\mathbf{M})^H\}, \tag{5.87}$$

where $\mathbf{vec}(\mathbf{M}) \in \mathcal{C}^{MN \times 1}$. This correlation matrix represents the correlation characteristics among different input ports of the transmit DN and the output ports of the receive DN, that is, by taking into account the propagation of the signals over the channel, the radiation characteristics of the transmit and receive MPAs, the characteristics of the DNs, and the termination characteristics of the source and drain. For the normalized version of the superextended channel matrix correlation matrix $\mathcal{R}_0^M \in \mathcal{C}^{MN \times MN}$, we can write

$$\mathcal{R}_0^M = E\{\mathbf{vec}(\mathbf{M}_0)\mathbf{vec}(\mathbf{M}_0)^H\}. \tag{5.88}$$

On the other hand, we can also define the superextended transmit antenna correlation matrix, $\mathbf{R}_M^T$, and the superextended receive antenna correlation matrix, $\mathbf{R}_M^R$, as follows:

$$\mathbf{R}_M^R = \frac{E\{\mathbf{M}(\mathbf{M})^H\}}{\sqrt{E\{\|\mathbf{M}\|_F^2\}}} \tag{5.89}$$

$$\mathbf{R}_M^T = \frac{E\{(\mathbf{M})^H\mathbf{H}\}}{\sqrt{E\{\|\mathbf{M}\|_F^2\}}}, \tag{5.90}$$

where $\mathbf{R}_{M_{ij}}^T$ is the correlation value between the ith and jth input ports of the transmit DN, whereas $\mathbf{R}_{M_{ij}}^R$ is the correlation value between the ith and jth output ports of the receive DN, including the effects of source and drain termination.

Similarly, as in Section 4.7.3, besides the fact that this separation is always possible, in general, it may not be possible to construct the resultant channel correlation matrix from the transmit and receive antenna correlation matrices. This is because the channel statistics at the transmitter are dependent on those at the receiver, and vice versa. In most of the rich scattered NLOS propagation scenarios, however, the random fading processes at the receiver are uncorrelated to those at the transmitter. In that case, a Kronecker channel model [18] can be used. As a result, it is much easier to evaluate how correlation may impact on the performance of a MIMO communication system. Let us assume a Kronecker channel model for now to provide an intuitive physical insight into the correlation concept. Then, a good approximation for $\mathcal{R}^M$ is given by

$$\mathcal{R}^M = \mathbf{R}_M^T \otimes \mathbf{R}_M^R, \tag{5.91}$$

and the superextended channel matrix $\mathbf{M}$ can be decomposed into

$$\mathbf{M} = (\mathbf{R}_M^R)^{\frac{1}{2}} \mathbf{W} ((\mathbf{R}_M^T)^{\frac{1}{2}})^\dagger, \tag{5.92}$$

where $\mathbf{W} \in \mathcal{C}^{N \times M}$ is a complex random matrix with all elements being iid Gaussian random variables with distribution $N(0, 1)$. Notice that $\mathbf{R}_M^T = (\mathbf{R}_M^T)^{\frac{1}{2}} ((\mathbf{R}_M^T)^{\frac{1}{2}})^H$ and $\mathbf{R}_M^R = (\mathbf{R}_M^R)^{\frac{1}{2}} ((\mathbf{R}_M^R)^{\frac{1}{2}})^H$.

5.6.3   Superextended Transmit and Receive Antenna Correlation

Let us first use the separated model for the transmitter and receiver described in Section 4.4.3 to define the covariance of the transmitted signals in the superextended channel, $\mathbf{X}_M^T$, as

$$\mathbf{X}_M^T = E\{\mathbf{U}_{\alpha_1}^{MT} (\mathbf{U}_{\alpha_1}^{MT})^H\} \tag{5.93}$$

and the covariance of the received signals in the superextended channel, $\mathbf{X}_M^R$, as

$$\mathbf{X}_M^R = E\{\mathbf{U}_{\alpha_2}^{MR} (\mathbf{U}_{\alpha_2}^{MR})^H\}. \tag{5.94}$$

The (i,j)th entry of the superextended transmit antenna correlation matrix $\mathbf{R}_M^T$ can be computed as

$$\mathbf{R}_{M_{ij}}^T = \frac{\mathbf{X}_{M_{ij}}^T}{\sqrt{\mathbf{X}_{M_{ii}}^T \mathbf{X}_{M_{jj}}^T}}, \tag{5.95}$$

whereas for the superextended receive antenna correlation matrix $\mathbf{R}_M^R$, it can be calculated using

$$\mathbf{R}_{M_{ij}}^R = \frac{\mathbf{X}_{M_{ij}}^R}{\sqrt{\mathbf{X}_{M_{ii}}^R \mathbf{X}_{M_{jj}}^R}}. \tag{5.96}$$

We derive now a general expression for $\mathbf{X}_M^R$ for a generic load termination scenario in which $\mathbf{\Gamma}_{\alpha_R}^D \neq \mathbf{0}$ and includes a specific lossless DN ($\mathbf{S}^{MR}$). To do this, we define first the following parameter $\mathbf{Q} = Z_{\alpha_2}^{0,MR \frac{1}{2}} \left(\mathbf{I} + \mathbf{\Gamma}_{\alpha_R}^D\right) \left(\mathbf{I} - \mathbf{S}_{22}^{MR} \mathbf{\Gamma}_{\alpha_R}^D\right)^{-1} \mathbf{S}_{21}^{MR} \left(\mathbf{I} - \mathbf{S}_{\alpha_R \alpha_R}^E \mathbf{S}_{11}^{MR}\right)^{-1}$. Then $\mathbf{X}_M^R$ is given by

$$\mathbf{X}_M^R = \mathbf{Q} \mathbf{X}_H^R (\mathbf{Q})^H. \tag{5.97}$$

A similar analysis can easily be derived for the extended transmit antenna correlation. We first derive a generic expression for $\mathbf{X}_M^T$ for a general source termination scenario in which $\mathbf{\Gamma}_{\alpha_T}^S \neq \mathbf{0}$. We begin considering once again the transmitter as a receiving structure and the fact that

$$\mathbf{U}_{\alpha_T}^{MT} = Z_{\alpha_2}^{0,MT \frac{1}{2}} \left(\mathbf{I} + \mathbf{\Gamma}_{\alpha_T}^S\right) \left(\mathbf{I} - \mathbf{S}_{11}^{MR} \mathbf{\Gamma}_{\alpha_T}^S\right)^{-1} \mathbf{S}_{12}^{MR} \left(\mathbf{I} - \mathbf{S}_{\alpha_T \alpha_T}^E \mathbf{S}_{22}^{MR}\right)^{-1} \mathbf{b}_0^T. \tag{5.98}$$

Let us define $\mathbf{P} = Z_{\alpha_2}^{0,MT\frac{1}{2}} \left(\mathbf{I} + \mathbf{\Gamma}_{\alpha_T}^S\right) \left(\mathbf{I} - \mathbf{S}_{11}^{MR}\mathbf{\Gamma}_{\alpha_T}^S\right)^{-1} \mathbf{S}_{12}^{MR} \left(\mathbf{I} - \mathbf{S}_{\alpha_T\alpha_T}^E\mathbf{S}_{22}^{MR}\right)^{-1}$. Then $\mathbf{X}_M^T$ is given by

$$\mathbf{X}_M^T = \mathbf{P}\mathbf{X}_H^T(\mathbf{P})^H. \tag{5.99}$$

5.6.4   Normalized Far-Field Radiation Patterns at the DN Ports

Using the finding from Section 5.4.2, we can find an expression for the resultant normalized far-field radiation patterns at the input ports of the transmit DN and at the output ports of the receive DN. To do this, we assume $\mathbf{\Gamma}_{\alpha_T}^S = \mathbf{0}$ and $\mathbf{\Gamma}_{\alpha_R}^D = \mathbf{0}$. The vector containing the radiation patterns observed at the output ports of the receive DN can be constructed as [75]

$$\mathbf{F}^{MR}(\theta,\phi) = \mathbf{S}_{21}^{MR}(\mathbf{I} - \mathbf{S}_{\alpha_R\alpha_R}^E\mathbf{S}_{11}^{MR})^{-1}\mathbf{F}^R(\theta,\phi). \tag{5.100}$$

At the input ports of the transmit DN, the patterns can be constructed from

$$\mathbf{F}^{MT}(\theta,\phi) = \mathbf{S}_{12}^{MR}(\mathbf{I} - \mathbf{S}_{\alpha_T\alpha_T}^E\mathbf{S}_{22}^{MR})^{-1}\mathbf{F}^T(\theta,\phi). \tag{5.101}$$

On the other hand, from the unitarity conditions of reciprocal lossless MPAs, the following condition is satisfied:

$$(\mathbf{S}_{\beta\alpha}^A)^H\mathbf{S}_{\beta\alpha}^A = \mathbf{I} - (\mathbf{S}_{\alpha\alpha}^A)^H\mathbf{S}_{\alpha\alpha}^A, \tag{5.102}$$

which clearly shows that any modification in the antenna input parameters, that is, in the subblock scattering matrix $\mathbf{S}_{\alpha\alpha}^A$, using an external network and others, implies also a modification in the radiation pattern given by the matrix $\mathbf{S}_{\beta\alpha}^A$. One should know that because of the presence of mutual coupling, the radiation patterns associated with the ports of the MPAs, $\mathbf{F}^R(\theta,\phi)$ and $\mathbf{F}^T(\theta,\phi)$, may not be orthogonal. However, one can use appropriately designed DNs, satisfying the optimal condition of multiport conjugate match, such that the radiation patterns seen at the ports of the receive DN $\mathbf{F}^{MR}(\theta,\phi)$ and transmit DN $\mathbf{F}^{MT}(\theta,\phi)$ will be orthogonal.

5.6.5   System Bandwidth

As shown in Ref. [69], although the optimal design goal for narrow-band system can always be achieved by means of antenna diversity techniques or the use of DNs, wideband matching has fundamental theoretical as well as practical limits. That is, the effective bandwidth seen at the ports of the MPAs is determined by the fundamental limitations of antennas on its Q factor, as explained in Section 4.8, as well as the Fano limit [56] in matching coupled antennas. For a preliminary study on the impact of wideband matching on the performance of a MIMO system, refer to Refs. [69, 76].

Following the approach on Ref. [76], to analyze the spectral efficiency of a volume, it is, however, essential to relate three classes of theories giving fundamental limitations in the disciplines of information theory, broadband matching, and antenna theory. The capacity depends on the number and gain of the orthogonal subchannels (degrees of freedom) and their signal-to-noise ratio. Of course, these channels depend on the choice of antenna, matching, transmission lines, and the scattering characteristics of the channel. The classical theory of broadband matching, developed by Fano, shows how much power can be transmitted over a certain bandwidth between a transmission line and a given load or antenna. In Ref. [76], the classical theory of radiation Q also uses the spherical vector modes to analyze the properties of a hypothetical antenna inside a sphere. An antenna with a high Q factor has electromagnetic fields with large amounts of stored energy around them, and hence, typically low bandwidth and high losses. It is known from Fano's broadband matching theory that the best reflection coefficient $S_{\alpha\alpha}^A$ that can be achieved over a frequency range is restricted by the integral:

$$\int_0^\infty \ln \frac{1}{|S_{\alpha\alpha}^A|} d\omega \leq \frac{\pi}{RC}. \qquad (5.103)$$

This expression expresses the fact that the area under the curve $\ln \frac{1}{|S_{\alpha\alpha}^A|}$ cannot be greater than $\frac{\pi}{RC}$. Therefore, if matching is required over a certain bandwidth, it can only be obtained at the expense of power transfer, that is, a reduction of the SNR.

Typically, in antenna theory, the Q factor of an antenna is estimated from the energy stored in the equivalent circuit and the power radiated and has been freely interpreted as the reciprocal of the fractional bandwidth. To be more accurate, however, one must define the bandwidth in terms of allowable impedance variation or the tolerable reflection coefficient over the band; typically, a VSWR < 2 is chosen. For a given antenna, the bandwidth can be increased by choosing a proper DN; however, this is at the expense of restrictions on the power transfer. This is known to impact the SNR and ultimately the channel capacity.

The spherical modes associated with a particular radiation pattern of the MPA have a circuit equivalent representing the impedance of the modes [76]. On them, the resistance R, the capacitance C, and inductance L are the circuit equivalents of the radiated field, the stored electric field, and the stored magnetic field, respectively. These circuits can be interpreted as high-pass filters, and the Fano theory can be used to get the fundamental limitations on the DN for each one of these spherical modes. In Ref. [76], an expression for the capacity as the summation of the contributions due to each spherical mode including the matching limitations is obtained, and from it, it is possible to estimate the overall channel capacity given the number and order of the spherical modes that are being used (either by the antenna array of the channel response). This analysis is not limited to electrically small antennas—that is, it can be used for large antenna arrays and uses a strict definition of the bandwidth based on the VSWR < 2.

The transmission lines or networks that connect the electrical signals on the input ports with the radiated electromagnetic waves outside the antenna are, in general, very complex, and they can be used for matching, decoupling, and decorrelation purposes, as well as to perform beam forming, for example, using a Butler matrix network. Therefore, most of the time, the spherical modes cannot be accessed individually from a port but through a linear combination of them. In such case, the limitation on the bandwidth is given by the combined effect of the bandwidth limitation of the spherical modes involved.

5.7   DECORRELATING NETWORKS

The design of DNs is based on the condition for optimum matching of multiport networks, known as multiport conjugate match. The multiport conjugate match is a well-known condition [14, 66], and its practical implementation is a subject of current interest [77, 78]. One can, for example, consider implementing the optimum matching condition following the approach in Ref. [78], which uses distributed elements (couplers, transmission lines, and open-circuited stubs), or on the other hand, one could also use lumped elements [77], among other possibilities.

Antenna engineers then use DNs to achieve a multiport conjugate match between the transmit antennas and the source or between the receive antennas and the drain. In this section, a design methodology is presented for two types of DNs:

- lossless DNs
- lossless and reciprocal DNs.

In both approaches, we assume the following source and load terminations: $\mathbf{\Gamma}^S_{\alpha_T} = \mathbf{0}$ and $\mathbf{\Gamma}^D_{\alpha_R} = \mathbf{0}$. The design methodology will show how the characteristics of the MPAs impact on the design of DNs.

5.7.1   Lossless DNs

Lossless DNs are of particular interest because they may be constructed from all passive components exhibiting low-noise figures.

5.7.1.1   Properties of Lossless DNs.   Lossless DNs networks must satisfy the unitarity condition given by

$$(\mathbf{S}^{MA})^H \mathbf{S}^{MA} = \mathbf{I}, \tag{5.104}$$

where the upper index MA refers to either the transmit (MT) or receive (MR) DN. The design of DNs undergoes the use of SVD to factorize the subblock matrices of $\mathbf{S}^{MA}$. The SVD is based on the following theorem: for a complex matrix $\mathbf{A}$, the SVD is a decomposition into the form $\mathbf{A} = \mathbf{U}\mathbf{\Lambda}^{\frac{1}{2}}\mathbf{V}$, where $\mathbf{U}$ and $\mathbf{V}$ are unitary matrices, and $\mathbf{\Lambda}^{\frac{1}{2}}$ is a diagonal matrix of real singular values of the original matrix. For a complex matrix $\mathbf{A}$, there always exists such a decomposition with positive singular values.

As shown in Ref. [66], during the design of a lossless DN, only the subblocks $\mathbf{S}_{11}^{MA}$ and $\mathbf{S}_{22}^{MA}$ need to be specified because the subblocks $\mathbf{S}_{12}^{MA}$ and $\mathbf{S}_{21}^{MA}$ can be obtained from the following set of equations:

$$(\mathbf{S}_{11}^{MA})^H \mathbf{S}_{11}^{MA} + (\mathbf{S}_{21}^{MA})^H \mathbf{S}_{21}^{MA} = \mathbf{I} \tag{5.105}$$

$$(\mathbf{S}_{12}^{MA})^H \mathbf{S}_{12}^{MA} + (\mathbf{S}_{22}^{MA})^H \mathbf{S}_{22}^{MA} = \mathbf{I} \tag{5.106}$$

$$(\mathbf{S}_{11}^{MA})^H \mathbf{S}_{12}^{MA} + (\mathbf{S}_{21}^{MA})^H \mathbf{S}_{22}^{MA} = \mathbf{0} \tag{5.107}$$

$$(\mathbf{S}_{12}^{MA})^H \mathbf{S}_{11}^{MA} + (\mathbf{S}_{22}^{MA})^H \mathbf{S}_{21}^{MA} = \mathbf{0}. \tag{5.108}$$

Notice that there is an entire family of DNs that satisfies these conditions. Substitution of the SVD of the subblocks $\mathbf{S}_{ij}^{MA} = \mathbf{U}_{ij}^{MA}(\mathbf{\Lambda}_{ij}^{MA})^{\frac{1}{2}}(\mathbf{V}_{ij}^{MA})^H$ into the Eqs. (5.105) and (5.106) yields

$$\mathbf{V}_{21}^{MA}\mathbf{\Theta}_{21}^{MA} = \mathbf{V}_{11}^{MA} \tag{5.109}$$

$$\mathbf{V}_{12}^{MA}\mathbf{\Theta}_{12}^{MA} = \mathbf{V}_{22}^{MA} \tag{5.110}$$

$$\mathbf{\Lambda}_{21}^{MA} = \mathbf{I} - \mathbf{\Lambda}_{11}^{MA} \tag{5.111}$$

$$\mathbf{\Lambda}_{12}^{MA} = \mathbf{I} - \mathbf{\Lambda}_{22}^{MA}, \tag{5.112}$$

where $\mathbf{\Theta}_{21}^{MA}$ and $\mathbf{\Theta}_{12}^{MA}$ are diagonal phase-shift matrices with arbitrary complex elements of unit magnitude. On the other hand, as first mentioned in Ref. [66], an additional condition results from the substitution of the SVD of the subblocks in Eq. (5.107), given by

$$\mathbf{\Theta}_{21}^{MA}(\mathbf{\Lambda}_{11}^{MA})^{\frac{1}{2}}(\mathbf{U}_{11}^{MA})^H\mathbf{U}_{12}^{MA}(\mathbf{I} - \mathbf{\Lambda}_{22}^{MA})^{\frac{1}{2}} = -(\mathbf{I} - \mathbf{\Lambda}_{11}^{MA})^{\frac{1}{2}}(\mathbf{U}_{21}^{MA})^H\mathbf{U}_{22}^{MA}(\mathbf{\Lambda}_{22}^{MA})^{\frac{1}{2}}(\mathbf{\Theta}_{12}^{MA})^H, \tag{5.113}$$

where we have used the fact that $((\mathbf{\Lambda}_{11}^{MA})^{\frac{1}{2}})^H = (\mathbf{\Lambda}_{11}^{MA})^{\frac{1}{2}}$ and $((\mathbf{I} - \mathbf{\Lambda}_{11}^{MA})^{\frac{1}{2}})^H = (\mathbf{I} - \mathbf{\Lambda}_{11}^{MA})^{\frac{1}{2}}$. The above is a procedure to find these subblock matrices based on the SVD, which does not impose reciprocity on the resultant network. Using the previous findings, the subblock matrices of $\mathbf{S}^{MA}$ can be written in the form

$$\mathbf{S}_{11}^{MA} = \mathbf{U}_{11}^{MA}(\mathbf{\Lambda}_{11}^{MA})^{\frac{1}{2}}(\mathbf{V}_{11}^{MA})^H \tag{5.114}$$

$$\mathbf{S}_{12}^{MA} = \mathbf{U}_{12}^{MA}(\mathbf{I} - \mathbf{\Lambda}_{22}^{MA})^{\frac{1}{2}}\Theta_{12}^{MA}(\mathbf{V}_{22}^{MA})^H \tag{5.115}$$

$$\mathbf{S}_{21}^{MA} = \mathbf{U}_{21}^{MA}(\mathbf{I} - \mathbf{\Lambda}_{11}^{MA})^{\frac{1}{2}}\Theta_{21}^{MA}(\mathbf{V}_{11}^{MA})^H \tag{5.116}$$

$$\mathbf{S}_{22}^{MA} = \mathbf{U}_{22}^{MA}(\mathbf{\Lambda}_{22}^{MA})^{\frac{1}{2}}(\mathbf{V}_{22}^{MA})^H, \tag{5.117}$$

where Θ_{12}^{MA} and Θ_{21}^{MA} are diagonal matrices with complex unit-magnitude entries, which must satisfy Eq. (5.113).

Because we are interested in finding one lossless DN that achieves the specified design goals, we can further simplify Eqs. (5.114) to (5.117). We normally choose $\mathbf{U}_{12}^{MA} = \mathbf{U}_{11}^{MA}$ and $\mathbf{U}_{21}^{MA} = \mathbf{U}_{22}^{MA}$. Then, according to Eq. (5.113), we obtain $\Theta_{21}^{MA} = -(\Theta_{12}^{MA})^H = \Theta^{MA}$ (therefore, $\Theta_{12}^{MA} = (\Theta^{MA})^H$) and $\mathbf{\Lambda}_{22}^{MA} = \mathbf{\Lambda}_{11}^{MA}$. The subblocks of $\mathbf{S}^{MA}$ can then be expressed as

$$\mathbf{S}_{11}^{MA} = \mathbf{U}_{11}^{MA}(\mathbf{\Lambda}_{11}^{MA})^{\frac{1}{2}}(\mathbf{V}_{11}^{MA})^H \tag{5.118}$$

$$\mathbf{S}_{12}^{MA} = -\mathbf{U}_{11}^{MA}(\Theta^{MA})^H(\mathbf{I} - \mathbf{\Lambda}_{11}^{MA})^{\frac{1}{2}}(\mathbf{V}_{22}^{MA})^H \tag{5.119}$$

$$\mathbf{S}_{21}^{MA} = \mathbf{U}_{22}^{MA}\Theta^{MA}(\mathbf{I} - \mathbf{\Lambda}_{11}^{MA})^{\frac{1}{2}}(\mathbf{V}_{11}^{MA})^H \tag{5.120}$$

$$\mathbf{S}_{22}^{MA} = \mathbf{U}_{22}^{MA}(\mathbf{\Lambda}_{11}^{MA})^{\frac{1}{2}}(\mathbf{V}_{22}^{MA})^H. \tag{5.121}$$

5.7.1.2 Decomposition of Lossless Networks.

As shown in Ref. [79], a lossless $2n$-port network can be considered to be the cascade of three subnetworks, the end ones being all-passes and the middle one being a transformer bank of n uncoupled 2-port ideal transformers (Fig. 5.9). The networks are interconnected by transmission lines of arbitrary length. The two $2n$-port all-pass networks, also named decoupling networks, have a scattering matrix of the form

$$\mathbf{S}^L = \begin{pmatrix} \mathbf{0} & \mathbf{U}_{11}^{MA} \\ (\mathbf{V}_{11}^{MA})^H & \mathbf{0} \end{pmatrix} \tag{5.122}$$

$$\mathbf{S}^R = \begin{pmatrix} \mathbf{0} & (\mathbf{V}_{22}^{MA})^H \\ \mathbf{U}_{22}^{MA} & \mathbf{0} \end{pmatrix}, \tag{5.123}$$

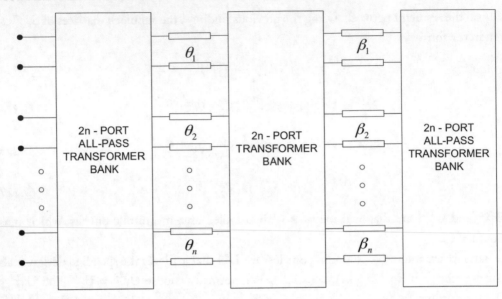

FIGURE 5.9: The cascade equivalent circuit of a lossless, reciprocal $2n$ ports in terms of $2n$ uncoupled lines and three real $2n$-port networks.

where superindices L and R refer to the decoupling networks located to the left and right sides, respectively, with respect to the bank of transformers. The bank of ideal uncoupled transformers has a scattering matrix of the form

$$\mathbf{S}^U = \begin{pmatrix} (\boldsymbol{\Lambda}_{11}^{MA})^{\frac{1}{2}} & -(\boldsymbol{\Theta}^{MA})^H (\mathbf{I} - \boldsymbol{\Lambda}_{11}^{MA})^{\frac{1}{2}} \\ \boldsymbol{\Theta}^{MA}(\mathbf{I} - \boldsymbol{\Lambda}_{11}^{MA})^{\frac{1}{2}} & (\boldsymbol{\Lambda}_{11}^{MA})^{\frac{1}{2}} \end{pmatrix}. \qquad (5.124)$$

Using the above subnetworks, the scattering matrix $\mathbf{S}^{MA}$ can be seen as the result of performing the following operation:

$$\mathbf{S}^{MA} = \mathbf{S}^L \odot \mathbf{S}^U \odot \mathbf{S}^R, \qquad (5.125)$$

where $\odot$ represents the concatenation of two scattering matrices as described in Section 4.4.1.

5.7.1.3 Implementation of Decorrelating Lossless Networks. From Section 5.5.2, let us recall the optimal multiport conjugate match condition at the transmit and receive side. We first assume $\boldsymbol{\Gamma}^S = \mathbf{0}$ and $\boldsymbol{\Gamma}^D = \mathbf{0}$. In that case, the conditions are

$$\mathbf{S}_{22}^{MT} = (\mathbf{S}_{\alpha_T \alpha_T}^E)^H \qquad (5.126)$$

$$\mathbf{S}_{11}^{MR} = (\mathbf{S}_{\alpha_R \alpha_R}^{E})^{H}. \tag{5.127}$$

It is important to realize that the multiport conjugate match condition only needs to be satisfied in one of the port groups of the DN. That is, only one of the subblock matrices of the DN may need to be specified. In that case, the complexity of the resultant DNs can be reduced. In fact, they can be made by using one decoupling subnetwork and one bank of transformers subnetwork, as shown:

$$\mathbf{S}^{MT} = \mathbf{S}^{U} \odot \mathbf{S}^{R} \tag{5.128}$$

$$\mathbf{S}^{MR} = \mathbf{S}^{L} \odot \mathbf{S}^{U}. \tag{5.129}$$

We can now proceed to implement the DNs such that the aforementioned conditions are satisfied. We first perform the SVD of $\mathbf{S}_{\alpha_T \alpha_T}^{E}$ and $\mathbf{S}_{\alpha_R \alpha_R}^{E}$, given by

$$\mathbf{S}_{\alpha_T \alpha_T}^{T} = \mathbf{U}^{T} \mathbf{\Lambda}_{T}^{\frac{1}{2}} (\mathbf{V}^{T})^{H} \tag{5.130}$$

$$\mathbf{S}_{\alpha_R \alpha_R}^{R} = \mathbf{U}^{R} \mathbf{\Lambda}_{R}^{\frac{1}{2}} (\mathbf{V}^{R})^{H}. \tag{5.131}$$

For the transmit side, one must pick

$$\mathbf{U}_{22}^{MT} = \mathbf{V}^{T} \tag{5.132}$$

$$\mathbf{V}_{22}^{MT} = \mathbf{U}^{T} \tag{5.133}$$

$$(\mathbf{\Lambda}_{22}^{MT})^{\frac{1}{2}} = (\mathbf{\Lambda}_{T}^{\frac{1}{2}})^{H} = \mathbf{\Lambda}_{T}^{\frac{1}{2}}. \tag{5.134}$$

Similarly, for the receiving side, one must pick

$$\mathbf{U}_{11}^{MR} = \mathbf{V}^{R} \tag{5.135}$$

$$\mathbf{V}_{11}^{MR} = \mathbf{U}^{R} \tag{5.136}$$

$$(\mathbf{\Lambda}_{11}^{MR})^{\frac{1}{2}} = (\mathbf{\Lambda}_{R}^{\frac{1}{2}})^{H} = \mathbf{\Lambda}_{R}^{\frac{1}{2}}. \tag{5.137}$$

Notice that in both cases, the remaining parameters can be chosen arbitrarily. In Ref. [66], it is shown that these degrees of freedom can be used to minimize the antenna correlation in environments without full angular spread or to match the antennas to other devices such as amplifiers [75] and others instead of loads terminated with the characteristic reference impedance.

5.7.2 Lossless Reciprocal DNs

Lossless DNs are of particular interest because they may be constructed from all passive reciprocal components exhibiting low-noise figures. In this section, we limit our study to the case of reciprocal MPAs for which $\mathbf{S}_{\alpha\alpha}^{A}$ is a complex symmetrical matrix.

5.7.2.1 Properties of Lossless and Reciprocal Decorrelating Networks.

Should the DN be lossless and reciprocal simultaneously, additional constraints need to be placed. If the microwave network is to be lossless and reciprocal, it must satisfy

$$(\mathbf{S}^{MA})^{H}\mathbf{S}^{MA} = \mathbf{I} \tag{5.138}$$

$$\mathbf{S}^{MA} = (\mathbf{S}^{MA})^{\dagger}. \tag{5.139}$$

Similarly as before, only the subblocks $\mathbf{S}_{11}^{MA}$ and $\mathbf{S}_{22}^{MA}$ need to be specified, because the subblocks $\mathbf{S}_{12}^{MA}$ and $\mathbf{S}_{21}^{MA}$ can be obtained from the following set of equations:

$$(\mathbf{S}_{11}^{MA})^{H}\mathbf{S}_{11}^{MA} + (\mathbf{S}_{21}^{MA})^{H}\mathbf{S}_{21}^{MA} = \mathbf{I} \tag{5.140}$$

$$(\mathbf{S}_{12}^{MA})^{H}\mathbf{S}_{12}^{MA} + (\mathbf{S}_{22}^{MA})^{H}\mathbf{S}_{22}^{MA} = \mathbf{I} \tag{5.141}$$

$$(\mathbf{S}_{11}^{MA})^{H}\mathbf{S}_{12}^{MA} + (\mathbf{S}_{21}^{MA})^{H}\mathbf{S}_{22}^{MA} = \mathbf{0} \tag{5.142}$$

$$(\mathbf{S}_{12}^{MA})^{H}\mathbf{S}_{11}^{MA} + (\mathbf{S}_{22}^{MA})^{H}\mathbf{S}_{21}^{MA} = \mathbf{0} \tag{5.143}$$

$$\mathbf{S}_{11}^{MA} = (\mathbf{S}_{11}^{MA})^{T} \tag{5.144}$$

$$\mathbf{S}_{12}^{MA} = (\mathbf{S}_{12}^{MA})^{T} \tag{5.145}$$

$$\mathbf{S}_{21}^{MA} = (\mathbf{S}_{21}^{MA})^{T} \tag{5.146}$$

$$\mathbf{S}_{22}^{MA} = (\mathbf{S}_{22}^{MA})^{T}. \tag{5.147}$$

By inspection of Eqs. (5.118) to (5.121), one realizes that for them to satisfy the additional requirements for reciprocity one must select $\mathbf{U}_{11}^{MA} = (\mathbf{V}_{11}^{MA})^{*}$, $\mathbf{U}_{22}^{MA} = (\mathbf{V}_{22}^{MA})^{*}$, and $\Theta^{MA} = j$. In this case, the decomposition of the subblock matrices of $\mathbf{S}^{MA}$ is given by

$$\mathbf{S}_{11}^{MA} = \mathbf{U}_{11}^{MA}(\mathbf{\Lambda}_{11}^{MA})^{\frac{1}{2}}(\mathbf{U}_{11}^{MA})^{T} \tag{5.148}$$

$$\mathbf{S}_{12}^{MA} = \mathbf{U}_{11}^{MA} j (\mathbf{I} - \mathbf{\Lambda}_{11}^{MA})^{\frac{1}{2}} (\mathbf{U}_{22}^{MA})^T \tag{5.149}$$

$$\mathbf{S}_{21}^{MA} = \mathbf{U}_{22}^{MA} j (\mathbf{I} - \mathbf{\Lambda}_{11}^{MA})^{\frac{1}{2}} (\mathbf{U}_{11}^{MA})^T \tag{5.150}$$

$$\mathbf{S}_{22}^{MA} = \mathbf{U}_{22}^{MA} (\mathbf{\Lambda}_{11}^{MA})^{\frac{1}{2}} (\mathbf{U}_{22}^{MA})^T. \tag{5.151}$$

On the other hand, the product of a unitary matrix $\mathbf{V}$ and a diagonal matrix with complex entries of unit magnitude Θ produces another unitary matrix $\mathbf{U} = \mathbf{V}\Theta$. Let us define $\mathbf{U}_{11}^{MA} = \mathbf{V}_{11}^{MA}\Theta_{11}^{MA}$ and $\mathbf{U}_{22}^{MA} = \mathbf{V}_{22}^{MA}\Theta_{22}^{MA}$. In that case, the previous equations can be expressed as

$$\mathbf{S}_{11}^{MA} = \mathbf{V}_{11}^{MA} (\Theta_{11}^{MA})^2 (\mathbf{\Lambda}_{11}^{MA})^{\frac{1}{2}} (\mathbf{V}_{11}^{MA})^T = \mathbf{V}_{11}^{MA} \mathbf{D}_{11}^{MA} (\mathbf{V}_{11}^{MA})^T \tag{5.152}$$

$$\mathbf{S}_{12}^{MA} = \mathbf{V}_{11}^{MA} j \Theta_{11}^{MA} (\mathbf{I} - \mathbf{\Lambda}_{11}^{MA})^{\frac{1}{2}} \Theta_{22}^{MA} (\mathbf{V}_{22}^{MA})^T = \mathbf{V}_{11}^{MA} \mathbf{D}^{MA} (\mathbf{V}_{22}^{MA})^T \tag{5.153}$$

$$\mathbf{S}_{21}^{MA} = \mathbf{V}_{22}^{MA} j \Theta_{22}^{MA} (\mathbf{I} - \mathbf{\Lambda}_{11}^{MA})^{\frac{1}{2}} \Theta_{11}^{MA} (\mathbf{V}_{11}^{MA})^T = \mathbf{V}_{22}^{MA} \mathbf{D}^{MA} (\mathbf{V}_{11}^{MA})^T \tag{5.154}$$

$$\mathbf{S}_{22}^{MA} = \mathbf{V}_{22}^{MA} (\Theta_{22}^{MA})^2 (\mathbf{\Lambda}_{11}^{MA})^{\frac{1}{2}} (\mathbf{V}_{22}^{MA})^T = \mathbf{V}_{22}^{MA} \mathbf{D}_{22}^{MA} (\mathbf{V}_{22}^{MA})^T, \tag{5.155}$$

where $\mathbf{D}_{11}^{MA} = (\Theta_{11}^{MA})^2 (\mathbf{\Lambda}_{11}^{MA})^{\frac{1}{2}}$, $\mathbf{D}_{22}^{MA} = (\Theta_{22}^{MA})^2 (\mathbf{\Lambda}_{11}^{MA})^{\frac{1}{2}}$ and $\mathbf{D}^{MA} = j\Theta_{11}^{MA}\Theta_{22}^{MA}(\mathbf{I} - \mathbf{\Lambda}_{11}^{MA})^{\frac{1}{2}}$. Eqs. (5.152) to (5.155) are equivalent expressions to those given from Eqs. (5.148) to (5.151), in which the diagonal inner matrices are allowed to contain complex numbers.

5.7.2.2 Decomposition of Lossless Reciprocal Networks. In this case, and associated with Eqs. (5.148) to (5.151), the two $2n$-port all-pass networks, also named decoupling networks, have a scattering matrix of the form

$$\mathbf{S}^L = \begin{pmatrix} \mathbf{0} & \mathbf{U}_{11}^{MA} \\ (\mathbf{U}_{11}^{MA})^\dagger & \mathbf{0} \end{pmatrix} \tag{5.156}$$

$$\mathbf{S}^R = \begin{pmatrix} \mathbf{0} & (\mathbf{U}_{22}^{MA})^\dagger \\ \mathbf{U}_{22}^{MA} & \mathbf{0} \end{pmatrix}, \tag{5.157}$$

where superindex L and R refer to the decoupling networks located to the left and right sides, respectively, of the bank of transformers. The bank of ideal uncoupled transformers has a scattering matrix of the form

$$\mathbf{S}^U = \begin{pmatrix} (\mathbf{\Lambda}_{11}^{MA})^{\frac{1}{2}} & j(\mathbf{I} - \mathbf{\Lambda}_{11}^{MA})^{\frac{1}{2}} \\ j(\mathbf{I} - \mathbf{\Lambda}_{11}^{MA})^{\frac{1}{2}} & (\mathbf{\Lambda}_{11}^{MA})^{\frac{1}{2}} \end{pmatrix}. \tag{5.158}$$

On the other hand, associated with Eqs. (5.152) to (5.155), the two $2n$-port all-pass networks, also called decoupling networks, have a scattering matrix of the form

$$S^L = \begin{pmatrix} \mathbf{0} & \mathbf{V}_{11}^{MA} \\ (\mathbf{V}_{11}^{MA})^\dagger & \mathbf{0} \end{pmatrix} \qquad (5.159)$$

$$S^R = \begin{pmatrix} \mathbf{0} & (\mathbf{V}_{22}^{MA})^\dagger \\ \mathbf{V}_{22}^{MA} & \mathbf{0} \end{pmatrix}. \qquad (5.160)$$

The bank of ideal uncoupled transformers has a scattering matrix of the form

$$S^U = \begin{pmatrix} \mathbf{D}_{11}^{MA} & \mathbf{D}^{MA} \\ \mathbf{D}^{MA} & \mathbf{D}_{22}^{MA} \end{pmatrix}. \qquad (5.161)$$

Using the above subnetworks, the scattering matrix $\mathbf{S}^{MA}$ can be seen as the result of performing the following operation:

$$\mathbf{S}^{MA} = \mathbf{S}^L \odot \mathbf{S}^U \odot \mathbf{S}^R, \qquad (5.162)$$

where $\odot$ represents the concatenation of two scattering matrices as described in Section 4.4.1.

5.7.2.3 Implementation of Lossless Reciprocal Decorrelating Networks.

Let us assume the same design conditions as those given in Section 5.7.1.3. Notice that because $\mathbf{S}_{\alpha\alpha}^A$ is the scattering matrix representations of a reciprocal lossless MPA, $\mathbf{S}_{\alpha\alpha}^A$ is a complex symmetrical matrix. In this case, the Takagi SVD (TSVD) can be used to factorize the subblock matrices of $\mathbf{S}_{\alpha\alpha}^A$. The Takagi factorization [17] is based on the following theorem: If $\mathbf{A}$ is a symmetric complex matrix, there exists a unitary matrix $\mathbf{U}$ and a real nonnegative diagonal matrix $\mathbf{\Lambda}^{\frac{1}{2}}$, such that $\mathbf{A} = \mathbf{U}\mathbf{\Lambda}^{\frac{1}{2}}\mathbf{U}^\dagger$. The columns of $\mathbf{U}$ are an orthonormal set of eigenvectors of $\mathbf{A}\mathbf{A}^*$ and the corresponding diagonal entries of $\mathbf{\Lambda}^{\frac{1}{2}}$ are the nonnegative square roots of the eigenvalues of $\mathbf{A}\mathbf{A}^*$. On the other hand, the product of a unitary matrix $\mathbf{V}$ by a diagonal matrix with complex unit magnitude entries Θ produces another unitary matrix, and for example, we can write $\mathbf{U} = \mathbf{V}\Theta$. In that case, $\mathbf{A}$ can be decomposed into $\mathbf{A} = \mathbf{V}\mathbf{D}\mathbf{V}^\dagger$, where $\mathbf{D}$ is a diagonal matrix with complex entries, given by $\mathbf{D} = (\Theta)^2 \mathbf{\Lambda}^{\frac{1}{2}}$. We refer to this decomposition as the generalized TSVD (GTSVD). Therefore, any complex symmetric matrix $\mathbf{A}$, such as $\mathbf{S}_{\alpha_T\alpha_T}^E$ and $\mathbf{S}_{\alpha_R\alpha_R}^E$, can also be decomposed in the form $\mathbf{A} = \mathbf{V}\mathbf{D}\mathbf{V}^\dagger$. We can now proceed to implement the DNs, such that the aforementioned conditions are satisfied. We first perform the GTSVDs of $\mathbf{S}_{\alpha_T\alpha_T}^E$ and $\mathbf{S}_{\alpha_R\alpha_R}^E$, which are given by

$$\mathbf{S}_{\alpha_T\alpha_T}^T = \mathbf{V}^T\mathbf{D}^T(\mathbf{V}^T)^\dagger \qquad (5.163)$$

$$\mathbf{S}^R_{\alpha_R \alpha_R} = \mathbf{V}^R \mathbf{D}^R (\mathbf{V}^R)^\dagger. \tag{5.164}$$

For the transmit side, one must pick

$$\mathbf{V}^{MT}_{22} = (\mathbf{V}^T)^* \tag{5.165}$$

$$\mathbf{D}^{MT}_{22} = (\mathbf{D}^T)^H. \tag{5.166}$$

Similarly, for the receiving side, one must pick

$$\mathbf{V}^{MR}_{11} = (\mathbf{V}^R)^* \tag{5.167}$$

$$\mathbf{D}^{MR}_{11} = (\mathbf{D}^R)^H. \tag{5.168}$$

Notice that in both cases, the remaining parameters can be chosen arbitrarily. In Ref. [66], it is shown that these degrees of freedom can be used to minimize the antenna correlation in environments without full angular spread or match the antennas to other devices such as amplifiers [75] instead of loads terminated with the characteristic reference impedance.

5.7.3 Proof of Optimal Superextended Antenna Correlation

To assess the benefits of these DN on the performance of MIMO communication systems in a more intuitive manner, we derive now the expressions for the optimal superextended transmit and receive antenna correlation matrices.

5.7.3.1 Optimal Superextended Receive Antenna Correlation.

Notice that if we design the receive DNs such that $\mathbf{S}^{MR}_{11} = (\mathbf{S}^E_{\alpha_R \alpha_R})^H$, with $\mathbf{\Gamma}^D_{\alpha_R} = \mathbf{0}$, one obtains

$$\mathbf{X}^R_M = \mathbf{Z}^{0,MR \frac{1}{2}}_{\alpha_2} \mathbf{S}^{MR}_{21} \left(\mathbf{I} - \mathbf{S}^E_{\alpha_R \alpha_R} (\mathbf{S}^E_{\alpha_R \alpha_R})^H \right)^{-1} \mathbf{X}^R_H$$
$$\cdot \left(\mathbf{I} - \mathbf{S}^E_{\alpha_R \alpha_R} (\mathbf{S}^E_{\alpha_R \alpha_R})^H \right)^{-1} (\mathbf{S}^{MR}_{21})^H \mathbf{Z}^{0,MR \frac{1}{2}}_{\alpha_2}. \tag{5.169}$$

Using the conditions for lossless DNs, the two relevant blocks of the DN are represented using

$$\mathbf{S}^{MR}_{11} = (\mathbf{S}^E_{\alpha_R \alpha_R})^H = \mathbf{V}^R \mathbf{\Lambda}^{\frac{1}{2}}_R (\mathbf{U}^R)^H \tag{5.170}$$

$$\mathbf{S}^{MR}_{21} = \mathbf{U}_{21} (\mathbf{I} - \mathbf{\Lambda}_R)^{\frac{1}{2}} \mathbf{\Theta}_{21} (\mathbf{U}^R)^H. \tag{5.171}$$

Using the fact that $\mathbf{S}^E_{\alpha_R \alpha_R} = \mathbf{U}^R \mathbf{\Lambda}^{\frac{1}{2}}_R (\mathbf{V}^R)^H$ and

$$\mathbf{I} - \mathbf{S}^E_{\alpha_R \alpha_R} (\mathbf{S}^E_{\alpha_R \alpha_R})^H = \mathbf{U}^R (\mathbf{I} - \mathbf{\Lambda}_R)(\mathbf{U}^R)^H \tag{5.172}$$

$$\left(\mathbf{I} - \mathbf{S}^E_{\alpha_R \alpha_R}(\mathbf{S}^E_{\alpha_R \alpha_R})^H\right)^{-1} = \mathbf{U}^R(\mathbf{I} - \mathbf{\Lambda}_R)^{-1}(\mathbf{U}^R)^H \tag{5.173}$$

and pluging them into Eq. (5.169), we obtain

$$\mathbf{X}^R_M = Z^{0,MR\frac{1}{2}}_{\alpha_2}\mathbf{U}_{21}\mathbf{\Theta}_{21}\mathbf{T}(\mathbf{\Theta}_{21})^H(\mathbf{U}_{21})^H Z^{0,MR\frac{1}{2}}_{\alpha_2}, \tag{5.174}$$

where $\mathbf{T} = (\mathbf{I} - \mathbf{\Lambda}_R)^{\frac{1}{2}}(\mathbf{U}^R)^H\mathbf{X}^R_H\mathbf{U}^R(\mathbf{I} - \mathbf{\Lambda}_R)^{\frac{1}{2}}$. Now, using the findings in Section 4.7.4.4, we can use the fact that in a full angular spread, $\mathbf{X}^R_H$ can be written as

$$\mathbf{X}^R_H = \frac{1}{8\pi}\left(\mathbf{I} - \mathbf{S}^E_{\alpha_R \alpha_R}(\mathbf{S}^E_{\alpha_R \alpha_R})^H\right) = \frac{1}{8\pi}\mathbf{U}^R(\mathbf{I} - \mathbf{\Lambda}_R)(\mathbf{U}^R)^H \tag{5.175}$$

to obtain a final expression of $\mathbf{X}^R_M$, given by

$$\mathbf{X}^R_M = Z^{0,MR}_{\alpha_2}\frac{1}{8\pi}\mathbf{I}. \tag{5.176}$$

Notice that with the above covariance term, the superextended receive antenna correlation matrix is equal to

$$\mathbf{R}^R_M = \mathbf{I}, \tag{5.177}$$

which clearly shows that we have accomplished our two objectives of maximum transmission power gain and minimum correlation.

When full angular spread does not exist, the optimal DN can be further specified to diagonalize the covariance matrix $\mathbf{X}^R_M$. Consider again

$$\mathbf{T} = (\mathbf{I} - \mathbf{\Lambda}_R)^{\frac{1}{2}}(\mathbf{U}^R)^H\mathbf{X}^R_H\mathbf{U}^R(\mathbf{I} - \mathbf{\Lambda}_R)^{\frac{1}{2}} = \mathbf{U}_{TT}\mathbf{\Lambda}_{TT}(\mathbf{U}_{TT})^H, \tag{5.178}$$

where the above expression represents the EVD of $\mathbf{T}$. Because the product $\mathbf{U}_{21}\mathbf{\Theta}_{21}$ is arbitrary as long as it is unitary, we can choose $\mathbf{U}_{21}\mathbf{\Theta}_{21} = (\mathbf{U}_{TT})^H$ to obtain

$$\mathbf{X}^R_M = \mathbf{\Lambda}_{TT}, \tag{5.179}$$

the diagonal matrix of eigenvalues of $\mathbf{T}$. Once again, minimum correlation is achieved. Finally, notice that the transmission power gain is independent of the characteristics of the channel, but correlation strongly depends on the channel.

5.7.3.2 Optimal Superextended Transmit Antenna Correlation.

Notice that if we design the receive DNs such that $\mathbf{S}^{MR}_{22} = (\mathbf{S}^E_{\alpha_T \alpha_T})^H$, with $\mathbf{\Gamma}^S_{\alpha_T} = \mathbf{0}$, one obtains

$$\mathbf{X}^T_M = Z^{0,MT\frac{1}{2}}_{\alpha_1}\mathbf{S}^{MT}_{12}\left(\mathbf{I} - \mathbf{S}^E_{\alpha_T \alpha_T}(\mathbf{S}^E_{\alpha_T \alpha_T})^H\right)^{-1}\mathbf{X}^T_H$$

$$\cdot \left(\mathbf{I} - \mathbf{S}^E_{\alpha_T \alpha_T}(\mathbf{S}^E_{\alpha_T \alpha_T})^H\right)^{-1}(\mathbf{S}^{MT}_{12})^H Z^{0,MT\frac{1}{2}}_{\alpha_1}. \tag{5.180}$$

Using the conditions for lossless DNs, the two relevant blocks of the DN are represented using

$$\mathbf{S}_{22}^{MT} = (\mathbf{S}_{\alpha_T \alpha_T}^{E})^H = \mathbf{V}^T \mathbf{\Lambda}_T^{\frac{1}{2}} (\mathbf{U}^T)^H \tag{5.181}$$

$$\mathbf{S}_{12}^{MT} = -\mathbf{U}_{12}(\mathbf{I} - \mathbf{\Lambda}_T)^{\frac{1}{2}} \mathbf{\Theta}_{12}(\mathbf{U}^T)^H. \tag{5.182}$$

Using the fact that $\mathbf{S}_{\alpha_T \alpha_T}^{E} = \mathbf{U}^T \mathbf{\Lambda}_T^{\frac{1}{2}} (\mathbf{V}^T)^H$ and

$$\mathbf{I} - \mathbf{S}_{\alpha_T \alpha_T}^{E}(\mathbf{S}_{\alpha_T \alpha_T}^{E})^H = \mathbf{U}^T(\mathbf{I} - \mathbf{\Lambda}_T)(\mathbf{U}^T)^H \tag{5.183}$$

$$\left(\mathbf{I} - \mathbf{S}_{\alpha_T \alpha_T}^{E}(\mathbf{S}_{\alpha_T \alpha_T}^{E})^H\right)^{-1} = \mathbf{U}^T(\mathbf{I} - \mathbf{\Lambda}_T)^{-1}(\mathbf{U}^T)^H \tag{5.184}$$

and plugging them into Eq. (5.180), we obtain

$$\mathbf{X}_M^T = Z_{\alpha_1}^{0,MT\frac{1}{2}} \mathbf{U}_{12}\mathbf{\Theta}_{12}\mathbf{T}(\mathbf{\Theta}_{12})^H(\mathbf{U}_{12})^H Z_{\alpha_1}^{0,MT\frac{1}{2}}, \tag{5.185}$$

where $\mathbf{T} = (\mathbf{I} - \mathbf{\Lambda}_T)^{\frac{1}{2}}(\mathbf{U}^T)^H \mathbf{X}_H^T \mathbf{U}^T(\mathbf{I} - \mathbf{\Lambda}_T)^{\frac{1}{2}}$. Now, using the findings in Section 4.7.4.4, we can use the fact that in a full angular spread, $\mathbf{X}_H^T$ can be written as

$$\mathbf{X}_H^T = \frac{1}{8\pi}\left(\mathbf{I} - \mathbf{S}_{\alpha_T \alpha_T}^{E}(\mathbf{S}_{\alpha_T \alpha_T}^{E})^H\right) = \frac{1}{8\pi}\mathbf{U}^T(\mathbf{I} - \mathbf{\Lambda}_T)(\mathbf{U}^T)^H \tag{5.186}$$

to obtain a final expression of $\mathbf{X}_M^T$, given by

$$\mathbf{X}_M^T = Z_{\alpha_1}^{0,MT} \frac{1}{8\pi}\mathbf{I}. \tag{5.187}$$

Notice that with the above covariance term, the superextended receive antenna correlation matrix is equal to

$$\mathbf{R}_M^T = \mathbf{I}, \tag{5.188}$$

which clearly shows that we have accomplished our two objectives of maximum transmission power gain and minimum correlation.

5.8 PRACTICAL IMPLEMENTATION ASPECTS OF DNs

In the previous sections, we have seen that DNs can be used to optimize the performance of MIMO communication system from a multiantenna system design perspective. It is important now to consider the practical implementation limitations of the DNs.

Along these lines, we should remark that DNs should be used in those cases in which the matching efficiency gains that can be recovered by using DNs are much larger than the insertion losses introduced by the DNs itself. Otherwise, the use of DNs will end up having a detrimental effect upon the system. The reader should also notice that besides the beneficial effects of DNs, especially in compact multiantenna systems in which the antennas are very closely spaced, these microwave networks will occupy some additional physical space. The resultant area/volume occupied by the multiantenna system (including the DNs) may sometimes be even bigger than the volume occupied by a multiantenna system having the same number of antennas separated by a larger distance and avoiding the use of DNs. Therefore, a tradeoff exists. It is, however, possible to implement decoupling networks with lumped components and thereby obtain very compact antenna designs with high matching efficiency. Finally, the use of DNs, in general, reduces the operative bandwidth of the resultant MPA; this needs to be taken into account, especially for wideband MIMO communication systems.

· · · ·

CHAPTER 6

Design Examples and Performance Analysis of Different MPAs

For a better understanding of the concepts explained in the previous chapters of the book, we now present some design examples of antenna diversity techniques and DNs. We conclude this chapter with a qualitative analysis of the impact of different antenna parameters, such as the topology, size, and polarization sensitivity of an MPA, on the performance within a MIMO wireless communication system.

6.1 PERFORMANCE ANALYSIS OF MPAs

In the following, we summarize the steps toward evaluating the performance of an MPA within a MIMO wireless communication system. We then present some basic examples. The performance analysis procedure is valid for any kind of MPA (including those using DNs), and the tools that we will use are extensively described in Sections 4.7 and 5.6.

The best way to investigate the performance of MPAs within a complete MIMO communication system is to conduct real-time measurements or use sophisticated channel models (such as those based on ray tracing, GTD, and others described in Chapter 2), which can take into account all the realistic electromagnetic propagation phenomena. If these resources are not available, then analytic MIMO channel models can be used. These models, despite their simplicity, can describe the impulse response (or equivalently, the transfer function) of the channel between the elements of the MPAs at both link ends by providing analytic expressions for the channel matrix, which later can be used for developing MIMO algorithms, in general, and the design of multiantenna systems.

For the sake of simplicity, we assume here the Kronecker channel model described in Section 1.6. Assuming such a channel, the first step toward evaluating the performance of an MPA, from a system-level perspective, is to compute its antenna correlation characteristics. Here, if we wish to investigate the MIMO performance without considering the effects of source and load terminations, we will use $\mathbf{R}_H^A$. On the other hand, if we wish to consider these terminations, we will use $\mathbf{R}_E^A$, and if our system is going to use DNs, then we should use $\mathbf{R}_M^A$. Notice that we use the upper-index A to refer to either the transmit side ($A = T$) or the receive side ($A = R$).

Expressions for these quantities are given in Sections 4.7 and 5.6, and to compute them, one will need to know the following two parameters: first, the radiation pattern at each one of the ports of the MPAs ($\mathbf{F}_i^A(\theta,\phi)$), and second, the propagation characteristics of the channel, that is, the transmit and receive power spectrum ($\mathbf{\Psi}^A(\theta,\phi)$). In the particular case that we can assume an ideal NLOS propagation scenario, the computation of these correlation matrix is simplified. This is because they can be computed from the scattering parameters of the antennas ($\mathbf{S}_{\alpha_R\alpha_R}^E$ and $\mathbf{S}_{\alpha_R\alpha_R}^E$), as shown in Chapter 4. Once the correlation properties are available, for example, they can be plugged into the Kronecker channel model, and with it, we can generate a large number of channel realizations with the desired correlation properties. Then, these channel realizations can be used to compute performance measures such as the capacity, diversity order, and other. (Such an approach is commonly known as Monte-Carlo methodology). Notice also that depending on the channel coding technique considered (see Chapter 3) for the MIMO system under investigation, the expression of the capacity and diversity order may change, as shown in Ref. [26]. Therefore, it is also important to identify the channel coding technique that will be used. However, any of these expressions can always be expressed in terms of the antenna correlation matrices and the characteristics of the propagation environment.

6.2 ANTENNA DIVERSITY DESIGN EXAMPLES

When talking about antenna diversity techniques, it is important to recall the fact that the different ports of an MPA may correspond to different antennas in an array or MEA, different polarizations in MPOAs, different radiating modes such as in MMAs or multibeam antennas or a combination of the above possibilities. These different configurations give rise to distinct antenna diversity techniques and designs. Some of them are discussed in what follows.

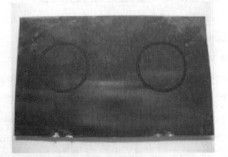 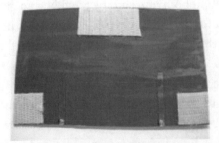

FIGURE 6.1: 2×1 Linear MEA in a space diversity configuration, using annular ring slot antennas.

6.2.1   Space Diversity Design

We begin by presenting a simple MEA exploiting space diversity techniques. Fig. 6.1 shows a picture of this particular 2×1 linear MEA using annular ring slot antennas, operating at 2.45 GHz. The antennas are separated by a distance of 0.5λ, where λ is the free-space wavelength associated with the operating frequency. Assuming a rich scattering environment, the correlation among these two antennas can be computed using (see Section 4.7)

$$\mathbf{R}^A_{H_{12}} = \frac{\frac{1}{8\pi}\int_{-\pi}^{\pi}\int_{0}^{\pi}(\mathbf{F}_1^A(\theta,\phi))^{\dagger}(\mathbf{F}_2^A(\theta,\phi))^*\sin(\theta)d\theta d\phi}{\sqrt{\left(\frac{1}{8\pi}\int_{-\pi}^{\pi}\int_{0}^{\pi}|\mathbf{F}_1^A(\theta,\phi)|^2\sin(\theta)d\theta d\phi\right)\left(\frac{1}{8\pi}\int_{-\pi}^{\pi}\int_{0}^{\pi}|\mathbf{F}_2^A(\theta,\phi)|^2\sin(\theta)d\theta d\phi\right)}}.$$

$$(6.1)$$

Using the above formula, it can be verified that this distance (0.5λ) is large enough for the radiation patterns associated with each one of the input ports of the MEA, $\mathbf{F}_1^A(\theta,\phi)$, and $\mathbf{F}_2^A(0,\phi)$, subject to a common coordinate system, to be orthogonal to each other. This idea is graphically shown in Fig. 6.2, where the correlation value among microstrip antennas is plotted versus the separation distance. Notice that for the case of an ideal NLOS propagation scenario (curve with uniform

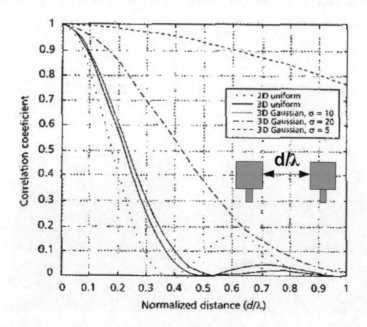

FIGURE 6.2: Correlation coefficients for different angle of arrival distribution and spreads versus antenna separation distance. Figure extracted from Ref. [80].

three-dimensional power spectrum), the correlation value is at its minimum at an antenna separation of approximately 0.5λ. We can conclude that spacing antennas apart can be used as a mechanism to decrease the mutual coupling and correlation among antennas. Notice also in Fig. 6.2 that if the channel cannot be assumed to be ideal NLOS, then the minimum antenna separation may be different. In particular, for example, on a channel with a single set of waves (cluster) arriving in a particular direction with an angular spread of $\sigma = 20°$, the required minimum antenna separation to satisfy the orthogonality condition would be around 1λ. In general, antenna separation in spatial diversity configurations depends on the characteristics of the channel, as shown also in Eqs (4.155) and (4.152).

6.2.2   Compact Dual-Polarized Antenna Design

On the other hand, Fig. 6.3 shows the design of a compact dual-polarized antenna that exploits two orthogonal polarizations. This antenna produces uncorrelated signals paths in scattered environments—that is, it achieves $\mathbf{R}^A_{H_{12}} \approx 0$. There exist two ways to study polarization diversity: on the one hand, by using one single antenna with multiple ports that can extract multiple polarizations, as with the antenna presented here, and on the other, by using several antennas, each one of them exploiting a different polarization. However, one must realize that this last approach uses, in fact, a combination of spatial and polarization diversity techniques. The shown MPOA operates as 2.45 GHz.

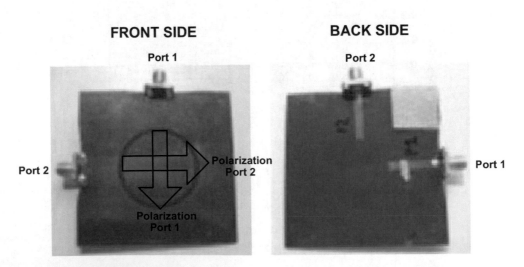

FIGURE 6.3: Dual-polarized antenna in a polarization diversity configuration, using an annular ring slot antenna.

6.2.3   Pattern Diversity Design Using Beam Forming

In early studies of linear arrays, it was shown that multiple independent beams formed by lossless networks are orthogonal [81], assuming an ideally rich scattering environment. Under the restriction of equal gain beams, there is a unique solution for the lossless network, the Butler matrix being its usual implementation [82]. Despite this particular conclusion, in general, one way of optimizing the design of MPAs through the use of pattern diversity techniques is by using external beam-forming networks. A particular design of a MEA using pattern diversity techniques is presented in Ref. [13]. The approach in Ref. [13] uses a linear MEA of four elements in combination with a Butler matrix [83, 84, 85] as an external beam-forming network, providing a twofold benefit. In an ideally scattering environment, this MMA provides orthogonal patterns. On the other hand, in poor scattered environments, the use of a Butler matrix arranges omnidirectional antenna patterns into directive beams, therefore providing additional array gain.

To clarify, an ideal Butler matrix is a $2n$-passive, reciprocal, and lossless network with n input and n output ports, such that a signal introduced at one of the inputs produces equal amplitude excitations at all the output ports but with a constant phase difference among them, resulting in radiation at a certain angle in space when combined with a linear MEA. A signal introduced at another input port results in radiation toward a different angle in space. A Butler matrix, in fact, performs a spatial fast Fourier transform, and up to n orthogonal directive beams can be generated when combined with a linear MEA.

In particular, as shown in Fig. 6.4, the proposed MMA in Ref. [13] is built by connecting the output ports of a Butler matrix to the input ports of a linear MEA, such that the input ports for the new MEA (including the linear array and the Butler matrix) are the input ports of the Butler matrix. This configuration simultaneously exploits the benefits of beam forming in low scattered environments, the benefits of spatial diversity in rich scattered environments or simultaneously

FIGURE 6.4: Linear MEA of four monopoles (left) and fabricated Butler matrix (right), as a particular design of pattern diversity using beam-forming networks.

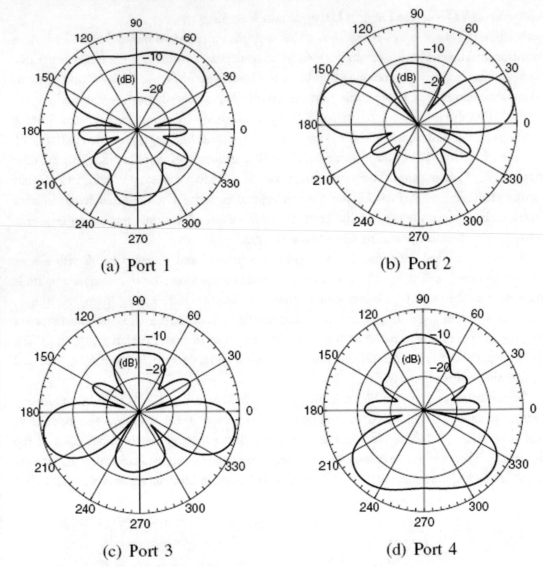

(a) Port 1 (b) Port 2

(c) Port 3 (d) Port 4

FIGURE 6.5: Radiation patterns associated with the proposed linear MEA in combination with a Butler matrix. (a) Port 1, (b) Port 2, (c) Port 3 and (3) Port 4.

takes advantage of both techniques in intermediate propagation conditions. The proposed linear MEA using a Butler beam-forming matrix is optimal in two ways. First, assuming that there are not electronically adjustable phase shifts components on the external circuitry, the Butler matrix is the only static microwave network that will not degrade the performance of a linear MEA in rich scattered environments because it produces orthogonal radiation patterns, as shown in Fig. 6.5. Second, the Butler matrix is a unique lossless reciprocal microwave network with which the full array gain of a linear MEA can be realized in each beam, as shown in Fig. 6.5.

6.2.4 Pattern Diversity Design Using MMAs

In Ref. [5], a novel way of exploiting pattern diversity by employing the higher-order modes of a biconical antenna is presented. Essentially, the biconical antenna offers characteristics similar to an antenna array through multiple modes using just a single antenna element with multiple accessible ports, one for each of the excited modes. In a multipath scenario, the radiation patterns of the higher order modes (see Fig. 6.6b) are different enough to result in low correlation between modes. The biconical antenna consists of two conical horns facing opposite directions, as illustrated in Fig. 6.6a.

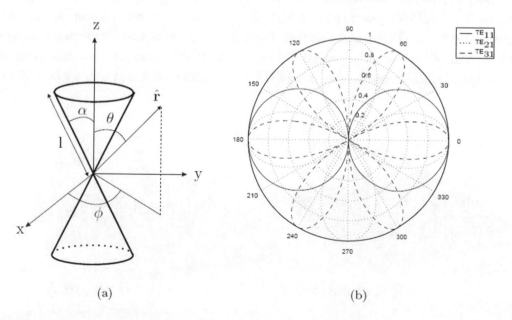

(a)   (b)

FIGURE 6.6: Pattern diversity concept using a multimode antenna. (a) representation of a biconical antenna, and (b) fundamental and some of the higher order modes excited by a biconical antenna.

FIGURE 6.7: Existing single antenna (SISO) wireless LAN product operating at 2.43 GHz.

6.2.5 Design Example of a Practical Antenna Diversity System

A company would like to improve the performance of their routers by adding MIMO capabilities into their baseline 802.11b router product. Their existing product only has a single antenna (SISO) to provide links at 2.43 GHz with existing devices, as shown in Fig. 6.7. The company would like to evaluate the performance improvements that they would obtain by going from their existing SISO system to a 2×1 MIMO system or 3×1 MIMO system, that is, a system using two and three antennas, respectively. The size of the router is fixed; therefore, the distance between the antennas can only be changed up to a certain extent. For the 2×1 MIMO system, the two monopoles will be 40 mm apart (0.32λ), whereas when using the 3×1 MIMO configuration, they will be 20 mm

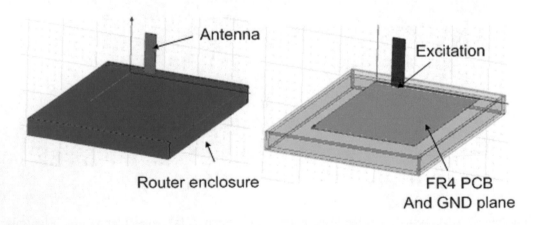

FIGURE 6.8: HFSS model of the router and the associated antenna operating at 2.43 GHz.

apart (0.161λ). Note that by adding a third dipole the mutual coupling among the antennas will also increase, and therefore, a boost in performance is not guaranteed. We know that the maximum DG can be achieved when the dipoles are uncorrelated. Assuming an ideal NLOS scenario, this happens when their radiation patterns are orthogonal to each other. If the antennas are too close to each other, this may not be guaranteed.

To evaluate the performance of these three systems, we first need to compute the correlation properties among the antennas involved. If we know the channel characteristics in which these systems are going to be operating, we can use Eq. (4.152). If we can assume an ideal NLOS propagation scenario (or one that is not ideal but is rich in scattering objects), then the correlation among the antennas can be simply computed from the mutual coupling coefficients. A simple model for the router and the antenna is created using a full-wave simulator (High Frequency Structure Simulator (HFSS) from Ansoft [86]) as shown in Fig. 6.8. If we use a quarter wave monopole on top of the ground plane, we can start by selecting the necessary length (l_m) to resonate at 2.4 GHz, given by

$$l_\text{m} = \frac{\lambda}{4} = \frac{c}{4f} = \frac{3 \times 10^8}{4 \times 3 \times 10^9} = 30.86 \times 10^{-3} \text{ m.} \qquad (6.2)$$

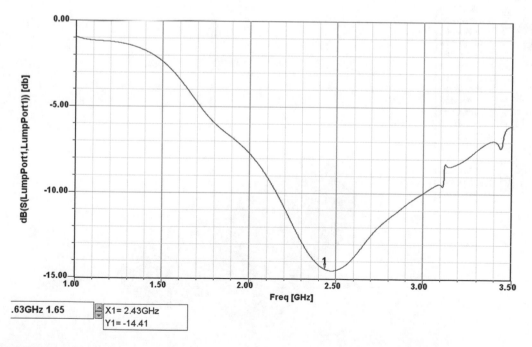

.63GHz 1.65   X1= 2.43GHz
               Y1= -14.41

FIGURE 6.9: Reflection coefficient for the model of the router represented in Fig. 6.8.

After tuning the antenna length to get the exact resonance, we obtain the following results for the reflection coefficient at the excitation port, shown in Fig. 6.9. Without loss of generality, assume the antenna in reception. The associated scattering parameter matrix at 2.43 GHz is equal to

$$\mathbf{S}^E_{\alpha_T \alpha_T} = 0.015 + j0.190. \tag{6.3}$$

On the other hand, the associated radiation pattern is shown in Figs. 6.10, 6.11, and 6.12. Notice that because of the asymmetry of the ground and the antenna location, the radiation pattern has some directivity with a maximum gain around 3 dBi. Based on the previous parameters, we can compute the capacity of the SISO system. To compute the antenna correlation, we may use the following expressions from Section 4.7.4.4, given by

$$\mathbf{X}^T_H = \frac{1}{8\pi} \left(\mathbf{I} - \mathbf{S}^E_{\alpha_T \alpha_T} (\mathbf{S}^E_{\alpha_T \alpha_T})^H \right) \tag{6.4}$$

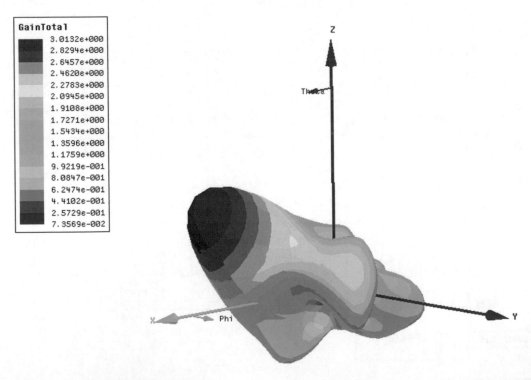

FIGURE 6.10: Simulated three-dimensional radiation pattern for the system depicted in Fig. 6.8.

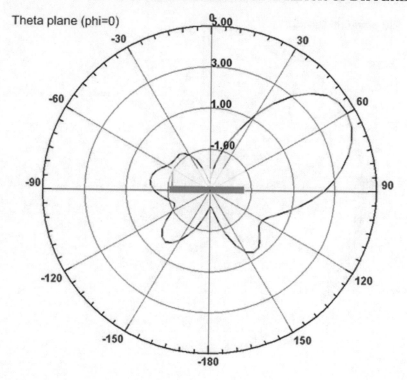

FIGURE 6.11: Simulated radiation pattern, at the $\phi = 0$ cut, for the system depicted in Fig. 6.8.

and also

$$\mathbf{R}_{H_{ij}}^T = \frac{\mathbf{X}_{H_{ij}}^T}{\sqrt{\mathbf{X}_{H_{ii}}^T \mathbf{X}_{H_{jj}}^T}}. \tag{6.5}$$

Notice that for this simple example, we are not considering the effects of source and load termination. However, in practical situations, these parameters must be taken into account. In this particular case, because the antenna system has only one port, notice that $\mathbf{R}_H^R = 1$. On the other hand, if the transmit antenna is connected to a generator with its internal impedance equal to the characteristic reference impedance of the system, then the transmit matching efficiency is given (from Section 4.7.1.2) by

$$\eta^{T,0} = \frac{E\{P^{S,0}\}}{E\{P^{S,0,\max}\}}, \tag{6.6}$$

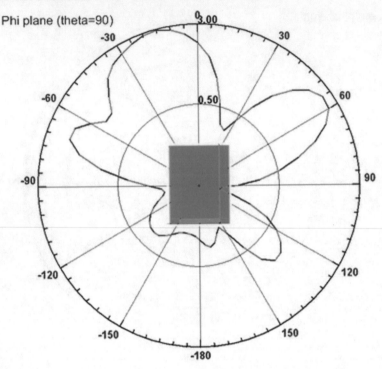

FIGURE 6.12: Simulated three-dimensional radiation pattern, at the $\phi = 0$ cut, for the system depicted in Fig. 6.8.

where $P^{S,0} = (\mathbf{b}_0^S)^H \left(\mathbf{I} - (\mathbf{S}_{\alpha_T \alpha_T}^E)^H \mathbf{S}_{\alpha_T \alpha_T}^E \right) \mathbf{b}_0^S$ and $P^{S,\max} = (\mathbf{b}_0^S)^H \mathbf{b}_0^S$. In this case, $\eta^{T,0} = 0.96$. Assuming a receive SNR = 10 dB, we can plot the capacity distribution for such system, as shown in Fig. 6.13, using Eqs. (1.21) and (1.22):

$$C = \log_2 \det \left(\mathbf{I} + \frac{\rho_0}{M} \mathbf{H}_0 \mathbf{H}_0^H \right) \tag{6.7}$$

$$\mathbf{H}_0^q = \mathbf{H}^q \sqrt{\frac{NMQ}{\sum_{q=1}^Q \|\mathbf{H}^q\|_F^2}}. \tag{6.8}$$

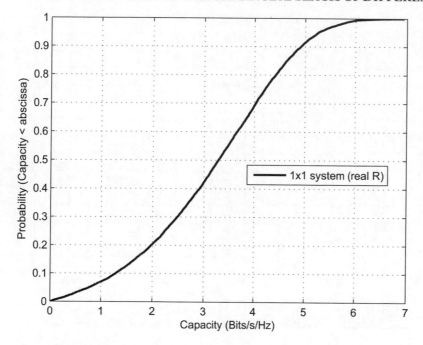

FIGURE 6.13: CDF of the capacity for the 1 × 1 SISO system.

For the 2 × 1 MIMO system, the antennas will be placed at 40-mm distance as shown in Fig. 6.14. As before, a simple model is used to extract the mutual coupling coefficients between the two antennas. The HFSS model for this case is the same as the one in Fig. 6.8 but has two antennas instead of one. The simulated results and coupling coefficients for this case are provided in Fig. 6.15.

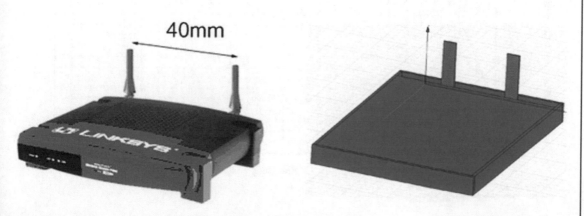

FIGURE 6.14: 2 × 1 MIMO system antenna arrangement and corresponding HFSS model.

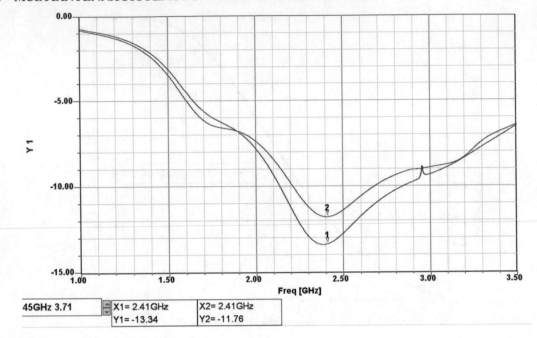

45GHz 3.71	X1= 2.41GHz	X2= 2.41GHz
	Y1= -13.34	Y2= -11.76

FIGURE 6.15: Simulation results of S_{ii} versus frequency for the system of Fig. 6.14.

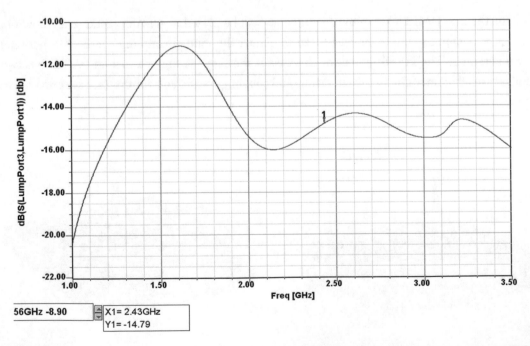

56GHz -8.90	X1= 2.43GHz
	Y1= -14.79

FIGURE 6.16: Simulation results of S_{ij} versus frequency for the system of Fig. 6.14.

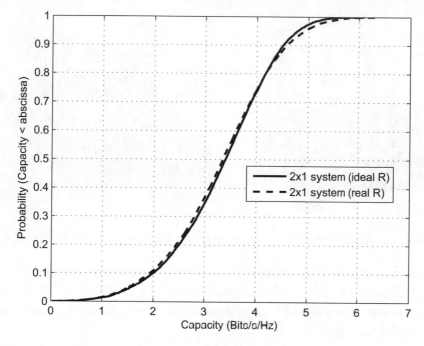

FIGURE 6.17: CDF of the capacity for the 2×1 MIMO system.

The associate S parameter matrix at 2.43 GHz is reported below:

$$\mathbf{S}^E_{\alpha_T\alpha_T} = \begin{pmatrix} 0.055 - j0.211 & -0.123 - j0.133 \\ -0.123 - j0.133 & 0.055 - j0.253 \end{pmatrix}. \tag{6.9}$$

We observe that because the two antenna are spaced around 0.32λ apart, a reasonable isolation of 14.79 dB is achieved, whereas the matching on both antennas is preserved (VSWR < 2.0). In fact, the matching efficiency is $\eta^T, 0 \approx 0.86$. We proceed now to compute the capacity of the 2×1 MIMO system and investigate the correlation properties of the system. We first compute the antenna correlation matrix as in the previous case, which produces

$$\mathbf{R}^T_H = \begin{pmatrix} 1 & -0.0530 + 0.0057j \\ -0.0530 - 0.0057j & 1 \end{pmatrix}. \tag{6.10}$$

Finally, assuming that $\mathbf{R}^R_H = \mathbf{I}$, the capacity distribution for an SNR = 10 dB is given in Fig. 6.17. Notice that the effect of a nonideal correlation matrix has very little impact on the capacity. This is because the influence of mutual coupling is negligible on this structure. We should remark that the

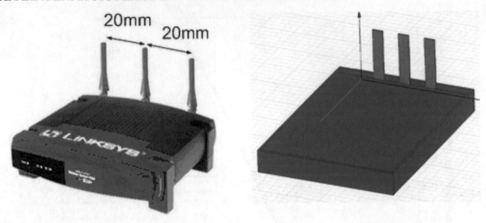

FIGURE 6.18: 3×1 MIMO system antenna arrangement and correspondent HFSS model.

SNR degradation due to lack of matching efficiency is not taken into account in Fig. 6.17 (thus, only the degradation effect on the correlation properties is being considered). This has to do with the fact that we have used normalized channel matrices. However, this can be taken into account using sophisticated channel models.

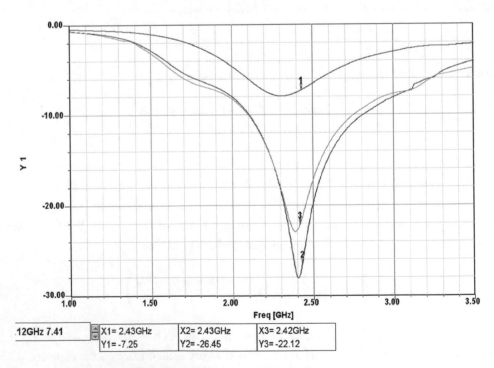

FIGURE 6.19: Scattering parameters (S_{ii}) for the 3×1 MIMO system.

Finally, for the 3×1 MIMO system, we will have the arrangement of Fig. 6.18. The corresponding simulated scattering parameters are now given in Figs. 6.19 and 6.20. The associate scattering parameter matrix at 2.43 GHz is reported below:

$$\mathbf{S}^E_{\alpha_T \alpha_T} = \begin{pmatrix} 0.033 - j0.033 & 0.105 - j0.450 & -0.122 + j0.015 \\ 0.105 - j0.450 & 0.167 + j0.400 & 0.086 - j0.468 \\ -0.122 + j0.015 & 0.086 - j0.468 & 0.059 - j0.061 \end{pmatrix}. \qquad (6.11)$$

Note that, in this case, because of the small distance between the elements, the isolation is degraded (around 6 dB between adjacent elements), and the return loss of the center element is also affected. The matching efficiency is now $\eta^{T,0} \approx 0.74$, thus, an important reduction on the received power level (and henceforth, on the SNR) is expected. Of course, the matching efficiency can be recovered using a matching network or some other matching techniques at the expense of bandwidth reduction, but this is beyond the scope of this exercise. We may now proceed to investigate the impact of having three closely spaced antennas, on the antenna correlation. To do

FIGURE 6.20: Scattering parameters (S_{ij} for $i \neq j$) for the 3×1 MIMO system.

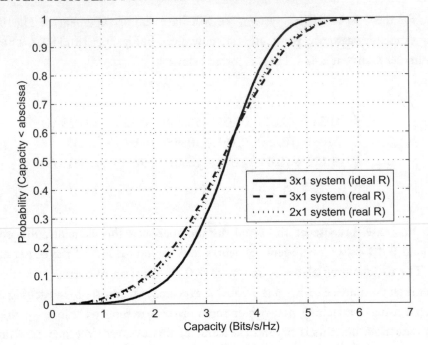

FIGURE 6.21: CDF of the capacity for the 3×1 MIMO system.

that, we can use the scattering matrix found above. Based on these parameters, we can compute the capacity of the 3×1 MIMO system. The antenna correlation matrix is now given by

$$\mathbf{R}_H^T = \begin{pmatrix} 1 & 0.3022 + 0.3020j & -0.2747 - 0.0293j \\ 0.3022 - 0.3020i & 1 & 0.4083 - 0.2516i \\ -0.2747 + 0.0293i & 0.4083 + 0.2516i & 1 \end{pmatrix}. \qquad (6.12)$$

In this case, with a receive SNR = 10 dB, the cumulative distribution function of the capacity is given in Fig. 6.21. Notice that now, the impact of the mutual coupling is not negligible. In fact, the capacity performance for the 3×1 is approximately the same as for the 2×1 case. This shows the importance of having the ports of an MPA antenna decorrelated and the necessity to evaluate the performance of any MPA from a system*level perspective. A summary of the average achievable capacity (ergodic capacity) for the three analyzed cases is provided in Table 6.1.

Following the aforementioned design procedure, many MIMO system architectures and different kinds of antennas can be easily investigated and compared. For example, one could now consider the problem of evaluating the performance of a 2×1 MIMO system for the two dipole

TABLE 6.1: Summary of the ergodic capacity for the three configurations mentioned above	
	Capacity (bits/s/Hz)
1×1	3.1848
2×1	3.3045
3×1	3.2869

configurations in the Fig. 6.22, which uses a combination of spatial diversity and polarization diversity. Also, one could find an answer to the following problem: A company would like to deploy a high-speed WLAN infrastructure based on a $N \times M$ MIMO system. They are considering two options that are practical to implement:

- 3×2 MIMO system with the transmitting site as the one of Fig. 6.18 and the receiver as the one of Fig. 6.14 in the same problem.
- 2×3 MIMO system with the receiving site as the one of Fig. 6.18 and the transmitter as the one of Fig. 6.14 in the same problem.

Following the design methodology laid out to this point, it is left to the reader to figure out which of the two systems would perform better.

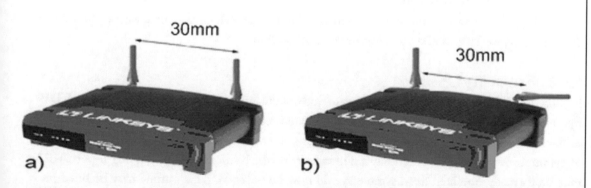

FIGURE 6.22: 2×1 MIMO system antenna arrangement: (a) aligned antenna and (b) orthogonal antenna.

6.3   DECORRELATING NETWORKS DESIGN EXAMPLES

In this section, we present some design examples of specific DNs for different MEA configurations. In particular, we provide the networks for a 2 × 1 linear MEA, a three-element circular MEA and a 2 × 2 planar MEA. We assume all the MEAs to be reciprocal components; therefore, its associate scattering matrix $\mathbf{S}_{\alpha\alpha}^A$ is given by a complex symmetric matrix. It is well-known from Chapter 5 and [17], that any complex symmetric matrix $\mathbf{A}$ can be factorized in the form:

$$\mathbf{A} = \mathbf{U}\Lambda^{\frac{1}{2}}\mathbf{V}^\dagger, \qquad (6.13)$$

where $\mathbf{U}$ is a unitary matrix and $\Lambda^{\frac{1}{2}}$ is a diagonal matrix with real elements on it. This factorization is called the TSVD. $\mathbf{A}$ can also be factorized in the form:

$$\mathbf{A} = \mathbf{V}\mathbf{D}\mathbf{V}^\dagger, \qquad (6.14)$$

where $\mathbf{V}$ is a unitary matrix and $\mathbf{D}$ is a diagonal matrix with complex elements on it. This factorization is called the GTSVD. As shown in Chapter 5, both decompositions are desirable for a straightforward design of reciprocal lossless matching networks, starting from the input ports scattering characteristics of the MPAs. There exist many techniques to find out these decompositions. We present here two approaches based on the

1. Takagi factorization
2. Symmetry operators.

The last approach is based on the GTSVD, whereas the former, as its name indicates, is based on the TSVD.

6.3.1   Takagi Factorization

The Takagi factorization method is based on the TSVD, which was extensively described in Section 5.7.2. Refer to that section for further information.

6.3.2   Symmetry Operators

It is also possible to use symmetry operators to find out the factors of the decomposition $\mathbf{A} = \mathbf{V}\mathbf{D}\mathbf{V}^\dagger$. Symmetry operators are mathematical tools that are used to describe the symmetry characteristics of the MPAs. A good reference on symmetry operators can be found in Refs. [60, 87] in the context of symmetrical microwave junctions. The manner in which information regarding n-port MPAs may be formally obtained from symmetry will now be outlined. This outline may be filled in, if necessary, by recourse to any of the abovementioned references. Because of symmetry, an MPA may have unusual properties. For example, a symmetrical MEA is characterized by the fact that it is left

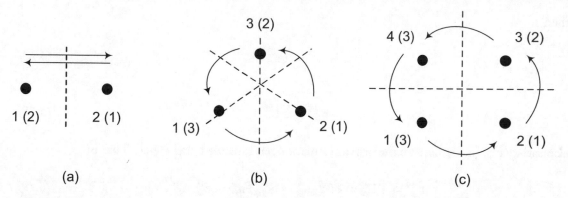

FIGURE 6.23: Symmetry schemes for (a) the 2×1 linear MEA, (b) the three-elements circular MEA, and (c) the 2×2 planar MEA, indicating the port numeration before (and after) applying a particular symmetry operator. The figure also shows some of the possible symmetry planes and rotations associated with these structures.

unchanged by a symmetry operation. For example, for a 2×1 linear MEA antenna configuration, shown in Fig. 6.23a, the symmetry rotation is a rotation of the structure by $180°$ about the symmetry axis. This operation turns the figure back into itself. The MEA is said to be invariant under this symmetry operation. A useful method for obtaining the properties of symmetrical MPA is found in the theory of eigenvalues, which is widely explained in the context of symmetrical junctions in Ref. [60]. The group of symmetry operations of an MPA are based on reflections or rotations that transform the MPA into itself and are normally represented by matrices $\overline{\mathbf{F}}$ to denote reflection with respect to a plane, $\overline{\mathbf{R}}$ to denote rotation with respect to an axis, and $\overline{\mathbf{P}}$ to refer to a reflection in the origin. Any of these matrices operate on the terminal quantities. Let us recall the notation of Chapter 4 to describe a particular MPA within our communication system, given by

$$\begin{pmatrix} \mathbf{b}_\alpha^A \\ \mathbf{b}_\beta^A \end{pmatrix} = \begin{pmatrix} \mathbf{S}_{\alpha\alpha}^A & \mathbf{S}_{\alpha\beta}^T \\ \mathbf{S}_{\beta\alpha}^T & \mathbf{S}_{\beta\beta}^T \end{pmatrix} \begin{pmatrix} \mathbf{a}_\alpha^A \\ \mathbf{a}_\beta^A \end{pmatrix}. \tag{6.15}$$

Thus, for example, the matrices $\overline{\mathbf{F}}$ (or $\overline{\mathbf{R}}$ or $\overline{\mathbf{P}}$) permute the components of the terminal vectors $\mathbf{a}_\alpha^A$ and $\mathbf{b}_\alpha^A$ so as to interchange terminal quantities of symmetrically disposed ports of the MPA. Such $\overline{\mathbf{F}}$ (or $\overline{\mathbf{R}}$ or $\overline{\mathbf{P}}$) matrices may commonly be written down by inspection from the conventional scattering matrix description or from physical inspection of the MPA or any other microwave component. Then if, for arbitrary $\mathbf{a}_\alpha^A$ of an MPA, the reflected wave $\mathbf{b}_\alpha^A$ satisfies

$$\mathbf{b}_\alpha^A = \mathbf{S}_{\alpha\alpha}^A \mathbf{a}_\alpha^A, \tag{6.16}$$

then,

$$\mathbf{a}_{\alpha}^{''A} = \overline{\mathbf{F}}\mathbf{a}_{\alpha}^{A} \qquad (6.17)$$

$$\mathbf{b}_{\alpha}^{''A} = \overline{\mathbf{F}}\mathbf{b}_{\alpha}^{A} \qquad (6.18)$$

also satisfy Eq. (6.16) because the physical situation has remained unchanged. That is,

$$\mathbf{b}_{\alpha}^{''A} = \overline{\mathbf{F}}\mathbf{b}_{\alpha}^{A} = \overline{\mathbf{F}}\mathbf{S}_{\alpha\alpha}^{A}\mathbf{a}_{\alpha}^{A} = \mathbf{S}_{\alpha\alpha}^{A}\overline{\mathbf{F}}\mathbf{a}_{\alpha}^{A} = \mathbf{S}_{\alpha\alpha}^{A}\mathbf{a}_{\alpha}^{''A}. \qquad (6.19)$$

We conclude from the third and fourth members of the previous equation that

$$\mathbf{S}_{\alpha\alpha}^{A}\overline{\mathbf{F}} = \overline{\mathbf{F}}\mathbf{S}_{\alpha\alpha}^{A}, \qquad (6.20)$$

which shows that the operators $\overline{\mathbf{F}}$ and $\mathbf{S}_{\alpha\alpha}^{A}$ commute. The same development applies for the $\overline{\mathbf{R}}$ and $\overline{\mathbf{P}}$ matrices. One now recalls the fact that two operators commute if, and only if, the operators possess a common set of eigenvectors (common invariant subspaces corresponding to different eigenvalues of $\overline{\mathbf{F}}$ (or $\overline{\mathbf{R}}$ or $\overline{\mathbf{P}}$). In other words, there exists a basis that simultaneously completely reduces $\overline{\mathbf{F}}$ (or $\overline{\mathbf{R}}$ or $\overline{\mathbf{P}}$) and $\mathbf{S}_{\alpha\alpha}^{A}$. A relation of the form given in Eq. (6.19) follows from each member of the set of independent generators of the symmetry group [60]. A transformation $\mathbf{V}$, which completely reduces the complicated scattering matrix $\mathbf{S}_{\alpha\alpha}^{A}$, is then deduced from the simple, known symmetry operators. That is, it is possible to have

$$\mathbf{S}_{\alpha\alpha}^{A} = \mathbf{V}\mathbf{D}\mathbf{V}^{-1}. \qquad (6.21)$$

Because the decomposition of $\mathbf{S}_{\alpha\alpha}^{A}$ from the symmetry operators procedure is based on the eigenvalue decomposition of the symmetry operator to find the $\mathbf{V}$ matrices, it is important that the final result is such that $\mathbf{V}$ is composed of pure real and orthogonal vectors. In that case, $\mathbf{V}^{-1} = \mathbf{V}^{\dagger}$, and therefore, we have a reciprocal design for the DN that verifies $\mathbf{S}_{\alpha\alpha}^{A} = \mathbf{V}\mathbf{D}\mathbf{V}^{-1} = \mathbf{V}\mathbf{D}\mathbf{V}^{\dagger}$. In Ref. [60], it is shown that it is possible, in general, to choose eigenvectors of $\mathbf{S}_{\alpha\alpha}^{A}$ that are real by doing linear combinations of the original ones found from the eigenvalue decomposition of the operators associated with the MPA.

6.3.3 The 2×1 Linear MEA

We describe now some specific examples of how DNs can be designed following the criteria for optimal MIMO performance given in Chapter 5 and using any of the factorization techniques described in Section 6.3. A 2×1 linear MEA, such as that shown in Fig. 6.23a, has a scattering matrix, $\mathbf{S}^A$, of the form

$$\mathbf{S}^A_{\alpha\alpha} = \begin{pmatrix} S^A_{11} & S^A_{12} \\ S^A_{12} & S^A_{11} \end{pmatrix}. \tag{6.22}$$

For this particular MEA, there exists only one independent symmetry operator, given by $\overline{\mathbf{F}}$ [60]. The transformation $\overline{\mathbf{F}}$ that interchanges components of the terminal vectors at symmetrically disposed ports of the MEA is evidently specified by the unitary matrix:

$$\overline{\mathbf{F}} = \begin{pmatrix} 0 & 1 \\ 1 & 0 \end{pmatrix} \tag{6.23}$$

found from inspection of Fig. 6.23a. Notice that after applying this transformation, the MEA structure is invariant with respect to the original structure. Notice that this transformation produces

$$\begin{pmatrix} a_2 \\ a_1 \end{pmatrix} = \begin{pmatrix} 0 & 1 \\ 1 & 0 \end{pmatrix} \begin{pmatrix} a_1 \\ a_2 \end{pmatrix} \tag{6.24}$$

$$\begin{pmatrix} b_2 \\ b_1 \end{pmatrix} = \begin{pmatrix} 0 & 1 \\ 1 & 0 \end{pmatrix} \begin{pmatrix} b_1 \\ b_2 \end{pmatrix}. \tag{6.25}$$

The symmetry group is of order 2, containing only $\overline{\mathbf{F}}$ as operator, satisfying $\overline{\mathbf{F}}^2 = \mathbf{I}$. We find the invariant subspaces in the usual fashion, that is, by solving the eigenvalue problem of $\overline{\mathbf{F}}$. The associated eigenvalues of the operator $\overline{\mathbf{F}}$ are $\lambda_1 = 1$ and $\lambda_2 = -1$. The invariant subspaces are, therefore, nondegenerated because all the eigenvalues are different, and the eigenvectors are given by

$$\mathbf{v}_1 = \frac{1}{\sqrt{2}} \begin{pmatrix} 1 \\ 1 \end{pmatrix} \tag{6.26}$$

$$\mathbf{v}_2 = \frac{1}{\sqrt{2}} \begin{pmatrix} 1 \\ -1 \end{pmatrix}. \tag{6.27}$$

Therefore, the transformation matrix $\mathbf{V}^A$ can be built by placing the previous eigenvectors in its columns and is thus given by

$$\mathbf{V}^A = \frac{1}{\sqrt{2}} \begin{pmatrix} 1 & 1 \\ 1 & -1 \end{pmatrix}. \tag{6.28}$$

We can finally decompose $\mathbf{S}_{\alpha\alpha}^A$ into

$$\mathbf{S}_{\alpha\alpha}^A = \mathbf{V}^A \mathbf{D}^A (\mathbf{V}^A)^\dagger, \qquad (6.29)$$

with

$$\mathbf{D}^A = \begin{pmatrix} S_{11}^A + S_{12}^A & 0 \\ 0 & S_{11}^A - S_{12}^A \end{pmatrix}, \qquad (6.30)$$

which is the desired solution. For simplicity, assume the design of the receiving matching network. A similar analysis can be followed for the transmit matching network. Let us recall from Chapter 5 that

$$\mathbf{S}^{MR} = \mathbf{S}^L \odot \mathbf{S}^U. \qquad (6.31)$$

Since we desire $\mathbf{S}_{11}^{MR} = (\mathbf{S}_{\alpha\alpha}^A)^H$, we need to set

$$\mathbf{V}_{11}^{MR} = (\mathbf{V}^A)^* \qquad (6.32)$$

$$\mathbf{D}_{11}^{MR} = (\mathbf{D}^A)^H. \qquad (6.33)$$

With the computed $\mathbf{V}^A$, one obtains that the lossless and reciprocal decoupling network must have the following parameters:

$$\mathbf{S}^L = \begin{pmatrix} \mathbf{0} & \mathbf{V}_{11}^{MR} \\ (\mathbf{V}_{11}^{MR})^\dagger & \mathbf{0} \end{pmatrix} = \begin{pmatrix} \mathbf{0} & (\mathbf{V}^A)^* \\ (\mathbf{V}^A)^H & \mathbf{0} \end{pmatrix} = \frac{1}{\sqrt{2}} \begin{pmatrix} 0 & 0 & 1 & 1 \\ 0 & 0 & 1 & -1 \\ 1 & 1 & 0 & 0 \\ 1 & -1 & 0 & 0 \end{pmatrix}, \qquad (6.34)$$

and that is readily identified as a 180° hybrid. The following matching network needed to match each one of the uncoupled ports at the output of the decoupling network can easily be found using Chapter 5 and Eq. (6.30). The matching network is given by

$$\mathbf{S}^U = \begin{pmatrix} (\mathbf{D}^A)^H & \mathbf{D}^{MR} \\ \mathbf{D}^{MR} & \mathbf{D}_{22}^{MR} \end{pmatrix}, \qquad (6.35)$$

where $(\mathbf{D}^A)^H = \mathbf{D}_{11}^{MR} = (\Theta_{11}^{MR})^2 (\Lambda_{11}^{MR})^{\frac{1}{2}}$, $\mathbf{D}_{22}^{MR} = (\Theta_{22}^{MR})^2 (\Lambda_{11}^{MR})^{\frac{1}{2}}$ and $\mathbf{D}^{MR} = j\Theta_{11}^{MR}\Theta_{22}^{MR}(\mathbf{I} - \Lambda_{11}^{MR})^{\frac{1}{2}}$ and where Θ_{22}^{MR} can be set to any arbitrary value and Θ_{11}^{MR} is such that $(\Lambda_{11}^{MR})^{\frac{1}{2}}$ contains real entries.

As remarked on before, there exist many other possible solutions to the problem of decorrelating the input ports of a 2×1 linear MEA. In fact, another interesting possible solution that is easy to synthesize with microwave components would be as follows. Notice that because $\mathbf{V}^A$

is a unitary matrix, one can observe that $\mathbf{V}^A = \mathbf{W}^A \Theta^A$, where Θ^A is a complex diagonal matrix, is another known solution. Henceforth, in the new decomposition of $\mathbf{S}^A_{\alpha\alpha}$, we have

$$\mathbf{S}^A_{\alpha\alpha} = \mathbf{W}^A \mathbf{T}^A (\mathbf{W}^A)^\dagger, \qquad (6.36)$$

where $\mathbf{T}^A = (\Theta^A)^2 \mathbf{D}^A$. In the particular case that

$$\Theta^A = \begin{pmatrix} 1 & 0 \\ 0 & j \end{pmatrix}, \qquad (6.37)$$

then

$$\mathbf{W}^A = \frac{1}{\sqrt{2}} \begin{pmatrix} 1 & j \\ 1 & -j \end{pmatrix} \qquad (6.38)$$

and

$$\mathbf{T}^A = \begin{pmatrix} S^A_{11} + S^A_{12} & 0 \\ 0 & S^A_{12} - S^A_{11} \end{pmatrix}. \qquad (6.39)$$

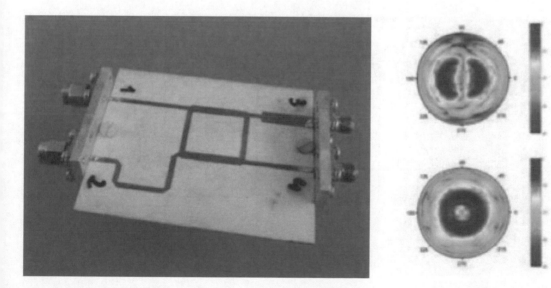

FIGURE 6.24: Picture of the fabricated decoupling network for a 2×1 linear MEA using a $180°$ hybrid. On the right, associated orthogonal three-dimensional normalized far-field radiation patterns at the output of the DN for two monopoles separated 0.21λ.

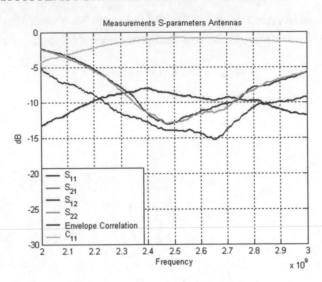

FIGURE 6.25: Measured scattering parameters for two monopoles separated 0.21λ.

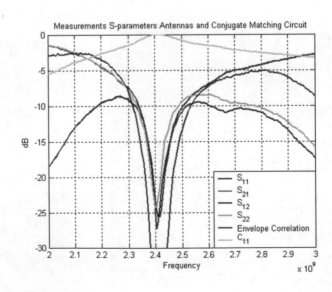

FIGURE 6.26: Measured scattering parameters at the output ports of the DN when connected to an antenna array consisting of two monopoles separated 0.21λ.

In this case, one obtains that the lossless and reciprocal decoupling network must have the following parameters:

$$\mathbf{S}^L = \frac{1}{\sqrt{2}} \begin{pmatrix} 0 & 0 & 1 & -j \\ 0 & 0 & 1 & j \\ 1 & 1 & 0 & 0 \\ -j & j & 0 & 0 \end{pmatrix}, \qquad (6.40)$$

which is readily identified as a 90° hybrid.

Finally, notice that the bank of uncoupled matching transformers is easily implemented with lumped components or transmission lines and stubs. When the matching section is added to the new structure, exciting one port will result in exciting the port with one eigensource, that is, with the magnitude and phase proportional to one of the eigenvectors, and alternatively, the other eigensource is excited when an incident wave is applied to the other port of the new MPA structure. As expected, when one adds a decorrelating matching network, these two eigensources will generate two orthogonal radiation patterns.

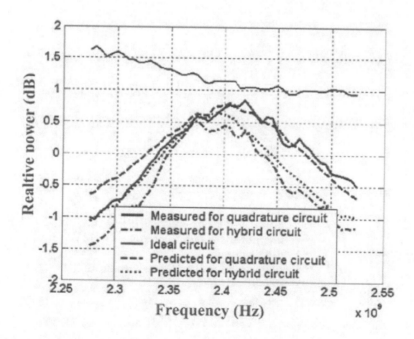

FIGURE 6.27: Measured improvement on the received power at the output ports of the DN (line corresponding to the hybrid circuit) when connected to an antenna array consisting of two monopoles separated 0.21λ, compared with the case without DN.

To conclude, we present some results and a real implementation of a DN for a 2×1 linear MEA consisting of two monopoles separated by a distance of 0.21λ. Fig. 6.24 shows a picture of the fabricated decoupling network noted above, for a 2×1 linear MEA, using the $180°$ hybrid as a decoupling network. The associated three-dimensional normalized far-field radiation patterns at the output of the DN are represented on the right of Fig. 6.24. To understand these three-dimensional radiation pattern plots, one must know that the radial axis (in the polar plot) represents the $\theta \in [0°, 180°]$ angle, whereas the polar angle represents the $\phi \in [0°, 360°]$ angle. That is, the center of the plot corresponds to $\theta = 0°$. As expected, using Eq. (6.1) and assuming an ideal NLOS scenario, the orthogonality among the radiation patterns was verified.

Figs. 6.25 and 6.26 show the measured scattering parameters for two monopoles separated 0.21λ without a DN and with the aforementioned DN (see Fig. 6.24), respectively. It is interesting to observe that the correlation is minimized after including the DN. This is done by decoupling ($S_{ij} \approx 0$) and matching ($S_{ii} \approx 0$) all the ports of the antenna array. Notice also that the operational bandwidth of the resultant antennas (including the DN) has been reduced, although the performance has been improved within that bandwidth range.

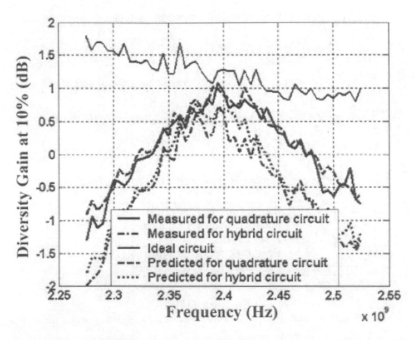

FIGURE 6.28: Measured improvement on the DG after selection among the output ports of the DN (line corresponding to the hybrid circuit) when connected to an antenna array consisting of two monopoles separated 0.21λ, compared with the case without DN.

Finally, Fig. 6.27 shows the improvement on the received power as a result of using the DN, compared with the case in which the coupled antennas are directly connected to the loads. On the other hand, Fig. 6.28 shows the improvement on the DG after selection among the output ports of the DN, compared with the case in which the coupled antennas are directly connected to the loads. Notice that for this particular case, using a DN gives a twofold benefit: matching efficiency and antenna decorrelation, thus producing larger DGs. The matching efficiency gains are in the order of 0.7 dB, whereas the diversity gains are in the order of 1 dB at the resonant frequency of the structure. Interestingly, notice that the performance of the DNs begins to degrade as we move away from the resonant frequency. Thus, these designs are interesting for narrowband MIMO systems. The receive power and DGs presented above correspond to channel measurements conducted in realistic NLOS propagation scenarios. Larger improvements would be expected as the scattering richness of the propagation increases. For the seek of completeness, Fig. 6.29 shows the improvements of using a DN in terms of the capacity of the system.

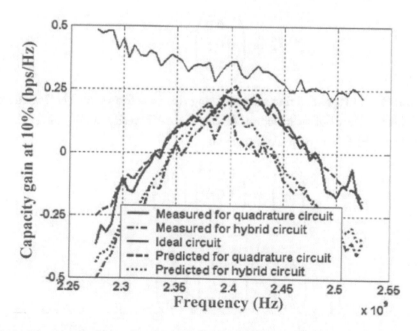

FIGURE 6.29: Measured improvement on the capacity at the output ports of the DN (line corresponding to the hybrid circuit) when connected to an antenna array consisting of two monopoles separated 0.21λ, compared with the case without DN.

6.3.4   Three-Element Circular MEA

We proceed now with another example. A three-element circular MEA, as that shown in Fig. 6.23b, has a scattering matrix, $\mathbf{S}^A$, of the form

$$\mathbf{S}^A_{\alpha\alpha} = \begin{pmatrix} S^A_{11} & S^A_{12} & S^A_{12} \\ S^A_{12} & S^A_{22} & S^A_{12} \\ S^A_{12} & S^A_{12} & S^A_{11} \end{pmatrix}. \tag{6.41}$$

As shown in Ref. [60], for this particular MEA, there exist two independent symmetry operators, given by the reflection $\overline{\mathbf{F}}$ and rotation $\overline{\mathbf{R}}$. The rotation transformation $\overline{\mathbf{R}}$, which interchanges components of the terminal vectors at symmetrically rotated ports of the MEA by $120°$, is clearly specified by the unitary matrix expressing a rotation:

$$\overline{\mathbf{R}} = \begin{pmatrix} 0 & 0 & 1 \\ 1 & 0 & 0 \\ 0 & 1 & 0 \end{pmatrix}. \tag{6.42}$$

On the other hand, the transformation $\overline{\mathbf{F}}$ is given by the following reflection operation:

$$\overline{\mathbf{F}} = \begin{pmatrix} 1 & 0 & 0 \\ 0 & 0 & 1 \\ 0 & 1 & 0 \end{pmatrix}. \tag{6.43}$$

For both operations $\overline{\mathbf{F}}$ and $\overline{\mathbf{R}}$, notice that after applying this transformation, the MEA structure is invariant with respect to the original structure. As one can easily verify in Fig. 6.23b, the particular $\overline{\mathbf{R}}$ operation, produces the following transformation:

$$\begin{pmatrix} a_3 \\ a_1 \\ a_2 \end{pmatrix} = \begin{pmatrix} 0 & 0 & 1 \\ 1 & 0 & 0 \\ 0 & 1 & 0 \end{pmatrix} \begin{pmatrix} a_1 \\ a_2 \\ a_3 \end{pmatrix} \tag{6.44}$$

$$\begin{pmatrix} b_3 \\ b_1 \\ b_2 \end{pmatrix} = \begin{pmatrix} 0 & 0 & 1 \\ 1 & 0 & 0 \\ 0 & 1 & 0 \end{pmatrix} \begin{pmatrix} b_1 \\ b_2 \\ b_3 \end{pmatrix}. \tag{6.45}$$

The rotation symmetry operator $\overline{\mathbf{R}}$ is of order 3, whereas the reflection symmetry operator $\overline{\mathbf{F}}$ is of order 2, that is, $\overline{\mathbf{R}}^3 = \mathbf{I}$ and $\overline{\mathbf{F}}^2 = \mathbf{I}$. We find the invariant subspaces in the usual fashion, for example, by solving the eigenvalue problem of $\overline{\mathbf{R}}$. The associated eigenvalues with operator

$\overline{\mathbf{R}}$ are $\lambda_1 = 1$ and $\lambda_2 = -\frac{1}{2} + \frac{\sqrt{3}}{2j}$ and $\lambda_3 = -\frac{1}{2} - \frac{\sqrt{3}}{2j}$. The invariant subspaces are therefore nondegenerated because all the eigenvalues are different, and the eigenvectors are given by

$$\mathbf{v}_1 = \frac{1}{\sqrt{2}} \begin{pmatrix} 1 \\ 1 \\ 1 \end{pmatrix} \tag{6.46}$$

$$\mathbf{v}_2 = \frac{1}{\sqrt{2}} \begin{pmatrix} 1 \\ \lambda_3 \\ \lambda_2 \end{pmatrix} \tag{6.47}$$

$$\mathbf{v}_3 = \frac{1}{\sqrt{2}} \begin{pmatrix} 1 \\ \lambda_2 \\ \lambda_3 \end{pmatrix}. \tag{6.48}$$

Notice that the above eigenvectors are not real. As noted in Section 6.3.2, because the symmetry operators procedure is based on the EVD of the symmetry operator to find the $\mathbf{V}^A$ matrices, it is important that the final result is such that $\mathbf{V}^A$ is composed of pure real and orthogonal vectors. In that case, $(\mathbf{V}^A)^{-1} = (\mathbf{V}^A)^\dagger$, and therefore, we have a reciprocal design for the DN that verifies

$$\mathbf{S}^A_{\alpha\alpha} = \mathbf{V}^A \mathbf{D}^A (\mathbf{V}^A)^{-1} = \mathbf{V}^A \mathbf{D}^A (\mathbf{V}^A)^\dagger. \tag{6.49}$$

In Ref. [60], it is shown that it is possible to choose eigenvectors of $\mathbf{S}^A_{\alpha\alpha}$ that are real by doing linear combinations of those above. This happens when we choose the following set of independent and orthonormal vectors, which span the same subspace as the previous ones, given by

$$\mathbf{w}_1 = \frac{1}{\sqrt{3}} \mathbf{v}_1 \tag{6.50}$$

$$\mathbf{w}_1 = \frac{1}{\sqrt{6}} (\mathbf{v}_2 + \mathbf{v}_3) \tag{6.51}$$

$$\mathbf{w}_1 = \frac{\mathbf{v}}{\sqrt{6}} j(\mathbf{v}_2 - \mathbf{v}_3), \tag{6.52}$$

that is,

$$\mathbf{w}_1 = \frac{1}{\sqrt{3}} \begin{pmatrix} 1 \\ 1 \\ 1 \end{pmatrix} \tag{6.53}$$

$$\mathbf{w}_1 = \frac{1}{\sqrt{6}} \begin{pmatrix} 2 \\ -1 \\ -1 \end{pmatrix} \tag{6.54}$$

$$\mathbf{w}_1 = \frac{1}{\sqrt{2}} \begin{pmatrix} 0 \\ 1 \\ -1 \end{pmatrix}. \tag{6.55}$$

Therefore, the transformation matrix $\mathbf{W}^A$ is given by arranging the previous eigenvectors in the following way:

$$\mathbf{W}^A = \begin{pmatrix} \frac{1}{\sqrt{3}} & \frac{2}{\sqrt{6}} & 0 \\ \frac{1}{\sqrt{3}} & -\frac{1}{\sqrt{6}} & \frac{1}{\sqrt{2}} \\ \frac{1}{\sqrt{3}} & -\frac{1}{\sqrt{6}} & -\frac{1}{\sqrt{2}} \end{pmatrix}. \tag{6.56}$$

We can finally decompose $\mathbf{S}_{\alpha\alpha}^A$ into

$$\mathbf{S}_{\alpha\alpha}^A = \mathbf{W}^A \mathbf{D}^A (\mathbf{W}^A)^\dagger, \tag{6.57}$$

with

$$\mathbf{D}^A = \begin{pmatrix} S_{11}^A + 2S_{12}^A & 0 & 0 \\ 0 & S_{11}^A - S_{12}^A & 0 \\ 0 & 0 & S_{11}^A - S_{12}^A \end{pmatrix}, \tag{6.58}$$

which is the desired solution. Following the same approach as in Section 6.3.3, with the computed $\mathbf{W}^A$, one obtains that the receiving lossless and reciprocal decoupling network must have the following parameters:

$$\mathbf{S}^L = \begin{pmatrix} 0 & 0 & 0 & \frac{1}{\sqrt{3}} & \frac{2}{\sqrt{6}} & 0 \\ 0 & 0 & 0 & \frac{1}{\sqrt{3}} & -\frac{1}{\sqrt{6}} & \frac{1}{\sqrt{2}} \\ 0 & 0 & 0 & \frac{1}{\sqrt{3}} & -\frac{1}{\sqrt{6}} & -\frac{1}{\sqrt{2}} \\ \frac{1}{\sqrt{3}} & \frac{1}{\sqrt{3}} & \frac{1}{\sqrt{3}} & 0 & 0 & 0 \\ \frac{2}{\sqrt{6}} & -\frac{1}{\sqrt{6}} & -\frac{1}{\sqrt{6}} & 0 & 0 & 0 \\ 0 & \frac{1}{\sqrt{2}} & -\frac{1}{\sqrt{2}} & 0 & 0 & 0 \end{pmatrix}. \tag{6.59}$$

Notice that the above decoupling network may not be easily implementable with microwave circuits; in such cases, one may use lumped components.

6.3.5   The 2 × 2 Planar MEA

Finally, we present the design for the DN of a 2 × 2 planar MEA. A 2 × 2 planar MEA, such as that shown in Fig. 6.23c, has a scattering matrix, $\mathbf{S}^A$, of the form

$$\mathbf{S}^A_{\alpha\alpha} = \begin{pmatrix} S^A_{11} & S^A_{12} & S^A_{13} & S^A_{12} \\ S^A_{12} & S^A_{11} & S^A_{12} & S^A_{13} \\ S^A_{13} & S^A_{12} & S^A_{11} & S^A_{12} \\ S^A_{12} & S^A_{13} & S^A_{12} & S^A_{11} \end{pmatrix}. \qquad (6.60)$$

As shown in Ref. [60], for this particular MEA, there exist several independent symmetry operators. In Ref. [60], it is shown that the symmetry operators associated with this MEA have degenerated eigenvalues. In the aforementioned reference, however, a mathematical procedure is given to find a set of orthogonal eigenvectors from a linear combination of some of the independent symmetry operators. Therefore, following the approach in Ref. [60], it is possible to find a matrix $\mathbf{V}^A$, such that

$$\mathbf{V}^A = \frac{1}{2} \begin{pmatrix} 1 & 1 & 1 & 1 \\ 1 & -1 & 1 & -1 \\ 1 & 1 & -1 & -1 \\ 1 & -1 & -1 & 1 \end{pmatrix}, \qquad (6.61)$$

with which we can decompose $\mathbf{S}^A_{\alpha\alpha}$ into

$$\mathbf{S}^A_{\alpha\alpha} = \mathbf{V}^A \mathbf{D}^A (\mathbf{V}^A)^\dagger \qquad (6.62)$$

and where

$$\mathbf{D}^A = \begin{pmatrix} S^A_{11} + 2S^A_{12} + S^A_{13} & 0 & 0 & 0 \\ 0 & S^A_{11} - 2S^A_{12} + S^A_{13} & 0 & 0 \\ 0 & 0 & S^A_{11} - S^A_{13} & 0 \\ 0 & 0 & 0 & S^A_{11} - S^A_{13} \end{pmatrix}. \qquad (6.63)$$

Following the same approach as in Section 6.3.3, with the computed $\mathbf{V}^A$, one obtains that the receiving lossless and reciprocal decoupling network must have the following parameters:

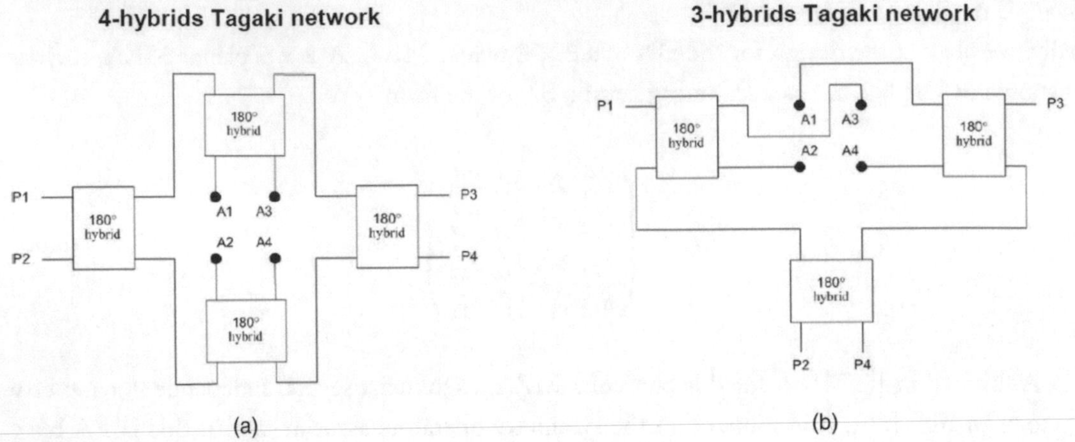

4-hybrids Tagaki network

3-hybrids Tagaki network

(a)

(b)

FIGURE 6.30: Schematic of two possible decoupling networks for a 2×2 planar MEA using $180°$ hybrids: (a) decoupling network using four hybrids and (b) another solution using three hybrids.

$$
\mathbf{S}^L = \frac{1}{2}
\begin{pmatrix}
0 & 0 & 0 & 0 & 1 & 1 & 1 & 1 \\
0 & 0 & 0 & 0 & 1 & -1 & 1 & -1 \\
0 & 0 & 0 & 0 & 1 & 1 & -1 & -1 \\
0 & 0 & 0 & 0 & 1 & -1 & -1 & 1 \\
1 & 1 & 1 & 1 & 0 & 0 & 0 & 0 \\
1 & -1 & 1 & -1 & 0 & 0 & 0 & 0 \\
1 & 1 & -1 & -1 & 0 & 0 & 0 & 0 \\
1 & -1 & -1 & 1 & 0 & 0 & 0 & 0
\end{pmatrix}.
\tag{6.64}
$$

As shown in Fig. 6.30a, the above decoupling network can be implemented using four $180°$ hybrids. The associated orthogonal radiation patterns at the output ports of the DN for a 2×2 MEA using monopoles separated 0.21λ are given in Fig. 6.31a.

Using another independent symmetry operator, one can find other solutions. As shown in Ref. [60], another solution is that given by the following transformation network $\mathbf{V}^B$, such that

$$
\mathbf{V}^B = \frac{1}{2}
\begin{pmatrix}
0 & -\sqrt{2} & -1 & 1 \\
-\sqrt{2} & 0 & 1 & 1 \\
0 & \sqrt{2} & -1 & 1 \\
\sqrt{2} & 0 & 1 & 1
\end{pmatrix}.
\tag{6.65}
$$

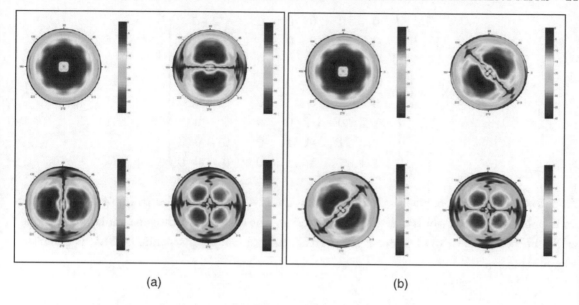

(a)   (b)

FIGURE 6.31: Associated orthogonal radiation patterns at the output of the DN for the 2×2 MEA using monopoles separated 0.21λ. Note in (a) those corresponding to the four-hybrid decoupling network and in (b) those associated with the three-hybrid solution.

We can finally decompose $\mathbf{S}_{\alpha\alpha}^{B}$ into

$$\mathbf{S}_{\alpha\alpha}^{B} = \mathbf{V}^{B}\mathbf{D}^{B}(\mathbf{V}^{B})^{\dagger}, \qquad (6.66)$$

with

$$\mathbf{D}^{B} = \begin{pmatrix} S_{11}^{A} - S_{13}^{A} & 0 & 0 & 0 \\ 0 & S_{11}^{A} - S_{13}^{A} & 0 & 0 \\ 0 & 0 & S_{11}^{A} - 2S_{12}^{A} + S_{13}^{A} & 0 \\ 0 & 0 & 0 & S_{11}^{A} + 2S_{12}^{A} + S_{13}^{A} \end{pmatrix}. \qquad (6.67)$$

With the computed $\mathbf{V}^{B}$, one obtains that the lossless and reciprocal decoupling network must have the following parameters:

$$\mathbf{S}^L = \frac{1}{2} \begin{pmatrix} 0 & 0 & 0 & 0 & 0 & -\sqrt{2} & -1 & 1 \\ 0 & 0 & 0 & 0 & -\sqrt{2} & 0 & 1 & 1 \\ 0 & 0 & 0 & 0 & 0 & \sqrt{2} & -1 & 1 \\ 0 & 0 & 0 & 0 & \sqrt{2} & 0 & 1 & 1 \\ 0 & -\sqrt{2} & 0 & \sqrt{2} & 0 & 0 & 0 & 0 \\ -\sqrt{2} & 0 & \sqrt{2} & 0 & 0 & 0 & 0 & 0 \\ -1 & 1 & -1 & 1 & 0 & 0 & 0 & 0 \\ 1 & 1 & 1 & 1 & 0 & 0 & 0 & 0 \end{pmatrix}. \tag{6.68}$$

As shown on Fig. 6.30b, a decoupling network for a 2×2 MEA can also be implemented based on the above network, which uses three $180°$ hybrids. The associated orthogonal radiation patterns at the output ports of the DN for a 2×2 MEA using monopoles separated 0.21λ, is given in Fig. 6.31b.

6.3.6   Performance Analysis of MPAs Using DNs

As discussed previously in Chapter 5, besides the design methodology described above, it is now important to consider the practical implementation limitation of the DNs. In this context, we should remark that DNs should be used in those cases in which the matching efficiency gains that can be recovered by using DNs are much larger than the insertion losses introduced by the DNs itself. Otherwise, the use of DNs will end up having a detrimental effect on the system. Also, the use of DNs in general reduces the operative bandwidth of the resultant MPA; thus, this needs to be taken into account, especially for wideband MIMO communication systems.

Once the design of the DN has been concluded, it is interesting to evaluate its performance within a complete MIMO communication system. The procedure to follow is based on the computation of the transmit and receive correlation matrix, as explained in 6.1. From these quantities, it is then possible to estimate the matching efficiency, capacity, and the diversity order, among other quantities, as shown in Chapter 4.

6.4   PERFORMANCE PREDICTION OF DIFFERENT MPAs

In this section, we comment on the impact of several MPA parameters on the performance of MIMO system. We can differentiate several of these parameters, such as the geometry and size of the transmit and receive MPAs, the polarization sensitivity of the antennas, and the operating frequency of the MPAs, among others. In this section, we will comment on the basic topology principles of multiantenna systems in an attempt to give a qualitative explanation of the impact of the aforementioned properties on the performance of a MIMO communication system.

6.4.1 Geometry and Size of an MPA

The size of an array determines the finesse with which different directions of propagation can be distinguished. Each one of the possible directions of propagation of a wave in an open space is described by $\mathbf{k}$. In spherical coordinates, it can be represented by all the linear combinations of the elevation and the azimuth direction vectors, $\widehat{\theta}$ and $\widehat{\phi}$, respectively. That is, $\mathbf{k} = k_\theta \widehat{\theta} + k_\phi \widehat{\phi}$. Then, for example, a larger antenna array is able to form narrower beams, and hence, it can point into space with smaller solid angles. This can intuitively be seen from the fact that the radiation pattern of an antenna is, in fact, the Fourier transform of the currents on the antenna. Therefore, the larger the array, the more directive the antenna can be. This allows the antenna array to be more directive and thus to distinguish more precisely among signals coming from different propagation directions.

Using the approach in Ref. [88], one can define $|\Omega_T|$ and $|\Omega_R|$ as the areas of projection of the scattering clusters onto the unit sphere enclosing the transmit and receive array respectively, as shown in Figs. 6.32 and 6.33. For example, in the ideal fully scattered environment, the channel solid angle is 4π, the whole surface of the unit sphere. To determine the degrees of freedom for a given channel response and array configuration, the decomposition in singular values approach or

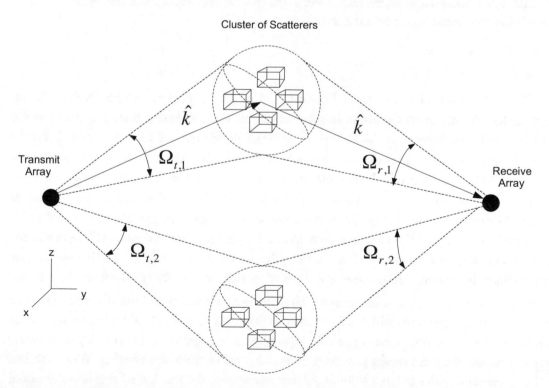

FIGURE 6.32: Graphical representation of the transmit and receive power spectra.

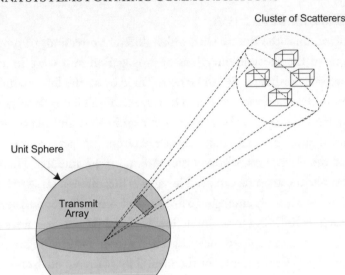

FIGURE 6.33: Shaded area on the unit sphere is the channel solid angle subtended by the scattering cluster being illuminated by the transmit array.

the solid angles approach can be used. This last gives us a more physical intuitive insight. Let us recall the fact that an array of a certain aperture size is able to create beams that illuminate a region in space, and those regions can also be represented with an effective solid angle $\frac{1}{A^T}$ and $\frac{1}{A^R}$ for the transmitting and receiving arrays, respectively, where A^T and A^R are the aperture size of the transmit and receive MPAs, respectively. Notice that the resolution directly depends on the dimensions of the array. On the other hand, the number of degrees of freedom of the channel is computed as the ratio of the channel solid angle $|\Omega|$ to the array resolution given by the inverse of the effective aperture of the array $\frac{1}{A}$. This is an equivalent approach to the number of degrees of freedom in the temporal domain, on a communication system with bandwidth B over which a transmission time of t is defined. In this case, the number of degrees of freedom is $\frac{B}{\frac{1}{T}} = BT$ (a waveform of duration t can fit BT times into a given bandwidth B). Similarly then, for the spatial domain, the degrees of freedom are, in general, $\min(|\Omega_t|A_t, |\Omega_r|A_r)$. An intuitive way to understand how many degrees of freedom exist is to count how many times the equivalent solid angle of the transmitting or receiving array fits within the solid angles generated by the channel because of scattering. When the size of the arrays and/or the channel solid angles are substantial, the number of degrees of freedom is large, due in part to a finer resolution in focusing the energy due to the arrays. Including the twofold benefits of polarization, the total number of degrees (spatial and temporal) of freedom of a multiantenna system for MIMO is, therefore, $2BTA|\Omega|$.

On the other hand, the geometry of the array determines the efficiency in resolving the two directions, $\widehat{\theta}$ and $\widehat{\phi}$ and the resolution uniformity, that is, to what degree the beam-width changes along those two directions. For example, a linear array can resolve the elevation direction, although its resolution is not uniform over it—that is, the beam width and gain are not constant depending on the elevation angle. A circular array can resolve either the azimuth or the elevation directions and has uniform resolution over the azimuth but nonuniform resolution over the elevation. On the other hand, a spherical array could resolve both the elevation and azimuth and have uniform resolution over them both.

As noted before, the spherical modes of decomposition can help us to link array properties such as its size and geometry to the concept of MIMO capacity and to help evaluate the performance of a MIMO system, as shown in Ref. [88]. From the spherical modes analysis [88], it can be seen that the array size and geometry determine how many of those spherical harmonics are significant or easy to excite. In particular, assuming a totally scattered NLOS environment, $\Omega = 4\pi$ and the number of these significant modes are $4L$ for linear arrays of length $2L$, $2\pi R$ for circular arrays of radius R and $4\pi A$ for spherical arrays of effective aperture A. Notice that L, R, and A are dimensions normalized to a wavelength. For example, for a linear array with $L = 1$, that is, with a size of $2L = 2$ wavelengths, the number of uncorrelated propagation modes is $4L = 4$. If we account for the polarization effects, then a twofold benefit exists, and the total number of degrees of freedom is 8. In Ref. [88], it is shown that this number matches with the number of antenna elements, four in this particular case, presumed in the conventional statistical fading model where the antennas are assumed to be separated 0.5λ, for similar capacity values. Given the size of the arrays, these numbers form the upper bound on the number of spatial degrees of freedom independent of the number of antennas. Packing in more antennas beyond these limits will not increase the channel capacity significantly. In fact, if one puts too many antennas together, the capacity may even get worse because of the detrimental effects of mutual coupling.

Finally, in Ref. [88], the $|\Omega|A$ product is used to determine, for a given channel and antenna characteristics, which of the two techniques, beam forming (or array gain) or multiplexing gain, is more beneficial. Using array gain, the transmitter focuses its energy toward the receiver using mainly one single propagation path or, in other words, using one of the solid angles offered by the channel. SM instead divides the power among several channel solid angles or among different beams within a given solid angle (typically large) and uses a parallel transmission scheme. When the channel solid angles are very few and very small, that is, when the $|\Omega|A$ product is small, such as in LOS propagation, the multiplexing gain approach performs worse than the array gain approach, and hence, it is better to send all the energy in one specific direction. On the other hand, in NLOS propagation, the $|\Omega|A$ product is large—that is, the number of degrees of freedom or uncorrelated propagation paths are large—and the SM/diversity approach is a better technique.

6.4.2 Polarization Sensitivity of an MPA

As noted in the previous section, the spatial degrees of freedom obtained from resolving distinct directions in the propagation space are determined by the size of the antenna array. Now, for each propagation direction, there exists an additional degree of freedom given by the polarization, that is, by the direction of oscillation of the propagating field.

Let us restate that the number of degrees of freedom of a system can be limited because of three reasons: the characteristics of the propagation channel, the resolution of the transmitter array, and the resolution of the receiver array. In Ref. [88], it is shown that there is, at most, a threefold increase in degrees of freedom, corresponding to the three orthogonal directions of polarization in space. Being more specific, in the intermediate and near-field regions, in Ref. [88], it is shown that a threefold increase is attainable. However, in the far-field region, the increase is limited to twofold, and the limitation comes from the inherent dependence of the electric and magnetic field on the transmitter and receiver array.

On the other hand, in Ref. [89], it is also shown how a tri-dipole antenna is used for tripling the capacity of a wireless communication using polarization techniques. This would conflict with the expected twofold increase in the number of degrees of freedom noted above. To explain this, in Ref. [89], the system is using a combination of different propagation directions and different polarizations—that is, a combination of polarization and pattern diversity techniques. Combining both, it is possible to achieve a threefold benefit. The approach followed in Ref. [88] instead makes a substantial effort to distinguish the impact of the propagation direction and the polarization of the electric field on the performance of a MIMO system.

6.4.3 Operating Frequency of an MPA

It is commonly believed that by operating at a higher frequency, it is possible to increase the degrees of freedom of a MIMO system by packing more antennas into the same wireless device. This idea can be easily seen in the following formula from Ref. [88] that expresses the number of degrees of freedom as follows:

$$\frac{1}{\lambda^2}\min\{2A^T|\Omega_T|, 2A^R|\Omega_R|\}, \tag{6.69}$$

where λ represents the wavelength associated with the operating frequency. However, as the frequency of operation increases, the channel solid angles Ω_T and Ω_R—that is, the existing propagable paths between the transmitter and receiver—decrease. There are two reasons that explain this phenomenon: first, the electromagnetic waves of higher frequency attenuate more after passing through or bouncing off channel objects, which reduces the number of scattering clusters; second, at high frequency, the wavelength is small relative to the feature size of typical objects on the channel, so scattering appears to be more specular in nature and reduces the EDOF available in the channel. In Ref. [88], based on the aperture size of the array and the solid angles of the channel, the degrees of freedom are estimated as a function of the frequency of operation. There, it is shown that the available degrees of freedom do not always increase with frequency, and in fact, it is possible to find an optimal operating frequency. Interestingly, the optimal frequencies for typical office and residential environments are between 5 and 6 GHz, which is where the IEEE 802.11a standard is located. Therefore, it is difficult to evaluate the performance of a MIMO system in different environments and frequency bands. It is believed that the performance on each band depends on the dimensions and dielectric properties of the scattering sources and the sensitivity of the transceivers.

·   ·   ·   ·

References

[1] G. Foschini and M. J. Gans, On limits of wireless communications in a fading environment when using multiples antennas, *Wireless Personal Communications* **6** (1998) 331–335.

[2] M. A. Jensen and J. W. Wallace, A review of antennas and propagation for MIMO wireless communications, *IEEE Transactions on Antennas and Propagation* **25**(11) (2004) 2810–2824. doi:10.1109/TAP.2004.835272

[3] C.-J. Ahn, Y. Kamio, S. Takahashi, and H. Harada, Reverse link performance improvement for an wideband OFDM using Alamouti coded heterogeneous polarization antennas, *Proceedings of the 9th IEEE Symposium on Computers and Communications 2004, ISCC 2004)* **2** (2004) 702–707.

[4] R. U. Nabar, H. Bolcskei, V. Erceg, D. Gesbert, and A. J. Paulraj, Performance of multi-antenna signaling techniques in the presence of polarization diversity, *Proceedings of the 9th IEEE Symposium on Computers and Communications 2004, ISCC 2004)* **50**(10) (2002) 2553–2562.

[5] T. Svantesson, On the potential of multimode antenna diversity, *IEEE 52nd Vehicular Technology Conference, 2000 (VTC 2000 Fall)* **5** (2000) 2368–2372. doi:10.1109/VETECF. 2000.883290

[6] C. Waldschmidt and W. Wiesbeck, Compact wide-band multimode antennas for MIMO and diversity, *IEEE Transactions on Antennas and Propagation* **52**(8) (2004) 1963–1969. doi:10.1109/TAP.2004.832495

[7] S. Verdu, Fifty years of Shannon theory, *IEEE Transactions on Information Theory* **44**(6) (1998) 2057–2078. doi:10.1109/18.720531

[8] H. Nyquist, Certain topics in telegraph transmission theory, *Transactions of the AIEE* **47**(1928) 617–644. Reprint as classic paper in: Proc. IEEE, Vol. 90, No. 2, Feb 2002. doi:10.1109/5.989875

[9] R. V. L. Hartley, Transmission of information, Bell System Technical Journal (July 1928).

[10] C. A. Balanis, *Antenna Theory, Analysis and Design*, 2nd ed., Wiley (1997). doi:10.1109/ 5.119564

[11] J. B. Keller, Geometrical theory of diffraction, *Journal of the Optical Society of America* **5**(2) (1962) 116–130.

[12]   A. Grau, J. Romeu, L. Jofre, and F. De Flaviis, On the polarization diversity gain using the ORIOL antenna in fading indoor environments, *Antennas and Propagation Society International Symposium, 2005 IEEE* **4B** (2005) 14–17.

[13]   A. Grau, J. Romeu, S. Blanch, L. Jofre, and F. De Flaviis, Optimization of linear multi-element antennas for selection combining by means of a Butler matrix in different MIMO environments, *IEEE Transactions on Antennas and Propagation* **54**(11, Pt. 1)(2006) 3251–3264.

[14]   J. W. Wallace and M. A. Jensen, Mutual coupling in MIMO wireless system: a rigorous network theory analysis, *IEEE Transactions on Wireless Communications* **3**(4) (2004) 1317–1325. doi:10.1109/TWC.2004.830854

[15]   C. Waldschmidt, S. Schulteis, and W. Wiesbeck, Complete RF system model for analysis of compact MIMO arrays, *IEEE Transactions on Vehicular Technology* **53**(3) (2004) 579–586.

[16]   A. Grau, H. Jafarkhani, and F. De Flaviis, A reconfigurable multiple-input multiple-output communication system, *IEEE Transactions on Wireless Communications*, submitted.

[17]   R. A. Horn and C. R. Johnson, *Matrix Analysis*, 1st ed., Cambridge University Press (1985).

[18]   J. P. Kermoal, L. Schumacher, K. I. Pedersen, P. E. Mogensen, and F. Frederiksen, A stochastic MIMO radio channel model with experimental validation, *IEEE Journal on Selected Areas in Communications* **20**(6) (2002) 1211–1226.

[19]   M. Ozcelik, N. Czink, and E. Bonek, What makes a good MIMO channel model?, *IEEE 62nd Semiannual Vehicular Technology Conference, 2005 (VTC 2005 Spring)* **1** (2005) 156–160. doi:10.1109/VETECS.2005.1543269

[20]   A. M. Sayeed, Deconstructing multiantenna fading channels, *IEEE Transactions on Signal Processing* **50**(10) (2002) 2563–2579 [see also IEEE Transactions on Acoustics, Speech, and Signal Processing]. doi:10.1109/TSP.2002.803324

[21]   TGn channel models, IEEE Technical Report 802.11-03/940r4, *Antennas and Propagation Society International Symposium, 2004 IEEE* (2004).

[22]   C. Waldschmidt, J. v. Hagen, and W. Wiesbeck, Influence and modelling of mutual coupling in MIMO and diversity systems, *Antennas and Propagation Society International Symposium 2002, IEEE* **3** (2002) 16–21. doi:10.1109/APS.2002.1018187

[23]   D. Chizhik, G. J. Foschini, M. J. Gans, and R. A. Valenzuela, Keyholes, correlations, and capacities of multielement transmit and receive antennas, *IEEE Transactions on Wireless Communications* **1**(2) (2002) 361–369. doi:10.1109/7693.994830

[24]   S. Lin and D. J. Costello, *Error Control Coding*, 2nd ed., Prentice-Hall (2004).

[25] A. Paulraj, R. Nabar, and D. Gore, *Introduction to Space–Time Wireless Communications*, 2nd ed., Cambridge University Press (2003).

[26] H. Jafarkhani, *Space–Time Coding: Theory and Practice*, 1st ed., Cambridge University Press (2005).

[27] A. F. Molisch, *Wireless Communications*, IEEE-Wiley (2005).

[28] S. R. Saunders *Antennas and Propagation for Wireless Communication Systems*, Wiley (1999).

[29] R. Vaughan and J. B. Andersen, *Channel, Propagation and Antennas for Mobile Communications*, IEEE (2003).

[30] T. Zwick, C. Fischer, D. Didascalou, and W. Wiesbeck, A stochastic spatial channel model based on wave-propagation modeling, *IEEE Transactions on Selected Areas in Communications* **18**(1) (2000) 6–15. doi:10.1109/49.821698

[31] A. Cardama, L. Jofre, J. M. Rius, J. Romeu, S. Blanch and M. Ferrando, *Antennas*, IEEE, Washington, DC (2003).

[32] M. Martone, *Multiantenna. Digital Radio Transmission*, Artech House (2002).

[33] W. C. Jakes, *Microwave Mobile Communications*, IEEE Press (1974).

[34] A. M. Saleh and R. A. Valenzuela, A statistical model for indoor multipath propagation, *IEEE Transactions on Selected Areas in Communications* **5**(2) (1987) 128–137. doi:10.1109/JSAC.1987.1146527

[35] A. M. Sayeed, Characterization of randomly time-variant linear channels, *IEEE Transactions on Communication Systems* **11** (1993) 967–978.

[36] M. Hata, Empirical formula for propagation loss in land mobile radio services, *IEEE Transactions on Vehicular Technology* **29** (1980) 317–325.

[37] International Telecommunication Union, *ITU-R Recommendation 1238: Propagation Data and Prediction Models for the Planning of Indoor Radiocommunication Systems and Radio Local Are Networks in the Frequency Range 900 MHz to 100 GHz*, International Telecommunication Union, Geneva (1997).

[38] T. Zwick, C. Fischer, and W. Wiesbeck, A stochastic multipath channel model including path directions for indoor environments, *IEEE Transactions on Selected Areas in Communications* **20**(6) (2002) 1178–1192. doi:10.1109/JSAC.2002.801218

[39] W. Braun and U. Dersch, A physical mobile radio channel model, *IEEE Transactions on Vehicular Technology* **40**(2) (1991) 472–482.

[40] D. Chizhik, F. Rashid-Farrokhi, J. Ling, and A. Lozano, Effect of antenna separation on the capacity of BLAST in correlated channels, *IEEE Communication Letters* **4**(11) (2000) 337–349. doi:10.1109/4234.892194

[41] S. Wang, A. Abdi, J. Salo, H. M. El Sallabi, J. W. Wallace, P. Vainikainen, and A. Jensen, Time-varying MIMO channels: parametric statistical modeling and experimental results, *IEEE Transactions on Vehicular Technology* **4**(56) (2007) 1949–1963.

[42] V. Tarokh, N. Seshadri, and A. R. Calderbank, Space–time codes for high data rate wireless communication: performance analysis and code construction, *IEEE Transactions on Information Theory* **44**(2) (1998) 744–765.

[43] V. Tarokh, A. Naguib, N. Seshadri, and A. R. Calderbank, Space–time codes for high data rate wireless communication: performance criteria in the presence of channel estimation errors, mobility, and multiple paths, *IEEE Transactions on Communications* **47**(2) (1999) 199–207.

[44] L. Zheng and D. N. C. Tse, Diversity and multiplexing: a fundamental tradeoff in multiple-antenna channels, *IEEE Transactions on Information Theory* **49**(5) (2003) 1073–1096.

[45] H. Bolcskei and A. J. Paulraj, Performance of space–time codes in the presence of spatial fading correlation, *Conference Record of the Thirty-Fourth Asilomar Conference on Signals, Systems and Computers, 2000* **1** (2000) 687–693. doi:10.1109/ACSSC.2000.911041

[46] R. E. Jaramillo, O. Fernandez, and R. P. Torres, Empirical analysis of a 2 × 2 MIMO channel in outdoor–indoor scenarios for BFWA applications, *IEEE Antennas and Propagation Magazine* **48**(6) (2006) 57–69.

[47] D. Samardzija and N. Mandayam, Pilot-assisted estimation of MIMO fading channel response and achievable data rates, *Transactions on Signal Processing, IEEE* **51**(11) (2003) 2882–2890. doi:10.1109/TSP.2003.818158

[48] J. Yang, Y. Sun, J. M. Senior, and N. Pem, Channel estimation for wireless communications using space–time block coding techniques, *Proceedings of the 2003 International Symposium on Circuits and Systems 2003, ISCAS 2003* **2** (2003) 220–223.

[49] Z. Chen, J. Yuan, and B. Vucetic, Analysis of transmit antenna selection/maximal-ratio combining in Rayleigh fading channels, *IEEE Transactions on Vehicular Technology* **54**(4) (2005) 1312–1321. doi:10.1109/TVT.2005.851319

[50] S. M. Alamouti, A simple transmit diversity technique for wireless communications, *IEEE Journal on Selected Areas in Communications* **16**(8) (1998) 1451–1458. doi:10.1109/49.730453

[51] V. Tarokh, H. Jafarkhani, and A. R. Calderbank, Space–time block codes from orthogonal designs, *IEEE Transactions on Information Theory* **45**(5) (1999) 1456–1467. doi:10.1109/49.730453

[52] H. Jafarkhani and N. Seshadri, Super-orthogonal space–time trellis codes, *IEEE Transactions on Information Theory* **49**(4) (2003) 937–950.

[53]   G. J. Foschini, Layered space–time architecture for wireless communication in a fading environment when using multi-element antennas, *Bell Laboratories Technical Journal* **1**(2) (1996) 41–59. doi:10.1002/bltj.2015

[54]   F. R. Farrokhi, A. Lozano, and G. J. Foschini and R. A. Valenzuela, Spectral efficiency of FDMA/TDMA wireless systems with transmit and receive antenna arrays, *IEEE Transactions on Wireless Communications* **1**(4) (2002) 591–599.

[55]   A. Lozano, F. R. Farrokhi, and R. A. Valenzuela, Asymptotically optimal open-loop space–time architecture adaptive to scattering conditions, *IEEE 54th Vehicular Technology Conference, 2001 (VTC 2001 Spring)* **1** (2001) 73–77. doi:10.1109/VETECS.2001.944806

[56]   D. M. Pozar, *Microwave Engineering*, 2nd ed., Wiley (1998). doi:10.1109/13.53636

[57]   W. Wasylkiwskyj, Response of an antenna to arbitrary incident fields, *Antennas and Propagation Society International Symposium, 2005 IEEE* **3B** (2005) 39–42. doi:10.1109/APS.2005.1552426

[58]   W. Wmylkiwskyj and W. K. Kahn, Theory of mutual coupling among minimum scattering antennas, *IEEE Transactions on Antennas and Propagation* **18**(2) (1970) 204–216.

[59]   J. Hansen, P. Clarricoats, and J. B. Andersen, *Spherical Near-Field Antenna Measurements*, IEEE (1988).

[60]   E. R. Beringer, N. Marcuvitz, C. G. Montgomery, R. H. Dicke, and E. M. Purcell, *Principles of Microwave Circuits, Radiation Laboratory Series*, 1st ed., McGraw-Hill, New York (1945).

[61]   N. Marcuvitz, *Waveguide Handbook*, Radiation Lab Series, Vol. 10, McGraw-Hill, New York (1951).

[62]   H. A. Abdallah and W. Wasylkiwskyj, A numerical technique for calculating mutual impedance and element patterns of antenna arrays based on the characteristics of an isolated element, *IEEE Transactions on Antennas and Propagation* **53**(10) (2005) 3293–3299.

[63]   S. Nordebo, M. Gustafsson, and G. Kristensson, On the capacity of the free space antenna channel, *IEEE Transactions on Antennas and Propagation* (2006) 3105–3108. doi:10.1109/APS.2006.1711266

[64]   M. Gustafsson and S. Nordebo, Characterization of MIMO antennas using spherical vector waves, *IEEE Transactions on Antennas and Propagation* **54**(9) (2006) 2679–2682. doi:10.1109/TAP.2006.880777

[65]   N. Geng and W. Wiesbeck, Simulation of radio relay link performance using a deterministic 3D wave propagation model, *IEEE Conference 1993* **386** (1993) 343–348.

[66]   J. W. Wallace and M. A. Jensen, Termination-dependent diversity performance of coupled antennas: network theory analysis, *IEEE Transactions on Antennas and Propagation* **52**(1) (2004) 98–105. doi:10.1109/TAP.2003.822444

[67] E. A. Jorswieck and H. Boche, On the impact of correlation on the capacity in MIMO systems without CSI at the transmitter, *Proceedings of the Conference on Information Sciences and System, John Hopkins University* (2003).

[68] S. Blanch, J. Romeu, and I. Corbella, Exact representation of antenna system diversity performance from input parameter description, *Electronics Letters* **39**(9) (2003) 705–707. doi:10.1049/el:20030495

[69] B. K. Lau, J. B. Andersen, G. Kristensson, and A. F. Molisch, Impact of matching network on bandwidth of compact antenna array, *IEEE Transactions on Antennas and Propagation* **54**(11) (2006) 3225–3238.

[70] L. Schumaker, K. I. Pedersen, and P. E. Morgensen, From antenna spacing to theoretical capacities—guidelines for simulating MIMO systems, *The 13th IEEE International Symposium on Personal, Indoor and Mobile Radio Communications, 2002* **2** (2002) 587–592. doi:10.1109/PIMRC.2002.1047289

[71] H. Shin and J. H. Lee, Performance analysis of space–time block codes over keyhole MIMO channels, *Proceedings of the 14th IEEE 2003 International Symposium on Personal, Indoor and Mobile Radio Communication , 2003)*, (2003) 2933–2937. doi:10.1109/TVT.2004.823540

[72] L. J. Chu, Physical limitations of omni-directional antennas, *Journal of Applied Physics* **19** (1948) 1163–1175. doi:10.1063/1.1715038

[73] G. A. Thiele, P. L. Detweiler, and R. P. Penno, On the lower bound of the radiation Q for electrically small antennas, *IEEE Transactions on Antennas and Propagation* **51**(6) (2003) 1263–1269.

[74] R. F. Harrington, *Time-Harmonic Electromagnetic Fields*, IEEE Press (1961).

[75] M. L. Morris and M. A. Jensen, Network model for MIMO systems with coupled antennas and noisy amplifiers, *IEEE Transactions on Antennas and Propagation* **53**(1) (2005) 545–552.

[76] M. Gustafsson and S. Nordebo, On the spectral efficiency of a sphere, *Progress in Electromagnetics Research*, **67** (2007) 275–296. doi:10.2528/PIER06091202

[77] H. J. Chaloupka and X. Wang, On the properties of small array with closely spaced antenna elements, *Proceedings of the IEEE Antennas Propagation Society International Symposium, Monterey, CA* **3** (2004) 2699–2702. doi:10.1109/APS.2004.1331931

[78] S. Dossche, S. Blanch, and J. Romeu, Optimum antenna matching to minimise signal correlation on a two-port antenna diversity system, *Electronics Letters* **40**(19) (2004) 1164–1165. doi:10.1049/el:20045737

[79] D. C. Youla, Weissfloch equivalents for lossless 2n-ports, *IEEE Transactions on Circuit Theory* **7**(3) (1960) 193–199.

[80] M. K. Ozdemir, E. Arvas, and H. Arslan, Dynamics of spatial correlation and implications on MIMO systems, *IEEE Communications Magazine* **42**(6) (2004) 14–19.

[81] J. L. Allen, A theoretical limitation on the formation of lossless multiple beams in linear arrays, *IEEE Transactions on Antennas and Propagation* **3**(7) (1960) 350–352. doi:10.1109/TAP.1961.1145014

[82] W. K. Kahn and H. Kurss, The uniqueness of the lossless feed network for a multibeam array, *IEEE Transactions on Antennas and Propagation* **10**(1)(1962) 100–101. doi:10.1109/TAP.1962.1137823

[83] J. Butler and R. Lowe, Beam forming matrix simplifies design of electronically scanned antennas, *IEEE Transactions on Applied Superconductivity* **9** (1961) 170–173.

[84] L. Accatino, A. Angelucci, and B. Piovani, Design of Butler matrices for multiport amplifier applications, *Microwave Engineering Europe* (1992) 45–50.

[85] H. J. Moody, The systematic design of the Butler matrix, *IEEE Transactions on Antennas and Propagation* **12**(6) (1964) 786–788. doi:10.1109/TAP.1964.1138319

[86] Ansoft Corporation, *HFFS 10*, www.ansoft.com.

[87] W. K. Kahn, Scattering equivalent circuits for common symmetrical junctions, *IEEE Transactions on Circuit Theory* **3**(2)(1956) 121–127.

[88] A. S. Y. Poon, R. W. Brodersen, and D. N. C. Tse, Degrees of freedom in multiple-antenna channels: a signal space approach, *IEEE Transactions on Information Theory* **51**(2) (2005) 523–536.

[89] M. R. Andrews, P. P. Mitra, and R. deCarvalho, Tripling the capacity of wireless communications using electromagnetic polarization, *Nature* **409** (2001) 316–318.

[90] M. Muraguchi, T. Yukitake, and Y. Naito, Optimum design of 3 dB branch-line couplers using microstrip lines, *IEEE Transactions on Microwave Theory and Techniques* **83**(8) (1983) 674–678. doi:10.1109/TMTT.1983.1131568

[91] T. Kawai, Y. Kokubo, and I. Ohta, Broadband lumped-element 180-degree hybrids utilizing lattice circuits, *Microwave Symposium Digest, 2001 IEEE MTT-S International* **1**(20–25) (2001) 47–50. doi:10.1109/MWSYM.2001.966836

[92] F. Ali and A. Podell, A wide-band GaAs monolithic spiral quadrature hybrid and its circuit applications, *IEEE Journal of Solid-State Circuits* **26**(10) (1991) 1394–1398.

[93] F. Giannini and L. Scucchia, A double frequency 180° lumped-element hybrid, *Microwave and Optical Technology Letters* **33**(4) (2002) 247–251. doi:10.1002/mop.10288

[94] F.-L. Wong and K.-K. M. Cheng, A novel planar branch-line coupler design for dual-band applications, *Microwave Symposium Digest, 2004 IEEE MTT-S International* **2** (2004) 903–906.

[95] K.-K. M. Cheng and F.-L. Wong, A novel approach to the design and implementation of dual-band compact planar 90° branch-line coupler, *IEEE Transactions on Microwave Theory and Techniques* **52**(11) (2004) 2458–2463. doi:10.1109/TMTT.2004.837151

[96] T. Hirota, A. Minakawa, and M. Muraguchi, Reduced-size branch-line and rat-race hybrids for uniplanar MMIC's, *IEEE Transactions on Microwave Theory and Techniques* **38**(3) (1990) 270–275. doi:10.1109/22.45344

[97] J. Reed and G. J. Wheeler, A method of analysis of symmetrical four-port networks, *IEEE Transactions on Microwave Theory and Techniques* **4** (1956) 246–252.

[98] I.-H. Lin, M. DeVincentis, C. Caloz, and T. Itoh, Arbitrary dual-band components using composite right/left-handed transmission lines, *IEEE Transactions on Microwave Theory and Techniques* **4** (1956) 246–252. doi:10.1109/TMTT.2004.825747

[99] A. Corona and M. J. Lancaster, A high-temperature superconducting Butler matrix, *IEEE Transactions on Applied Superconductivity* **13**(4) (2003) 3867–3872. doi:10.1109/TASC.2003.820507

[100] Agilent Technologies (2004), http://www.agilent.com.

[101] M. Lisi, An introduction to multiport power amplifiers, *Microwave Engineering Europe* (1992) 43–48.

[102] T. A. Denidni and T. E. Libar, Wide band four-port Butler matrix for switched multi-beam antenna arrays, *Personal, Indoor and Mobile Radio Communications* **3** (2003) 2461–2464.

[103] C. Collado, A. Grau, and F. De Flaviis, A dual-band planar quadrature hybrid with enhanced bandwidth, *IEEE Transactions on Microwave Theory and Techniques* (submitted).

[104] M. Kavehrad and P. J. Mclane, Performance of low-complexity channel coding and diversity for spread spectrum in indoor, wireless communications, *ATT Technical Journal* **64**(8) (1985) 1927–1965.

[105] P. Petrus, J. H. Reed, and T. S. Rappaport, Geometrical-based statistical macrocell channel model for mobile environments, *IEEE Transactions on Communications* **50**(3) (2002) 495–502. doi:10.1109/26.990911

[106] S. Wang, K. Raghukumar, A. Abdi, J. Wallace, and M. Jensen, Indoor MIMO channel: a parametric correlation model and experiments results, *IEEE Transactions on Communications*.

[107] H. B. Andersen and H. H. Rasmussen, Decoupling and descattering networks for antennas, *IEEE Transactions on Antennas and Propagation* **24**(6) (1976) 841–846.

[108] A. Grau, S. Liu, B. A. Cetiner, and F. De Flaviis, Investigation of the influence of antenna array parameters on adaptive MIMO performance, *IEEE Transactions on Antennas and Propagation* **22** (2004) 1704–1707. doi:10.1109/APS.2004.1330524

[109] C. W. Jung and F. De Flaviis, RF-MEMS capacitive series switches for reconfigurable antenna application, *IEEE Transactions on Antennas and Propagation* (submitted).

[110] E. Shin and S. Safavi-Nacini, A simple theoretical model for polarization diversity reception in wireless mobile environments, *Antennas and Propagation Society International Symposium IEEE* **2** (1999) 1332–1335.

[111] A. Grau, J. Romeu, M.-J. Lee, S. Blanch, L. Jofre, and F. De Flaviis, Octagonal reconfigurable isolated orthogonal element antenna for enhanced receiving polarization diversity, *IEEE Transactions of Antennas and Propagation* (in preparation).

[112] C. Degen and W. Keusgen, Performance of polarization multiplexing in mobile radio systems, *IEEE Electronic Letters* **38**(25) (2002) 1730–1732.

[113] I. E. Telatar, Capacity of multi-antenna Gaussian channels, 1995 Technical Memorandum, Bell Laboratories, Lucent Technologies, *European Transactions on Telecommunications* **10**(6) (1999) 585–595.

[114] H. J. Chaloupka, X. Wang, and J. C. Coetzee, Performance enhancement of smart antennas with reduced element spacing, *Wireless Communications and Networking 2003, WCNC 2003, 2003 IEEE* **1**(16–20) (2003) 425–430. doi:10.1109/WCNC.2003.1200387

[115] C. W. Jung, M.-J. Lee, G. P. Li, and F. De Flaviis, Monolithic integrated re-configurable antenna with RF-MEMS switches fabricated on printed circuit board, *Industrial Electronics Society 2005, IECON 2005, 32nd Annual Conference of IEEE* (2005) 2329–2334.

[116] A. Ghrayeb and T. M. Duman, Performance analaysis of MIMO systems with antenna selection over quasi-static fading channels, *IEEE Transaction on Vehicular Technology* **52** (2003) 281–288.

[117] J. G. Proakis, *Digital Communications*, 3rd ed., McGraw-Hill, New York (1995).

[118] P. Panaia, C. Luxey, G. Jacquemod, R. Staraj, G. Kossiavas, L. Dussopt, F. Vacherand, and C. Billard, EMS-based reconfigurable antennas, *2004 IEEE International Symposium on Industrial Electronics* **1** (2004) 175–179.

[119] D. E. Anagnostou, G. Zheng, M. T. Chryssomallis, J. C. Lyke, G. E. Ponchak, J. Papapolymerou, and C. G. Christodoulou, Design, fabrication, and measurements of an RF-MEMS-based self-similar reconfigurable antenna, *Proceedings of the 2003 International Symposium on Circuits and Systems 2003, ISCAS 2003* **54** (2006) 422–432. doi:10.1109/TAP.2005.863399

[120] S. Sandhu and A. Paulraj, Space–time block codes: a capacity perspective, *Communications Letters, IEEE* **4**(12) (2000) 384–386. doi:10.1109/4234.898716

[121] C. W. Jung, M.-J. Lee, G. P. Li, and F. De Flaviis, Reconfigurable scan-beam single-arm spiral antenna integrated with RF-MEMS switches, *IEEE Transactions on Antennas and Propagation* **54**(2) (2006) 455–463. doi:10.1109/TAP.2005.863407

[122] R. G. Vaughan, Two-port higher mode circular microstrip antennas, *IEEE Transactions on Antennas and Propagation* **36**(3) (1988) 309–321. doi:10.1109/8.192112

[123] N. Michishita and H. Arai and Y. Ebine, Mutual coupling characteristics of patch array antenna with choke for repeater systems, *Antennas and Propagation 2003, ICAP 2003* **1**(31) (2003) 229–232.

[124] W. T. Slingsby and J. P. McGeehan, Antenna isolation measurements for on-frequency radio repeaters, *Antennas and Propagation 1995, ICAP 1995* **1**(4–7) (1995) 239–243. doi:10.1049/cp:19950300

[125] C. W. Jung and F. De Flaviis, RF-MEMS capacitive series switches of CPW and MSL configurations for reconfigurable antenna application, *Antennas and Propagation Society International Symposium, 2005 IEEE* **2** (2005) 425–428.

[126] A. Grau, M.-J. Lee, J. Romeu, H. Jafarkhani, L. Jofre, and Franco De Flaviis, A multi-functional MEMS-reconfigurable pixel antenna for narrowband MIMO Communications, *IEEE Antennas and Propagation Symposium 2007* (submitted).

[127] A. Grau, J. Romeu, L. Jofre, and F. De Flaviis, Back-to-back high-isolation iso-frequency repeater antenna using MEMS-reconfigurable-parasitics, *IEEE Antennas and Propagation Symposium 2007* (submitted).

[128] H.-P. Chang, J. Qian, B. A. Cetiner, F. De Flaviis, M. Bachman, G. P. Li, Design and process considerations for fabricating RF MEMS switches on printed circuit boards, *Journal of Microelectromechanical Systems* **14**(6) (2005) 1311–1322.

[129] M. DiZazzo, M. D. Migliore, F. Schettino, V. Patriarca, and D. Pinchera, A novel parasitic-MIMO antenna, *IEEE Antennas and Propagation Society International Symposium, 2006* (2006) 4447–4450.

[130] D. Esser and H. J. Chaloupka, Compact reactively reconfigurable multi-port antennas, *IEEE Antennas and Propagation Society International Symposium, 2006* (2006) 2309–2312.

[131] J. Weber, C. Volmer, K. Blau, R. Stephan, and M. A. Hein, Miniaturized antenna arrays using decoupling networks with realistic elements, *IEEE Transactions on Microwave Theory and Techniques* **54**(6) (2006) 2733–2740.

List of Acronyms

AWGN	additive white Gaussian noise
BER	bit error rate
BB	base band
BLAST	Bell Labs layered space–time
BP	band pass
BPSK	binary phase-shift keying
CDF	cumulative distribution function
CGD	coding gain distance
CIR	channel impulse response
CMSA	canonical minimum scattering antenna
CSI	channel state information
DG	diversity gain
DoA	direction of arrival
DoD	direction of departure
DN	decorrelating network
EDOF	effective degrees of freedom
EGC	equal gain combining
EVD	eigenvector decomposition
FER	frame error rate
FF	fast fading
GTD	geometric theory of diffraction
GTSVD	generalized Takagi singular value decomposition
iid	independent and identically distributed
LOS	line-of-sight
LTI	linear time invariant
LTV	linear time variant
MEA	multielement antenna
MIMO	multiple-input multiple-output
ML	maximum likelihood
MMA	multimode antenna

MRC	maximal ratio combining
MPA	multiport antenna
MPC	multipath component
MPOA	multipolarized antenna
MSA	minimum scattering antenna
NLOS	non-line-of-sight
OFDM	orthogonal frequency division multiplexing
OSTBC	orthogonal space–time block code
PCM	pulse coded modulation
PEP	pairwise error probability
PL	Propagation Loss
QOSTBC	quasi-orthogonal space–time block code
QPSK	quadrature phase-shift keying
RF	radio frequency
SC	selection combining
SF	slow fading
SISO	single-input single-output
SM	spatial multiplexing
SMG	spatial multiplexing gain
SNR	signal-to-noise ratio
SOSTTC	superorthogonal space–time trellis code
SPA	single-port antenna
STBC	space–time block code
ST	space–time
STC	space–time code
STTC	space–time trellis code
SVD	singular value decomposition
SWC	switching combining
TSVD	Takagi singular value decomposition
US	uncorrelated scattering
VSWR	voltage standing wave ratio
WSS	wide sense stationary
WSSUS	wide sense stationary uncorrelated scattering
XPR	cross-polarization ratio

List of Symbols

M	number of transmit antennas
N	number of receive antennas
C	capacity
C_E	ergodic capacity
$C_{\text{out},q}$	q% outage capacity
$\mathbf{H}$	channel matrix (single-frequency transfer function of a narrowband LTI channel)
$\mathbf{H}_{nm}$	(n,m)th entry of the channel matrix $\mathbf{H}$
$\mathbf{H}_0$	normalized channel matrix
$\widehat{\mathbf{H}_0}$	estimation of the normalized channel matrix
B	bandwidth
B_c	coherence bandwidth
B_d	Doppler spread
f_s	sampling rate
f_c	center frequency within the bandwidth
f_d^p	Doppler frequency associated with the pth MPC
$f_{\max}$	maximum frequency within the bandwidth
$f_{\min}$	minimum frequency within the bandwidth
P^S	transmitted power ($P^S = P_{\alpha_T}^S = P_{\alpha_T}^T$)
P^D	received power ($P^D = P_{\alpha_R}^D = P_{\alpha_R}^R$)
N_0	noise power
ρ	SNR at the transmitter
ρ_0	SNR at the receiver
$\rho_{0,MRC}$	SNR at the receiver after MRC combining
$\rho_{0,EGC}$	SNR at the receiver after EGC combining
$\rho_{0,SC}$	SNR at the receiver after SC combining
G^T	maximum gain of the transmit SPA
G^R	maximum gain of the receive SPA
$G^A(\theta,\phi)$	gain pattern of an SPA ($A = T,R$)
T^A	temperature of antenna of an SPA ($A = T,R$)

$T^A(\theta,\phi)$	temperature of antenna distribution seen by an SPA ($A = T,R$)
$\mathbf{s} = \mathbf{s}(t)$	transmitted symbols in the time domain (MIMO system)
$\mathbf{S}(t,f)$	transmitted symbols in the frequency domain (MIMO system)
$s = s(t)$	transmitted symbols in the time domain (SISO system)
$S(t,f)$	transmitted symbols in the frequency domain (SISO system)
$\mathbf{n} = \mathbf{n}(t)$	noise vector in the time domain (SISO system)
$\mathbf{r} = \mathbf{r}(t)$	received symbols in the time domain (MIMO system)
$\mathbf{R}(t,f)$	received symbols in the frequency domain (MIMO system)
$r = r(t)$	received symbols in the time domain (SISO system)
$R(t,f)$	received symbols in the frequency domain (SISO system)
$\mathbf{y}$	received symbols after combining
$\mathbf{\Gamma}_{nm}^{p}$	polarimetric transfer matrix
P_{e}	error probability
R	rate of an STC
G_{d}	diversity gain or diversity order
G_{sm}	spatial multiplexing gain
G_{tp}	transmission power gain
t	code length
T_{c}	channel coherence time
$\mathbf{W}$	channel matrix with iid Gaussian random entries with distribution $N(0,1)$
$\mathbf{G}$	STC-generating matrix
ξ	correlation level among two generic entries of the channel matrix
$\mathbf{C}$	transmitted STBC codeword
$\mathbf{C}^i$	ith codeword of an STBC
$\mathbf{R}$	received STBC codeword
$\mathbf{N}$	noise matrix in an STBC architecture
$\mathbf{P}$	beam-forming matrix
$\overline{\mathbf{C}}$	transmitted STBC codeword with beam forming
$I(\mathbf{r};\mathbf{s})$	mutual information between the transmitted and received symbols
$\overline{\mathbf{F}}$	plane reflection symmetry operator
$\overline{\mathbf{R}}$	rotation symmetry operator
$\overline{\mathbf{P}}$	point reflection symmetry operator
$\mathbf{R}_{H}^{A}$	antenna correlation ($A = T,R$)
$\mathbf{R}_{E}^{A}$	extended antenna correlation ($A = T,R$)
$\mathbf{R}_{M}^{A}$	superextended antenna correlation ($A = T,R$)

$\mathbf{J}^A_{V',i}$	volumetric current density phasor associated with port i of the MPA
$\mathbf{l}^T_{ef}(\theta,\phi)$	directional effective transmission length
$\mathbf{l}^R_{ef}(\theta,\phi)$	directional effective receive length
n_{PL}	PL exponent
P	number of MPCs
$\mathbf{h}_{ij}(\tau)$	(i,j)th entry of the CIR (LTI channel)
$\mathbf{h}_{ij}(t,\tau)$	(i,j)th entry of the CIR (LTV channel)
$\mathbf{h}_{ST_{ij}}(t,\tau)$	(i,j)th entry of the space−time CIR (LTV channel)
$\mathbf{H}_{ij}$	(i,j)th entry of the transfer function (narrowband LTI channel)
$\mathbf{H}_{ij}(f)$	(i,j)th entry of the transfer function (LTI channel)
$\mathbf{H}_{ij}(t,f)$	(i,j)th entry of the transfer function (LTV channel)
$\mathbf{H}_{ST_{ij}}(t,f)$	(i,j)th entry of the space−time transfer function (LTV channel)
$\mathbf{r}'_{T_i}$	position of the ith transmit antenna
$\mathbf{r}'_{R_i}$	position of the ith receive antenna
$\mathbf{v}'_{T_i}$	velocity of the ith transmit antenna
$\mathbf{v}'_{R_i}$	velocity of the ith receive antenna
$\Omega_R = (\theta_R, \phi_R)$	direction of arrival at the receive antenna
$\Omega_T = (\theta_T, \phi_T)$	DoD at the transmit antenna
$\Omega^p_{R_j} = (\theta^p_{R_j}, \phi^p_{R_j})$	DoA of the pth MPC at the jth receive antenna
$\Omega^p_{T_i} = (\theta^p_{T_i}, \phi^p_{T_i})$	DoD of the pth MPC at the ith transmit antenna
$\mathbf{Z}^0_k$	characteristic reference impedance matrix at the kth port group
$Z^{0,A}_{k_i}$	characteristic reference impedance at the ith port, kth port group, Ath
$Z^{0,A}_i$	characteristic reference impedance at the ith if A is a single-port group component ($A = T, R$)
Z^0	characteristic reference impedance of the system
$\mathbf{a}^B_{k_i}$	incident wave at the ith port, kth port group, Bth component
$\mathbf{b}^B_{k_i}$	reflected wave from the ith port, kth port group, Bth component
$P^B_{k_i}$	instantaneous real power delivered to the ith port, kth port group, Bth component
$P^{B,\mathrm{av}}_{k_i}$	average real power delivered to the ith port, kth port group, Bth component
$P^{B,\mathrm{max}}_{k_i}$	maximum real power delivered to the ith port, kth port group, Bth component
$\mathbf{U}^B_k$	voltages at the port group k, Bth component
$\mathbf{U}^B_{k_i}$	voltages at the ith port, kth port group, Bth

$\mathbf{I}_k^B$	currents at the port group k, Bth component
$\mathbf{I}_{k_i}^B$	currents at the ith port, kth port group, Bth component
$\mathbf{S}^A$	scattering matrix of the Ath component
$\mathbf{S}_{kw}^A$	scattering matrix of the Ath component between port groups k, and w
$\mathbf{S}_{kw_{ij}}^A$	(i,j)th entry of $\mathbf{S}_{kw}^A$
$\mathbf{Z}^A$	impedance matrix of the Ath component (computed using $\mathbf{S}_{\alpha\alpha}^A$)
$\mathbf{Z}_{k_{ij}}^A$	(i,j)th entry of $\mathbf{Z}^A$ (computed using $\mathbf{S}_{kk}^A$)
$\mathbf{Z}_{ij}^A$	$\mathbf{Z}_{k_{ij}}^A$ if A is a single-port group component ($A = T,R$)
$\mathbf{Z}_i^A$	$\mathbf{Z}_{k_{ii}}^A$ if A is a single-port group component ($A = T,R$)
$\mathbf{Y}^A$	admittance matrix of the Ath component
$\mathbf{F}^A(\theta,\phi)$	normalized far-field radiation pattern of the SPA. ($A = T,R$)
$\mathbf{F}_i^A(\theta,\phi)$	normalized far-field radiation pattern at the ith port of an MPA ($A = T,R$)
$\mathbf{F}_{i,\theta}^A(\theta,\phi)$	θ component of $\mathbf{F}_i^A(\theta,\phi)$ ($A = T,R$)
$\mathbf{F}_{i,\phi}^A(\theta,\phi)$	ϕ component of $\mathbf{F}_i^A(\theta,\phi)$ ($A = T,R$)
$\mathbf{F}_i^T(\theta,\phi)$	$\mathbf{F}_i^A(\theta,\phi)$ for the transmit MPA
$\mathbf{F}_i^R(\theta,\phi)$	$\mathbf{F}_i^A(\theta,\phi)$ for the receive MPA
$\mathbf{F}_i^{A,0}(\theta,\phi)$	unit gain far-field radiation pattern at the ith port of the MPA ($A = T,R$)
$\mathbf{F}_i^{A,\text{iso}}(\theta,\phi)$	isolated far-field radiation pattern at the ith port of the MPA ($A = T,R$)
$\mathbf{F}_i^{A,\text{oc}}(\theta,\phi)$	open-circuit far-field radiation pattern at the ith port of the MPA ($A = T,R$)
$\mathbf{E}_i^T(r,\theta,\phi)$	radiated electric field at the ith port of the transmit MPA
$\mathbf{H}_i^T(r,\theta,\phi)$	radiated magnetic field at the ith port of the transmit MPA
$\mathbf{E}^T(r,\theta,\phi)$	total radiated electric field from the transmit MPA
$\mathbf{W}_i^T(r,\theta,\phi)$	average power density associated with the ith port of the transmit MPA
G_i^A	active gain associated with the ith port of the MPA ($A = T,R$)
$G_i^A(\theta,\phi)$	active gain radiation pattern associated with the ith port of the MPA ($A = T,R$)
$\mathbf{\Gamma}_{\alpha_T}^S$	signal source reflection matrix
$\mathbf{\Gamma}_{\alpha_R}^D$	signal drain reflection matrix
$P^{S,\text{av}}$	average transmitted power from the source
$P^{D,\text{av}}$	average received power to the loads
$P^{S,0}$	transmitted power from the source when $\mathbf{\Gamma}_{\alpha_T}^S = \mathbf{0}$
$P^{D,0}$	average received power to the loads when $\mathbf{\Gamma}_{\alpha_R}^D = \mathbf{0}$

$P^{S,\max}$	maximum transmitted power from the source
$P^{D,\max}$	maximum received power to the loads
η^{T}	transmit matching efficiency
η_{0}^{T}	transmit matching efficiency when $\Gamma_{\alpha_{T}}^{S} = \mathbf{0}$
η_{Ω}^{T}	transmit ohmic efficiency
η^{R}	transmit matching efficiency
η_{0}^{R}	transmit matching efficiency when $\Gamma_{\alpha_{R}}^{D} = \mathbf{0}$
η_{Ω}^{R}	transmit ohmic efficiency
$G_{i}^{A,\mathrm{MEG}} (= G_{\alpha_{i}}^{A,\mathrm{MEG}})$	mean effective gain associated with the ith port of the MPA ($A = T, R$)
$G^{A,\mathrm{MEAG}}$	mean effective array gain of the MPA ($A = T, R$)
$\mathcal{R}^{H_{0}}$	channel correlation matrix
$\mathcal{R}^{E_{0}}$	extended channel correlation matrix
$\mathcal{R}^{M_{0}}$	superextended channel correlation matrix
$\mathbf{E}$	extended channel matrix
$\mathbf{E}_{0}$	normalize extended channel matrix
$\mathbf{M}$	superextended channel matrix
$\mathbf{M}_{0}$	normalize superextended channel matrix
$\mathbf{S}^{T}$	scattering matrix of the transmit MPA
$\mathbf{S}^{R}$	scattering matrix of the receive MPA
$\mathbf{S}^{C}$	scattering matrix of the propagation channel
$\mathbf{S}^{MT}$	scattering matrix of the transmit DN
$\mathbf{S}^{MR}$	scattering matrix of the receive DN
$\mathbf{S}^{E}$	scattering matrix of the extended channel matrix
$\mathbf{S}^{M}$	scattering matrix of the superextended channel matrix
$\mathbf{\Psi}^{T}(\theta, \phi)$	transmit power spectrum
$\mathbf{\Psi}^{R}(\theta, \phi)$	receive power spectrum
$\mathbf{C}^{A}$	coupling matrix associated with the Ath component ($A = T, R$)
$\mathbf{Z}^{D}$	load impedance matrix of the signal drain
$\mathbf{Z}^{S}$	internal source impedance matrix of the signal source
$\mathbf{Z}_{i}^{D}$	load impedance at the ith port of the signal drain
$\mathbf{Z}_{i}^{S}$	internal source impedance at the ith port of the signal source
$\mathbf{\Gamma}_{\mathrm{in}}^{T}$	reflection matrix at the input of the transmit DN
$\mathbf{\Gamma}_{\mathrm{out}}^{T}$	reflection matrix at the output of the transmit DN
$\mathbf{\Gamma}_{\mathrm{in}}^{R}$	reflection matrix at the input of the receive DN
$\mathbf{\Gamma}_{\mathrm{out}}^{R}$	reflection matrix at the output of the receive DN
$\mathbf{b}_{0}^{S}$	incident waves from the source into the transmitter side

$\mathbf{b}_0^R$	incident waves from the channel into the receiver side
Q	Q factor
K	K factor
k	wave number
$\mathbf{k}$	wave propagation vector
a	radius of the virtual sphere enclosing an MPA
λ	free-space wavelength
η_0	free-space characteristic impedance
A^T	aperture size of the transmitter
A^R	aperture size of the receiver
τ^p	time delay associated with the pth MPC
τ_{ji}^p	time delay associated with the pth MPC between the ith transmit and jth receive antennas
$\tau_{\max}$	maximum time delay
$\Delta\tau_{\max}$	maximum time excess delay
$\mathbf{C}_{TR_{ji}}^p$	normalized channel polarimetric matrix
$\boldsymbol{\Gamma}_{ji}^p$	channel polarimetric matrix
$\boldsymbol{\Gamma}_{ji}^p$	channel polarimetric matrix
$S(v,\tau)$	Doppler spreading function
$B(v,f)$	Doppler-variant transfer function
ω	frequency variable
t	time variable

Operators and Mathematical Symbols

In this book, we use the spherical coordinate system, defined as a coordinate system for representing geometric figures in three dimensions using the following three coordinates: the radial distance (r) of a point from a fixed origin, the zenith angle (θ) from the positive z axis, and the azimuth angle (ϕ) from the positive x axis, as shown in Fig. 1.

Finally, in many practical systems involving electromagnetic waves, the time variations are of cosinusoidal form and are referred to as time-harmonic. In general, such time variations can be represented by $e^{j\omega t}$, and the instantaneous electromagnetic field vectors can be related to their

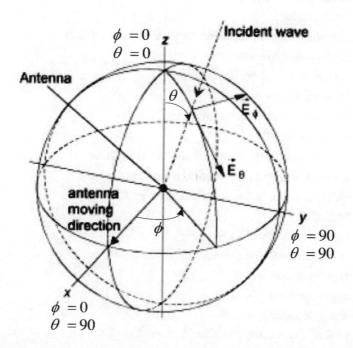

FIGURE 1: Graphical representation of the spherical coordinate system.

complex phasor forms in a very simple manner. For example, if $\mathbf{A}$ is the phasor of a particular electromagnetic field or current (which is a function of the position only), then the instantaneous field is given by $\mathbf{A}(t) = \Re(\mathbf{A})e^{j\omega t}$.

$\mathbf{A}$	matrix
$\mathbf{A} \in \mathcal{C}^{n \times n}$	matrix with dimensions $n \times n$
$\mathbf{A}_{ij}$	(i,j)th entry of matrix $\mathbf{A}$
$\mathbf{A}_{*j}$	jth column of matrix $\mathbf{A}$
$\mathbf{A}_{i*}$	ith row of matrix $\mathbf{A}$
$\mathbf{a}$	vector
$\mathbf{a} \in \mathcal{C}^{n \times 1}$	vector with dimensions $n \times 1$
$\mathbf{a}_i$	ith entry of vector $\mathbf{a}$
$(\cdot)^\dagger$	transpose operation
$(\cdot)^*$	complex conjugate operation
$(\cdot)^H$	Hermitian operation
$\mathbf{vec}(\cdot)$	vectorization operator
$\langle \cdot, \cdot \rangle$	inner product
$\otimes$	Kronecker product
$\times$	cross-product
$\odot$	operation that concatenates two scattering matrices
$\det(\cdot)$	determinant operator
$\log_2(\cdot)$	logarithmic (in base 2) operator
$\|\cdot\|_F$	Frobenius norm
$\Re(\cdot)$	real part of a complex number
$(\cdot)^+$	$(\cdot)^+ = \max(\cdot, 0)$
θ	zenith angle in the spherical coordinate system
ϕ	azimuth angle in the spherical coordinate system
r	radial distance in the spherical coordinate system
$\|$	such that
$\max(\cdot)$	maximum entry of a vector
$\min(\cdot)$	minimum entry of a vector
$\arg(\cdot)$	argument operator
$\angle(\cdot)$	angle operator
$abs(\cdot)$	absolute value operator
$p_x(\cdot)$	probability density function/distribution of x
$P(\cdot)$	probability function

$E\{\cdot\}$	expectation operator
$Var\{\cdot\}$	variance operator
$diag(\cdot)$	operator that retains only the diagonal elements of the matrix operand
$\delta(\cdot)$	delta function
$u_{\text{step}}(\cdot)$	step function
$\propto$	proportional to

Author Biography

Franco De Flaviis was born in Teramo, Italy, in 1963. He received his Italian degree (laurea) in electronics engineering from the University of Ancona (Italy) in 1990. In 1991, he was an engineer employee at Alcatel as researcher specialized in the area of microwave mixer design. In 1992, he was a visiting researcher at the University of California at Los Angeles (UCLA) working on low intermodulation mixers. He received his M.S. and Ph.D. degrees in electrical engineering from the Department of Electrical Engineering at UCLA in 1994 and 1997, respectively. Currently, he is an associate professor at the Department of Electrical and Computer Engineering at the University of California at Irvine. Dr. De Flaviis' research interests are computer-aided electromagnetics for high-speed digital circuits and antennas and microelectromechanical systems for RF applications fabricated on unconventional substrates such as printed circuit board and microwave laminates.

Lluís Jofre was born in Mataró, Spain, in 1956. He received his M.Sc. (Ing) and Ph.D. (Doctor Ing.) degrees in electrical engineering from the Technical University of Catalonia (UPC), Barcelona, Spain, in 1978 and 1982, respectively. From 1981 to 1982, he joined the École Normale Supérieure d'Electricite, Paris, France, where he was involved in microwave antenna design and imaging techniques for medical and industrial applications. In 1982, he was appointed associate professor at the Communications Department of the Telecommunication Engineering School at the UPC, where he became full professor in 1989. From 1986 to 1987, he was a Visiting Fulbright Scholar at the Georgia Institute of Technology, Atlanta, working on antennas and electromagnetic imaging and visualization. From 1989 to 1994, he served as the director of the Telecommunication Engineering School (UPC), and from 1994 to 2000, as UPC vice-rector for academic planning. From 2000 to 2001, he was a visiting professor at the Electrical and Computer Engineering Department, Henry Samueli School of Engineering, University of California, working on multiantenna systems for communications and imaging. From 2002 to 2004, he served as director of the Catalan Research Foundation, and since 2003, as director of the UPC-Telefonica chair. His research interests include antennas, electromagnetic scattering and imaging, and system miniaturization for wireless and sensing industrial and bio applications. He has published more than 100 scientific and technical papers, reports, and chapters in specialized volumes.

Jordi Romeu was born in Barcelona, Spain in 1962. He received the Ingeniero de Telecomunicación and Doctor Ingeniero de Telecomunicación, both from the Universitat Politècnica de Catalunya (UPC) in 1986 and 1991, respectively. In 1985, he joined the Antennalab at the Signal Theory and Communications Department, UPC. Currently, he is a full professor there, where he is engaged in research in antenna near-field measurements, antenna diagnostics, and antenna design. He was visiting scholar at the Antenna Laboratory, University of California at Los Angeles, in 1999, on a NATO Scientific Program Scholarship and in 2004 at University of California at Irvine. He holds several patents and has published 35 refereed papers in international journals and 50 conference proceedings. Dr. Romeu was the grand winner of the European IT Prize, awarded by the European Comission, for his contributions in the development of fractal antennas in 1998.

Alfred Grau was born in Barcelona, Spain, in 1977. He received his telecommunications engineering degree from the Universitat Politècnica de Catalunya (UPC), Barcelona, Spain, in 2001. He received his M.S. and Ph.D. degrees in electrical engineering from the Department of Electrical and Computer Engineering at the University of California at Irvine (UCI) in 2004 and 2007, respectively. Dr. Grau's research interest are in the field of miniature and integrated antennas, multiport antenna systems, MIMO wireless communication systems, software-defined antennas, reconfigurable and adaptive antennas, channel coding techniques, and microelectromechanical systems for RF applications.

Printed in the United States
by Baker & Taylor Publisher Services